International Workshop on Transportation and Supply Chain Engineering (IWTSCE'18)

MATEC Web of Conferences Volume 200 (2018)

Rabat, Morocco
8-9 May 2018

Editors:

B. Abou El Majd
H. El Ghazi

ISBN: 978-1-5108-7116-8

Printed from e-media with permission by:

Curran Associates, Inc.
57 Morehouse Lane
Red Hook, NY 12571

proceedings.com

Some format issues inherent in the e-media version may also appear in this print version.

Printed by Curran Associates, Inc. (2018)

For additional information, please contact EDP Sciences – Web of Conferences
at the address below.

EDP Sciences – Web of Conferences
17, Avenue du Hoggar
Parc d'Activité de Courtabœuf
BP 112
F-91944 Les Ulis Cedex A
France

Phone: +33 (0) 1 69 18 75 75
Fax: +33 (0) 1 69 28 84 91

contact-edps@webofconferences.org

Additional copies of this publication are available from:

Curran Associates, Inc.
57 Morehouse Lane
Red Hook, NY 12571 USA
Phone: 845-758-0400
Fax: 845-758-2633
Email: curran@proceedings.com
Web: www.proceedings.com

TABLE OF CONTENTS

Preface:

International Workshop on Transportation and Supply Chain Engineering (IWTSCE'18) is organized by Faculty of Science of Rabat (FSR-UM5) and National Institute of Posts and Telecommunications (INPT). The purpose of this meeting is to provide an international forum for recent developments in selected fields of research in transportation systems and supply chain engineering by bringing together researchers, practitioners and research scholars from all over the world to exchange and share their experiences and research results. The workshop is intended to provide chances for intensive discussions about their own selected topics, practical challenges encountered and identify future directions for the development of related modeling strategies and computational techniques. Innovative presentations are all invited, and participants for auditing only are also welcomed.

In order to make IWTSCE conference proceedings widely usable by the scientific community, we are proud having a partnership with MATEC as one of reputable partner with highly flexible publishing solutions. All accepted paper will be published by EDP Sciences on EI Serial Journal of MATEC web of Conferences {ISSN (Electronic Edition):2261-236X} as a proceeding. MATEC Web of Conferences is indexed in SCOPUS, Thomson Reuters (web of science), Chemical Abstracts Service (CAS), DOAJ, Engineering Village, Google Scholar, and Inspec. For further information about IWTSCE, please visit the web site at: https://tsc-eng.org.

We express our deepest gratitude to all members of the scientific committee, invited speakers, reviewers, sponsors, participants in the conference and members of the IWTSCE organizing committee for their valuable contribution to the successful organization of IWTSCE'18.

Badr Abou El Majd

General Chair
International Workshop on Transportation and Supply Chain Engineering

Statement of Peer review

In submitting conference proceedings to *Web of Conferences*, the editors of the proceedings certify to the Publisher that

1. They adhere to its **Policy on Publishing Integrity** in order to safeguard good scientific practice in publishing.
2. All articles have been subjected to peer review administered by the proceedings editors.
3. Reviews have been conducted by expert referees, who have been requested to provide unbiased and constructive comments aimed, whenever possible, at improving the work.
4. Proceedings editors have taken all reasonable steps to ensure the quality of the materials they publish and their decision to accept or reject a paper for publication has been based only on the merits of the work and the relevance to the journal.

Title, date and place of the conference

International workshop in Transportation and Supply Chain Engineering (IWTSCE'18)
May 08 - 09, 2018
National Institute of Posts and Telecommunications (INPT)
Rabat, Morocco

Proceedings editor(s):

Editor: Badr Abou El Majd, Mohammed V Universty-Rabat
Associate Editor: Hassan El Ghazi, INPT-Rabat

Date and editor's signature

August 11, 2018

MATEC Web of Conferences **200**, 00001 (2018) https://doi.org/10.1051/matecconf/201820000001
IWTSCE'18

Greedy heuristics for identical parallel machine scheduling problem with single server to minimize the makespan

Abdelhak El idrissi[1,3,⋆], *Mohammed* Benbrahim[1], *Rachid* Benmansour[2], and *David* Duvivier[3]

[1]*Laboratory MASI, Research team MOAD6, Ecole Mohammadia d'Ingénieurs, Rabat, Morocco*
[2]*Laboratory SI2M, Institut National de Statistique et d'Economie Appliquée , Rabat, Morocco*
[3]*LAMIH UMR CNRS 8201, Université de Valenciennes et du Hainaut Cambrésis, Valenciennes, France*

Abstract. This paper considers the problem of scheduling a set of n independent jobs on m identical parallel machines with setup times constraints. Immediately before processing, each job must be loaded on a common server to perform the setup operation. During the loading operation, both the machine and the server are occupied. Two greedy heuristics are developed for ($m \geq 2$) in order to minimize respectively the server waiting time and the machine idle time. These heuristics generalize those proposed in the literature for the case of two machines. The computational results show the efficiency of the proposed heuristics.

1 Introduction

In this paper, we consider a parallel machine scheduling problem with a single server (PSS) and the minimization of the makespan. Throughout this paper we consider the static version of the problem were all jobs are available at the beginning of the schedule. Scheduling problem under server constraints are intensively studied, only for special cases such as equal setup time, equal processing time and equal length (the sum of setup time and processing time), regular job constraints or the case of two machines. In this paper, the case of an arbitrary number of machines for the PSS problem is studied.

The PSS problem is denoted by $Pm, S1|s_j, p_j|C_{max}$ using the standard scheduling notation, where m represents the number of identical parallel machines, $S1$ represents the only available server, and s_j and p_j are respectively the setup time of job j and its processing time. The objective function considered in this problem is the minimization of the makespan C_{max}. The complexity has been studied by Kravchenko and Werner [1]. The authors showed that the problem $Pm, S1|s_j = 1|C_{max}$ is unary NP-hard and analyze several list scheduling heuristics. The complexity for the case of two machines and also an arbitrary number of machines is discussed by Bruker *et al.* [2] for all of the well-known objective functions C_{max}; $\sum C_j$; $\sum w_jC_j$; $\sum U_j$, $\sum w_jU_j$; $\sum T_j$; $\sum w_jT_j$ and L_{max}.

In 2006 Abdelkhodaee *et al.* [5] extended their previous study in [4],[3] dealing with the regular case where $p_i - p_j \leq s_j \; \forall i, j$ and the case of equal setup time and equal processing (see also [13]) for the $P2, S1|s_j, p_j|C_{max}$. The authors proposed two greedy heuristics, a genetic algorithm, and the Gilmore-Gomory algorithm for the general case of the $P2, S1||C_{max}$. The analysis started in [5] has been extended by Gan *et al.* [6], in which two mixed integer linear programming formulations and two variants of a branch-and-price scheme were developed.

Moreover, Hasani *et al.* [7] developed two metaheuristics namely simulated annealing (SA) and genetic algorithm (GA), for the problem $P2, S1||C_{max}$. The results obtained are much better than all the previous algorithms and models proposed in [5], [6]. Hasani *et al.* [8] proposed also two algorithms to solve very large instances for $P2, S1||C_{max}$. The computational results showed that the two algorithms outperformed the precedent methods existing in the literature, however, for small-sized and medium-sized instances, the results obtained were not promising.

Recently, Arnaout [9] proposed an ant colony optimization (ACO) metaheuristic for $P2, S1||C_{max}$. The computational results showed that the proposed method outperformed the metaheuristics (SA) and (GA) proposed in [7] for large instances of the problem. The problem with several machines ($m \geq 3$) was only studied by Kim and Lee [10]. The authors proposed two mixed integer programming formulations. The first one is developed using assignment and positional date variables, and the second one is developed by adding the concept of the server waiting time. A hybrid heuristic algorithm combining simulated annealing and tabu search is also suggested.

The remainder of this paper is organized as follows. In Section 2, we present the problem formally. In Section 3, two greedy heuristics algorithms for an arbitrary number of machines are proposed with two lower bounds. We conduct computational experiments in Section 4 to show the performance of those heuristics in comparison with the literature. The conclusion and future works end this paper.

⋆e-mail: abdelhakelidrissi@research.emi.ac.ma

2 Problem description

The aim of this section is to give a detailed description of the PSS problem. It can be described as follows. Suppose there are m identical parallel machines which must process n independent jobs. Each job i has a known integer processing time p_i. Immediately before its processing, job i must be loaded on a machine by the server. This duration, which can be also considered as a setup operation has a known integer value s_i. During the setup operation, both the machine and the server are occupied and after loading a job on a particular machine the server becomes available for loading the next job. The processing operation starts immediately after the end of setup. The preemption during setup and processing operations of jobs is not allowed. The objective is to find a feasible schedule that minimizes the makespan.

The following notations are used to define the problem throughout the paper:

- n: number of jobs.
- m: number of machines.
- $N = \{1, \ldots, n\}$: set of jobs.
- $M = \{1, \ldots, m\}$: set of machines.
- p_i: the processing time of the job i.
- s_i: the setup time of the job i.
- $A = \{A_1, A_2, \ldots, A_n\}$: the list of jobs to be scheduled.
- $B = \{B_1, B_2, \ldots, B_n\}$: the scheduled list sequence obtained by the heuristic.
- $St(J_i)$: the starting time of the setup operation of job i.
- $C(J_i)$: the completion time of the processing operation of job i.
- $C(M_k)$: the completion time of the last scheduled job on machine k.
- R_1: Sort the jobs in increasing order of their processing times (SPT).
- R_2: Sort the jobs in decreasing order of their processing times (LPT).
- R_3: Sort the jobs in increasing order of their setup times (SST).
- R_4: Sort the jobs in increasing order of their processing times (LST).
- R_5: Sort the jobs in increasing order of the sum of their processing and setup times (SPST).
- R_6: Sort the jobs in decreasing order of the sum of their processing and setup times (LPST).

3 Two Greedy heuristics

In this section, we propose two complementary greedy heuristics called HS1 and HS2 for the PSS problem under consideration in order to minimize respectively the server waiting time and the machine idle time. Indeed, it has been shown that the minimization respectively of the total server waiting time (i.e. the gaps between the loading of two jobs) and the minimization of the machine idle time is equivalent to the minimization of the makespan (see Kim and Lee [10]). The basic aim at each step of the two heuristics is, if possible, not to generate any machine idle time or any server waiting time.

3.1 Greedy heuristic HS1

This heuristic HS1 aims, to minimize machine idle time (the time machines are idle due to unavailability of the server). The pseudo-code of the proposed heuristic is as follows (see Algorithm 1).

In the first step jobs are arranged in a list in increasing order of their setup times. Then the first available $m-1$ jobs of the list are sequenced on the first available machines. After that, we sort the available unsequenced jobs according to some criterion and then choose an eligible job (i.e. one that would not create any machine idle time). If no eligible jobs exist we select the first job in the list. The list may be arranged in decreasing or increasing order of job lengths, setup times or processing times.

3.1.1 Illustrative example

Given an instance of $n = 10$ jobs and $m = 3$ machines. The processing times p_i and setup times of the jobs s_i are given in Table 1. The makespan value can be obtained by applying the heuristic HS1 as given above. It takes 0.002 seconds to solve the problem with the heuristics HS1. All the heuristics have been implemented using C++ language on a personal computer Intel(R) Core(TM) i5-2410M CPU @ 2.30 GHz with 8 GB of RAM memory. The feasible sequence is shown in Figure 1. With a value of the objective function (makespan) of 53.

Job	1	2	3	4	5	6	7	8	9	10
p_i	9	11	13	6	8	14	9	8	7	9
s_i	2	4	6	1	3	10	5	4	3	7

Table 1: Example instance with ($n = 10, m = 3$)

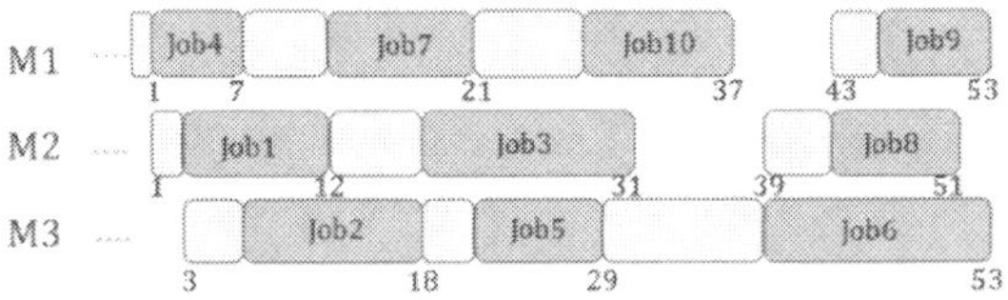

Figure 1: Feasible schedule for the ($n = 10, m = 3$) problem instance obtained by the heuristic HS1

In this illustrative example, we sort the initial list of jobs in increasing order of their setup times, then we schedule the first $m-1$ jobs on the first available machines, after this step, we sort for the second time the remaining jobs with the R_2 rule (longest processing time) and we start the job selection. Thus job $J4$ and job $J1$ are considered

Algorithm 1: HS1 Heuristic

```
Sort the initial list A = {A_1, A_2, ..., A_n} in increasing
  order of setup times
B ← ∅
for k = 1  to  m − 1 do
    B_k ← A_k,   A_k ← ∅
    Execute B_k on Machine M_k
    C(B_k) = St(B_k) + s(B_k) +p(B_k)
    St(B_{k+1}) = St(B_k) + s(B_k)
    C(M_k) = C(B_k)
    B ← B ∪ {B_k}
Sort the remaining list A according to one of the
  scheduling rule R_i
i ← m,   t ← m
j ← argmin_{h∈M}(C(M_h))
while i ≤ n − 1 do
    test=0
    for k = m  to  n − 1 do
        if (C(M_j) ≥
          max{St(B_{i−1}) + s(B_{i−1}); C(M_t)} + s(A_k))
          then
            test ← 1
            B_i ← A_k
            Execute B_i on Machine M_t
            C(B_i) = St(B_i) + s(B_i) +p(B_i)
            A ← A \ {A_k},   B ← B ∪ {B_i}
            C(M_t) = C(B_i)
            t ← argmin_{h∈M}(C(M_h))
            j ← argmin_{v∈M\M_t}(C(M_v))
            St(B_{i+1}) = max{St(B_i) + s(B_i); C(M_t)}
            i ← i + 1
            break
    if test=0 then
        for k = m  to  n − 1 do
            if (C(M_j) ≤
              max{St(B_{i−1}) + s(B_{i−1}); C(M_t)} + s(A_k))
              then
                test ← 1
                B_i ← A_k
                Execute B_i on Machine M_t
                C(B_i) = St(B_i) + s(B_i) +p(B_i)
                A ← A \ {A_k},   B ← B ∪ {B_i}
                C(M_t) = C(B_i)
                t ← argmin_{h∈M}(C(M_h))
                j ← argmin_{v∈M\M_t}(C(M_v))
                St(B_{i+1}) =
                  max{St(B_i) + s(B_i); C(M_t)}
                i ← i + 1
                break
Execute B_n on Machine M_t
C_max ← max{C(M_1), ..., C(M_m)}
```

as the first jobs. Then, the third job must be scheduled on the first available machine (machine $M3$) and a decision must be made according to the end of the setup time of the last scheduled job $J1$, the completion time of the last job scheduled on the next available machine, in this step the machine $M1$ was the next available one and the completion time of last job scheduled on the current available machine. Indeed, to minimize the machine idle time as much as possible, the best way is to choose a job whose loading time is less than or equal to the completion time of the next available machine minus the maximum value of the end of setup time of the last scheduled job in the finale list B and the completion time of the last job scheduled on the current available machine. Thus, the job $J2$ has to be selected as the third job. The fourth unscheduled job, which has to be determined, must be scheduled on the first available machine (machine $M1$) and must have a set-up time less than or equal to the completion time of the next available machine (Machine $M2$) minus the maximum value of the end of setup time of the last scheduled job (job $J2$) and the completion time of the last job scheduled on the current available machine (job $J4$). Among the remaining unscheduled jobs, job $J7$ will be chosen. And so on until scheduling all of the remaining jobs.

3.2 Greedy Heustic HS2

In this section, we present the second greedy heuristic HS2 to minimize the gap between the completion time of the setup operation of a job and the start of setup operation of the next job. Contrarily to the heuristic Min-loagap proposed for the case of two machines by Hasani *et al.* [8] where the jobs were scheduled in staggered order on machines (i.e. jobs alternate between machines), in this heuristic the jobs are scheduled according to the availability of the machines. And the job with the minimal processing time is considered as the last job to be scheduled in the final sequence.

3.2.1 Illustrative example

Given an instance of $n = 10$ jobs and $m = 4$ machines. The processing times p_i and setup times of the jobs s_i are given in Table 2. The makespan value can be obtained by applying the heuristic HS2 as given above. It takes 0.002 seconds to solve the problem with the heuristics HS2. The feasible sequence is shown in Figure 2. With a value of the objective function (makespan) of 90. In this example equality $C_{max} = LB$ holds.

Job	1	2	3	4	5	6	7	8	9	10
p_i	2	4	8	3	5	6	7	9	7	2
s_i	8	7	12	4	8	9	10	14	11	5

Table 2: Example instance with ($n = 10, m = 4$)

In this illustrative example, we sort the initial list of jobs according to the longest processing time rule (R_2), then we schedule the first $m - 1$ jobs on the first available machines, and the job with the minimal processing time

Algorithm 2: HS2 Heuristic

Sort the initial list $A = \{A_1, A_2, \ldots, A_n\}$ according to one of the scheduling rules R_i
$B \leftarrow \emptyset$
Find the job $A_k \in A$, $k \in \{1, .., n\}$ with the smallest processing time
$B_n \leftarrow A_k$
$A_t \leftarrow A_{t+1} \quad \forall t \in \{k, .., n-1\}$
for $k = 1$ *to* $m - 1$ **do**
  $B_k \leftarrow A_k, \quad A_k \leftarrow \emptyset$
  Execute B_k on Machine M_k
  $C(B_k) = St(B_k) + s(B_k) + p(B_k)$
  $St(B_{k+1}) = St(B_k) + s(B_k)$
  $C(M_k) = C(B_k)$
  $B \leftarrow B \cup \{B_k\}$
$i \leftarrow m, \quad t \leftarrow m$
$j \leftarrow \text{argmin}_{h \in M}(C(M_h))$
while $i \leq n - 1$ **do**
  test=0
  for $k = m$ *to* $n - 1$ **do**
    if $(C(M_j) \leq \max\{St(B_{i-1}) + s(B_{i-1}); C(M_t)\} + s(A_k))$ **then**
      $test \leftarrow 1$
      $B_i \leftarrow A_k$
      Execute B_i on Machine M_t
      $C(B_i) = St(B_i) + s(B_i) + p(B_i)$
      $A \leftarrow A \setminus \{A_k\}$, $B \leftarrow B \cup \{B_i\}$
      $C(M_t) = C(B_i)$
      $t \leftarrow \text{argmin}_{h \in M}(C(M_h))$
      $j \leftarrow \text{argmin}_{\upsilon \in M \setminus M_t}(C(M_\upsilon))$
      $St(B_{i+1}) = \max\{St(B_i) + s(B_i); C(M_t)\}$
      $i \leftarrow i + 1$
      break
  if *test=0* **then**
    for $k = m$ *to* $n - 1$ **do**
      if $(C(M_j) \geq \max\{St(B_{i-1}) + s(B_{i-1}); C(M_t)\} + s(A_k))$ **then**
        $test \leftarrow 1$
        $B_i \leftarrow A_k$
        Execute B_i on Machine M_t
        $C(B_i) = St(B_i) + s(B_i) + p(B_i)$
        $A \leftarrow A \setminus \{A_k\}$, $B \leftarrow B \cup \{B_i\}$
        $C(M_t) = C(B_i)$
        $t \leftarrow \text{argmin}_{h \in M}(C(M_h))$
        $j \leftarrow \text{argmin}_{\upsilon \in M \setminus M_t}(C(M_\upsilon))$
        $St(B_{i+1}) = \max\{St(B_i) + s(B_i); C(M_t)\}$
        $i \leftarrow i + 1$
        break
Execute B_n on Machine M_t
$C_{max} \leftarrow \max\{C(M_1), ..., C(M_m)\}$

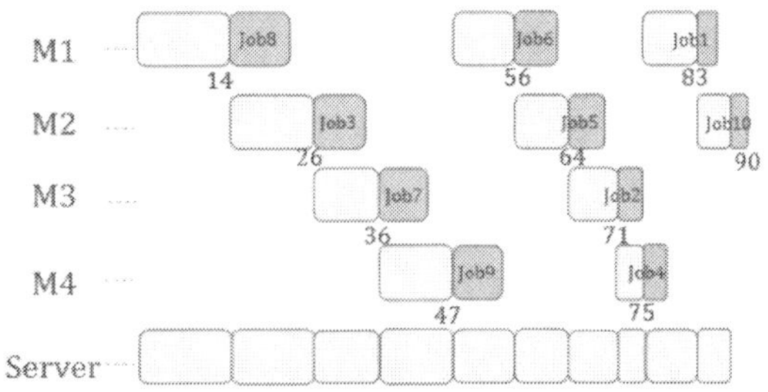

Figure 2: Optimal schedule for the ($n = 10, m = 4$) problem instance obtained by the heuristic HS2

is scheduled in the last position of the schedule. Thus job $J8$, job $J3$ and job $J7$ are considered as the first jobs. After this step, we start the job selection. The fourth job must be scheduled on the first available machine and a decision must be made according to the end of the setup time of the last scheduled job $J7$, the completion time of the last job scheduled on the next available machine, in our case the machine $M1$ was the next available one and the completion time of last job scheduled on the current available machine. Indeed, to minimize the server waiting time as much as possible, the best way is to choose a job whose loading time is greater than or equal to the completion time of the next available machine minus the maximum value of the end of setup time of the last scheduled job in finale list B and the completion time of last job scheduled on the current available machine. Thus, the job $J9$ has to be selected as the fourth job. The fifth unscheduled job, which has to be determined, must be scheduled on the first available machine (machine $M1$) and must have a set-up time greater or equal than the completion time of the next available machine (machine $M2$) minus the maximum value of the end of setup time of the last scheduled job and the completion time of last scheduled on the current machine (job $J8$). Among the remaining unscheduled jobs, job $J6$ will be chosen. And so on until scheduling all of the remaining jobs.

3.3 Lower bounds

The investigation of lower bounds is useful for benchmarking heuristic solutions. Thus, to evaluate the quality of the solutions, the following lower bounds are proposed.

Proposition 1.

$$LB1 = \frac{1}{m} \sum_{1 \leq i \leq n} \{p_i + s_i\}$$

is a lower bound for the problem of $P, S||C_{max}$.

Proof. by contradiction:
if $LB1 \leq \frac{1}{m} \sum_{1 \leq i \leq n} \{p_i + s_i\}$

In contradiction with the fact that the makespan cannot take a value less than the sum of the processing times and the setup times of all jobs divided by the number of machines. □

Proposition 2.

$$LB2 = \sum_{1\leq i\leq n} s_i + \min_{1\leq i\leq n} p_i$$

is a a lower bound for the problem of $P, S||C_{max}$.

Proof. In an optimal schedule, the solution will have a value of $\sum_{1\leq j\leq n} s_j$ plus the processing time of the last scheduled job. Indeed, the makespan is equal to the sum of setup times plus the total server waiting times as shown by Kim et al. [10] i.e. $C_{max} = \sum_{1\leq j\leq n} s_j + SWTT$ (with: $SWTT$ is the total server waiting time) and in an optimal schedule the total server waiting time will be equal to the processing time of the last executed job. So to obtain a good lower bound, the last scheduled job in the sequence should have the smallest processing time. □

Combining the two lower bounds $LB1$ and $LB2$, we obtain an overall lower bound $LB = \max\{LB1; LB2\}$ on the value of the makespan solution C_{max}. To show the effectiveness of the lower bound, we have compared it with the linear relaxation of the MIP models proposed by El idrissi *et al.* [11] for several instances. We have found that our LB is approximately equal to the linear relaxation for the majority of the cases, which shows its effectiveness.

4 Computational experiments

4.1 Description of data

In this section, we conduct a computational experiment where the above, heuristics HS1, HS2 and lower bounds are implemented. The data was generated in the same way as described firstly by Koulamas [12], and also described lately by Abdekhodaee and Wirth [3] and Hasani *et al.* [8]. The data are generated in the uncorrelated case, where the processing time values p_j are generated from a discrete uniform distribution $U(0, 100)$ and setup time values s_j are generated from a discrete uniform distribution $U(0, 100L)$ where $L = E(s_j)/E(p_j)$ is the server load and $E(x)$ denotes the mean of x.

HS1 was used for instances with $LB = LB1$, and HS2 for the instances with $LB = LB2$. From the experiments, it turned out that for all instances with $L \in \{0.1, 0.5, 0.8\}$, equality $LB = LB1$ always holds and for all instances with $L \in \{1.5, 1.8, 2\}$, equality $LB = LB2$ always holds. However, for the instances with $L = 1$, both lower bounds have to be taken into account. In the following, if $L \in \{0.1, 0.5, 0.8\}$ HS1 is only used and if $L \in \{1.5, 1.8, 2\}$ HS2 is only used.

In order to compare the performance of our proposed heuristics with the literature, three factors are used: a fixed number of machines, the problem size fixed to three levels: small-size instances with $n \in (8, 20)$, medium-size instances with $n \in (30, 40, 50, 100)$ and large-size instance with $n \in (200, 250, 300, 350)$. The server load is also fixed to $L \in \{0.1, 0.5, 0.8, 1.5, 1.8, 2\}$, for each value of the server load 5 instances are generated randomly for $n \in \{8, 20\}$ and 10 instances were generated randomly for each of the other value of n, m and L.

4.2 Performances

To evaluate the efficiency of the proposed heuristics, computational experiments were designed and conducted. The computational results with HS1 for small, medium and large-sizes instances are given in Table 3 and Table 4 and the computational results with HS2 for small, medium and large sizes instances are given in Table 5 and Table 6. In Tables 3, 4, 5, 6 , column 1 gives the number of jobs, column 2 gives the number of machines, column 4 until the last column give the values R_{avg}, denoting the average value of the ratio C_{max}/LB, and R_{max}, denoting the maximum value of the ratio C_{max}/LB among all instances for a particular value of L and for a particular heuristic.

Comparing the obtained results for HS1 with the results in Hasani *et al.* [8] for the Min-idle algorithm which we denote as (Mit) and the results in Abdekhodaee and Wirth [3] for forward heuristic which we denote as (FH) and also comparing the obtained results for HS2 with the results in Hasani *et al.* [8] for the Min-loadgap algorithm which we denote as (Mlg) and the results in Abdekhodaee and Wirth [3] for backward heuristic which we denote as (BH) when using the same instances, the following summary can be given:

In Table 3, HS1 is compared with FH and Mit for $L \in \{0.1, 0.5, 0.8\}$ and for $m = 2$ since FH and Mit can be applied to only the case of two machines. It must be noted that FH and Mit provide the same result for all proposed instances. With R_2 (LPT) rule and for the case of $L = 0.1$ HS1 outperform by far the two heuristics FH and Mit for the majority of cases. For $L = 0.5$ and $L = 0.8$ FH and Mit are better than HS1 for the two proposed scheduling rules R_2 and R_4.

In Table 4, HS1 is compared with FH in the case of $L = 0.1$ for medium/large instances. It is observed that HS1 with the LPST rule is better than FH in term of deviation from the lower bound in the most of the data sets $n \in \{50, 100, 150, 200\}$. In some cases of $n \in \{250, 300, 350\}$, HS1 with the LST rule gives the same results in comparison with FH.

In Table 5, HS2 is compared with BH and Mlg for small size instances. The symbol $*$ is used to specify that no solution can be found since Mlg was proposed only for the case of two machines and for BH we adapted the algorithm for the case of arbitrary number of machines. With R_2 (LPT) rule and for the case of $m \in \{2, 3, 4, 5\}$ HS2 is better than BH in term of deviation from the lower bound for some cases for $m \geq 2$ and also it gives less precise results in comparison with Mlg.

In Table 6, HS2 is compared with BH for medium/large instances. With LPT rule and for $m \in \{3, 4, 5\}$, HS2 outperform BH in term of deviation from the lower bound for the most cases and for all values of the server load $L \in \{1.5, 1.8, 2\}$. Therefore, with LST rule the heuristic HS2 gave less precise results in comparison with BH and with the R_2 rule also. It must be noted that the solution time with all proposed heuristics (HS1, HS2, BH, FH, Mit, Mlg) and for all proposed instances don't exceed 0,006 seconds.

n	m		$L = 0.1$			$L = 0.5$			$L = 0.8$		
			HS1-LPT	HS1-LST	FH / (Mit)	HS1-LPT	HS1-LST	FH / (Mit)	HS1-LPT	HS1-LST	FH / (Mit)
8	2	R_{avg}	1.01413	1.03591	1.03591	1.11347	1.17261	1.08607	1.17999	1.20681	1.13497
		R_{max}	1.03659	1.11385	1.11385	1.19561	1.30882	1.13388	1.34456	1.35988	1.17917
20	2	R_{avg}	1.0132	1.02269	1.02464	1.08481	1.06107	1.04708	1.19439	1.08417	1.1065
		R_{max}	1.02339	1.03768	1.04288	1.12479	1.13828	1.09338	1.26943	1.17402	1.27185
50	2	R_{avg}	1.00423	1.01033	1.01119	1.05972	1.0205	1.01431	1.14087	1.06299	1.03064
		R_{max}	1.00779	1.03451	1.0308	1.077	1.03686	1.02459	1.17022	1.12111	1.0761
100	2	R_{avg}	1.00341	1.00611	1.00611	1.06456	1.01405	1.0091	1.15299	1.04793	1.02527
		R_{max}	1.00821	1.01296	1.01296	1.08852	1.04478	1.03046	1.18127	1.08919	1.04048
150	2	R_{avg}	1.00539	1.00459	1.00459	1.06214	1.00572	1.00429	1.1301	1.01687	1.00722
		R_{max}	1.00953	1.00745	1.00745	1.08458	1.01021	1.00822	1.1466	1.03687	1.01504
200	2	R_{avg}	1.004	1.00484	1.00484	1.06572	1.00443	1.00259	1.13859	1.01611	1.00844
		R_{max}	1.00564	1.00825	1.00825	1.07941	1.00985	1.00518	1.15395	1.02805	1.01666
250	2	R_{avg}	1.00425	1.00251	1.00251	1.06451	1.0037	1.0023	1.13322	1.01129	1.00498
		R_{max}	1.00534	1.00614	1.00614	1.08621	1.00883	1.0061	1.15714	1.0228	1.01124
300	2	R_{avg}	1.00478	1.00182	1.00182	1.06345	1.00144	1.00098	1.14009	1.02038	1.00678
		R_{max}	1.00699	1.0047	1.0047	1.07094	1.00509	1.00262	1.14519	1.03069	1.01391
350	2	R_{avg}	1.00404	1.0009	1.0009	1.06402	1.00122	1.00098	1.13749	1.01518	1.00646
		R_{max}	1.00532	1.00248	1.00248	1.07527	1.00395	1.00167	1.17283	1.03719	1.02185

Table 3: Computational results with HS1 heuristic for small/medium and large-sizes instances

n	m		$L = 0.1$						
			HS1-SST	HS1-LST	HS1-SPT	HS1-LPT	HS1-SPST	HS1-LPST	FH
50	2	R_{avg}	1.01833	1.01033	1.02	1.00423	1.02074	1.00391	1.01119
		R_{max}	1.03776	1.03451	1.03127	1.00779	1.03253	1.00928	1.0308
100	2	R_{avg}	1.01191	1.00611	1.00931	1.00341	1.01007	1.00235	1.00611
		R_{max}	1.02391	1.01296	1.01621	1.00821	1.0139	1.00602	1.01296
150	2	R_{avg}	1.01008	1.00459	1.00787	1.00539	1.0065	1.0045	1.00459
		R_{max}	1.0144	1.00745	1.01156	1.00953	1.00798	1.008	1.00745
200	2	R_{avg}	1.0109	1.00484	1.0057	1.004	1.00459	1.00337	1.00484
		R_{max}	1.01831	1.00825	1.00857	1.00564	1.00582	1.0044	1.00825
250	2	R_{avg}	1.00831	1.00251	1.00391	1.00425	1.00377	1.00341	1.00251
		R_{max}	1.01483	1.00614	1.00686	1.00534	1.00494	1.00463	1.00614
300	2	R_{avg}	1.00829	1.00182	1.00303	1.00478	1.00318	1.00404	1.00182
		R_{max}	1.01161	1.0047	1.00527	1.00699	1.00431	1.00624	1.0047
350	2	R_{avg}	1.00723	1.0009	1.00339	1.00404	1.0027	1.00327	1.0009
		R_{max}	1.00994	1.00248	1.00546	1.00532	1.00374	1.00433	1.00248

Table 4: Computational results with HS1 heuristic for medium and large-sizes instances for $L = 0.1$

n	m		$L = 1.5$			$L = 1.8$			$L = 2$		
			HS2-LPT	BH	Mlg	HS2-LPT	BH	Mlg	HS2-LPT	BH	Mlg
8	2	R_{avg}	1.04044	1.02051	1.03388	1.03848	1.0274	1.03959	1.03309	1.02827	1.0348
		R_{max}	1.16529	1.08058	1.16942	1.11372	1.11372	1.11045	1.09957	1.09668	1.13131
	3	R_{avg}	1	1	*	1.00103	1.01303	*	1.00101	1.0206	*
		R_{max}	1	1	*	1.00266	1.0625	*	1.00504	1.05804	*
20	2	R_{avg}	1.02533	1.02085	1.01949	1.044	1.03147	1.05137	1.01054	1.00295	1.00536
		R_{max}	1.039	1.03874	1.04212	1.09224	1.07453	1.09835	1.03893	1.01416	1.02679
	3	R_{avg}	1.00045	1.00076	*	1.00011	1.00011	*	1.0001	1.0001	*
		R_{max}	1.00226	1.00226	*	1.00057	1.00057	*	1.0005	1.0005	*
	4	R_{avg}	1.00083	1.00547	*	1	1	*	1	1	*
		R_{max}	1.00413	1.02324	*	1	1	*	1	1	*
	5	R_{avg}	1	1	*	1	1	*	1.00011	1.00011	*
		R_{max}	1	1	*	1	1	*	1.00054	1.00054	*

Table 5: Computational results with HS2 heuristic for small-sizes instances

5 Conclusion

In this contribution, the problem $Pm, S1|s_j, p_j|C_{max}$ was considered. Two greedy heuristics were proposed to solve this problem for small, medium and large instances. The first one was developed for instances with $LB = LB1$, and the second one for instances with $LB = LB2$. The heuristics performed extremely well in terms of the deviation from the proposed lower bounds for small/medium and large instances and also in terms of the low computational times. A performance comparison was performed between these two heuristics and other heuristics proposed in the literature for the case of two machines and also an arbitrary number of machines. The comparison shows that our heuristic HS1 outperform the literature in term of deviation from the lower bound in some cases. And the heuristic HS2 outperforms by far the literature for the case of arbitrary number of machines.

In future work, it is intended to investigate several types of metaheuritics for the parallel machine scheduling problem with single server, and also take into consideration other types of objective functions such as the total completion time or the total tardiness.

References

[1] S.A. Kravchenko and F. Werner, Parallel machine scheduling problems with a single server, Mathemati-

n	m		$L = 1.5$			$L = 1.8$			$L = 2$		
			HS2-LST	HS2-LPT	BH	HS2-LST	HS2-LPT	BH	HS2-LST	HS2-LPT	BH
30	3	R_{avg}	1.03675	1	1.00264	1.01881	1	1.00026	1.01463	1	1
		R_{max}	1.0521	1	1.01634	1.03632	1	1.0026	1.03206	1	1
	4	R_{avg}	1.01587	1.00014	1.00014	1.01111	1.00018	1.00018	1.00667	1	1
		R_{max}	1.0427	1.00141	1.00141	1.03298	1.00181	1.00181	1.01518	1	1
40	3	R_{avg}	1.03217	1	1.0025	1.01793	1	1	1.02601	1.00034	1.0007
		R_{max}	1.06977	1	1.0184	1.04368	1	1	1.04756	1.00336	1.00336
	4	R_{avg}	1.01507	1	1	1.01129	1	1	1.00706	1	1
		R_{max}	1.02889	1	1	1.03012	1	1	1.02292	1	1
	5	R_{avg}	1.0104	1	1	1.00475	1	1	1.00434	1	1
		R_{max}	1.04315	1	1	1.01673	1	1	1.01656	1	1
50	3	R_{avg}	1.03099	1	1.00145	1.01656	1	1.00007	1.01139	1	1
		R_{max}	1.06379	1	1.00848	1.03549	1	1.00065	1.02464	1	1
	4	R_{avg}	1.01589	1	1	1.00885	1	1	1.01021	1	1
		R_{max}	1.03005	1	1	1.01821	1	1	1.02041	1	1
60	3	R_{avg}	1.03488	1.00002	1.00027	1.02079	1	1.00029	1.01353	1	1
		R_{max}	1.05075	1.00021	1.00142	1.03311	1	1.00235	1.0275	1	1
	5	R_{avg}	1.00866	1.00002	1.00002	1.00591	1	1	1.00388	1	1
		R_{max}	1.01784	1.00021	1.00021	1.01442	1	1	1.01184	1	1
100	3	R_{avg}	1.02453	1	1	1.0201	1	1.0001	1.01582	1	1
		R_{max}	1.03763	1	1	1.03738	1	1.00102	1.02496	1	1
	5	R_{avg}	1.00743	1	1	1.00481	1	1	1.00416	1	1
		R_{max}	1.01704	1	1	1.00885	1	1	1.01072	1	1
200	3	R_{avg}	1.02827	1	1.00002	1.01729	1	1	1.01626	1	1
		R_{max}	1.0397	1	1.00014	1.02506	1	1	1.02066	1	1.00005
	4	R_{avg}	1.01364	1	1	1.00945	1	1	1.00784	1	1
		R_{max}	1.02029	1	1	1.01339	1	1	1.01123	1	1
250	3	R_{avg}	1.02587	1	1.00006	1.01829	1	1.00001	1.01581	1	1
		R_{max}	1.04181	1	1.00051	1.03103	1	1.00009	1.0214	1	1
	5	R_{avg}	1.0069	1	1	1.00549	1	1	1.00513	1	1
		R_{max}	1.01086	1	1	1.00847	1	1	1.00819	1	1

Table 6: Computational results with HS2 heuristic for medium and large-sizes instances

cal and Computer Modeling, 26, 1-11, 1997.

[2] P. Brucker, C. Dhaenens-Flipo, S. Knust, S. Kravchenko, A. Svetlana and F. Werner, Complexity results for parallel machine problems with a single server, Journal of Scheduling, 5, 429-457, 2002.

[3] A.H. Abdekhodaee and A. Wirth, Scheduling parallel machines with a single server: some solvable cases and heuristics, Computers & Operations Research, 29, 295-315, 2002.

[4] A.H. Abdekhodaee, A. Wirth and H.S. Gan, Equal processing and equal setup time cases of scheduling parallel machines with a single server, Computers & Operations Research, 31, 1867-1889, 2004.

[5] A.H. Abdekhodaee, A. Wirth and H.S. Gan, Scheduling two parallel machines with a single server: the general case, Computers & Operations Research, 33, 994-1009, 2006.

[6] H. S. Gan, A. Wirth and A.H. Abdekhodaee, A branch-and-price algorithm for the general case of scheduling parallel machines with a single server, Computers & Operations Research, 39 (9), 2242-2247, 2012.

[7] K. Hasani, S. A. Kravchenko and F. Werner, Simulated annealing and genetic algorithms for the two-machine scheduling problem with a single server, International Journal of Production Research, 52 (13), 3778-3792, 2014.

[8] K. Hasani, S. A. Kravchenko and F. Werner, Minimizing the makespan for the two-machine scheduling problem with a single server: Two algorithms for very large instances, Engineering Optimization, 48, 173-183, 2016.

[9] J. P. Arnaout, Heuristics for the two machine scheduling problem with a single server, International Transactions in Operational Research, 24 (6), 1347-1355, 2017.

[10] M.Y. Kim and Y.H. Lee, MIP models and hybrid algorithm for minimizing the makespan of parallel machines scheduling problem with a single server, Computers & Operations Research, 39, 2457-2468, 2012.

[11] A. El idrissi, R. Benmansour, M. Benbrahim, D. Duviver, MIP formulations for identical parallel machine scheduling problem with single server, Proceedings of 4th IEEE International Conference on Optimization and Applications (ICOA), 2018.

[12] C.P. Koulamas, M.L. Smith, Look-ahead scheduling for minimizing machine interference, International Journal of Production Research, 26 (9), 1523-1533, 1988.

[13] R. Benmansour, O. Braun, H. Allaoui, Modeling the single processor scheduling problem with time restrictions as a parallel machine scheduling problem, Proceeding of 7th Multidisciplinary International Conference on Scheduling: Theory and Applications (MISTA), 2015.

A Game Theoretic Approach to Analyse Security between Smart Vehicles and Parcels in Smart Cities.

El Arbi ABDELLAOUI ALAOUI[1,2,*] and *Mustapha* EL MOUDDEN[3,**]

[1]EIGSI, 282 Route of the Oasis, Mâarif, 20140 Casablanca, Morocco.
[2]E3MI Team , Department of Computer Science, FST Errachidia - Moulay Ismail University, Morocco.
[3]Department of Mathematics , Faculty of Sciences, Meknes - Moulay Ismail University, Morocco.

Abstract. The Internet of Things (IoT) is considered as a modern concept that will revolutionize the near future. Its interest is to create an environment of combined intelligent devices and systems, communicating with each other through wireless networks. Urban logistics are an applicative field of this new technology, especially for smart parcels and vehicles. Actually, in the context of economy development, the competitiveness between companies and territories necessarily involves an improvement of logistics services. Although these gains offered by IoT, there are significant obstacles to counter. One of the important obstacles to consider is the security. In this paper, we will analyze the interaction between selfish smart vehicles/parcels and malicious smart vehicles/parcels, that was formulated as a game model. As a result, we have calculated the Nash equilibrium and the utilities for the both selfish smart vehicles/parcels and malicious smart vehicles/parcels, evaluated the parameters that can maximize the selfish smart vehicles/parcels's utility when the smart parcels are transported by vehicles between different centers (shops, supermarket, etc) was planned and identify the potential malicious smart vehicles/parcels.

1 Introduction

The term "Intelligent Transportation Systems (ITS) " is generally used to refer to technology, infrastructure and services of tomorrow, as well as the planning and the control methods used for the persons and freight transportation. In addition, the intelligent transport system (ITS) came as a solution to different transportation problems, such as traffic congestion; related to the world motorization, the urbanization, the population growth and the change in population density [1].

The goal of Intelligent Transportation Systems (ITS) is to integrate individual transportation elements and join them through use of information and communication technologies into a single system. ITS provide the opportunity to increase the use of existing transportation system and generate additional capacity from the existing physical infrastructure. Other benefits of freight ITS include, but not limited to, Increase safety and security, decrease negative environmental impacts of freight transportation.

The core of ITS consists in obtaining, processing and distributing information for a better use of the transportation system, infrastructure and services. Indeed, as data and information transmission become increasingly shared thanks to ITS, there is a need to establish more open data/information formats, interfaces and communication protocols. Harmonization and standardization of collected and processed data, as well as the information delivered, form part of this overall ITS standardization strategy. This can be achieved through examining and identifying the internationally adopted standards or best-practices and adapting them to the local context. Another area of importance is the communication security, especially when sensitive or confidential information is being transmitted. In this era of heightened security, some measures need to be taken to ensure that overall ITS ecosystem remain robust and secured. Likewise, the importance of privacy and anonymity should not be overlooked and proper measures need to be in place to maintain data anonymity. All these will enhance users' confidence and acceptance of ITS.

In this paper, we are exploring the applicability of game theoretic approaches to address the ITS security issues and some of these approaches look promising, such as freight transport in cities. The goal of the research is to design a solution of the security between the vehicles and parcels in smart cities using game theory. The challenge of this work is to manage the communication beteween the selfish and malicious smart vehicles/parcels in smart city in order to mitigate attacks based on malicious smart vehicles/parcels. We have modeled the selfish and malicious strategies using Game Theory approach in a normal form so as to find the Nash equilibrium to evaluate the benefits and costs for each strategie. We are exploring the applicability of game theoretic approach to protect the the communication beteween the smart vehicles and parcels in smart city [10] [9] [3]. So, the main contributions of this paper can be summarized as follows:

*e-mail: abdellaoui.e@gmail.com
**e-mail: muelmoudden@gmail.com

- to analyze of the security smart vehicle and parcel with the game theory;
- to model interaction between the selfish smart vehicles/parcels and the malicious smart vehicles/parcels in smart city;
- to determine the payoff of the selfish smart vehicles/parcels and the malicious smart vehicles/parcels;
- to characterize the Nash equilibrium (NE) of each strategy.

The rest of the paper is organized as follow. In Section 2 we describe our problematic. Then, the Section 3 provides an analysis of interactions between the selfish smart vehicles/parcels and the malicious smart vehicles/parcels, with a special focus on game theory approach in smart cities. In Section 4, we analyze the game equilibria for selfish smart vehicles/parcels and malicious smart vehicles/parcels. Finally, We conclude the paper in Section 5.

2 Problem and statement

In our study, we consider a smart city with a selfish and malicious smart parcels, and a set of smart vehicles (selfish and malicious) implicated in the inter-centers transportation process (shops,supermarket ...); and we want to know the action/interaction between the selfish and malicious smart vehicles/parcels, in order to secure the communication between them.

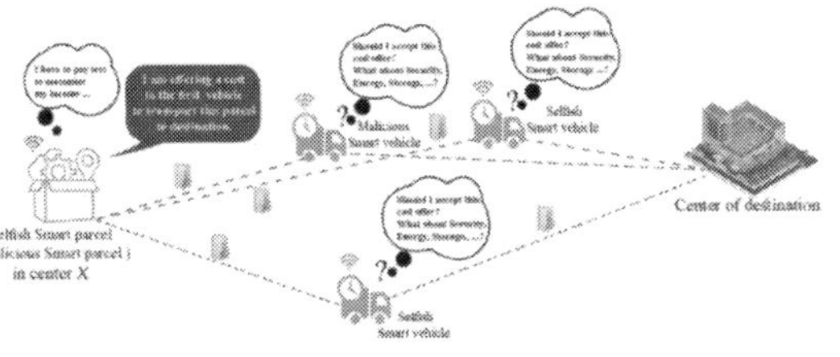

Figure 1. Diagram explaining our problematic

3 Game formulation

In this work, to formally analyze the security issue in smart cities, we model the dynamic interactions between the selfish smart vehicles/parcels and the malicious smart vehicles/parcels as a non-cooperative game [9] [10]. Our security non-cooperative game model is inspired by the work in [4].

The non-cooperative game model to be described has the following three components :

Players The set of the legitimate smart vehicles/parcels in an smart cities, which is composed of the malicious smart vehicles/parcels who is the abstraction of one or multiple smart vehicles/parcels with malicious intent to compromise the smart cities and the selfish smart vehicles/parcels defending them, $\{P^D_{nj}\}^m_{j=1}$ and $\{P^C_{mj}\}^m_{j=1}$ denote the set of the malicious smart vehicles/parcels and the selfish ones, respectively.

Strategy space Assume that an smart cities consisting of smart vehicles/parcels that are capable of deciding whether to collaborate with one another or defect. On the one hand, most of them are selfish and want to preserve their resource. On the other hand, they have to transport the smart parcels to the different centers in smart cities. At the same time, for some malicious smart vehicles/parcels, they may advertise false delivery information, not transport smart parcels correctly, fabricate, modify, or simply drop smart parcels, the action set of the smart vehicles/parcels is defined as follows :

$$S = \{C, D\} \tag{1}$$

where C represents collaboration and D denotes defection.

In this work, $x(s)$ denotes the probability that selfish smart vehicles/parcels discover the malicious smart vehicles/parcels at time s. For the selfish smart vehicles/parcels, the bigger $x(s)$, the more gains they will obtain. For simplicity, the gain $g_{mi}(x(s))$ of the selfish smart vehicle/parcel P^C_{mj} is given as follows

$$g_{mi}(x(s)) = q_{mi}(x(t)) \tag{2}$$

where q_{mi} is a positive parameter.

$f_{mi}(t)$ denotes the number of smart parcels that has been scheduled to transport and have successfully arrived at their centers of destination at time t by the selfish smart vehicle P^C_{mj}. For a collaborate smart vehicle/parcel P^C_{mj}, if a smart parcel originated from a selfish smart vehicle can be successfully delivered to its center of destination, then the smart vehicle can get gain g_{mi}. For simplicity, $r_{mi}(f_{mi}(t))$ represents the smart parcels that have been scheduled to transport and have successfully arrived at their centers of destination, which is given as follows :

$$r_{mi}(f_{mi}(t)) = f_{mi}(t)g_{mi} \tag{3}$$

where g_{mi} is a positive parameter.

$h_{nj}(i, t)$ denotes that a smart vehicle P^D_{nj} drops or injects the number of smart parcels, which are transported by the smart vehicle P^C_{mi} at time t. It is natural to assure that the smart vehicles/parcels are resource limited in smart cities, and they take into account the amount of energy spent. If the smart parcels are dropped or injected by the malicious smart vehicles and are not discovered by the selfish smart vehicles, the malicious smart vehicle can obtain gains $w_{nj}(h_{nj}(i, t))$. For simplicity, the gains $w_{nj}(h_{nj}(i, t))$ of the malicious smart vehicle P^D_{nj} is defined as follows :

$$w_{nj}(h_{nj}(i, t)) = (1 - x(s))h_{nj}(i, t)c \tag{4}$$

where c is a positive constant.

In what follows, $u_i(s)$ denotes the resource consume rate of a smart vehicle/parcel i (i.e., amount of resources) that is devoted to securing that information at time s, which is a ratio, so its unit is 1.

$$u_i(s) = \frac{\text{resource consume at time } s}{\text{the total of smart vehicle/parcel } i} \tag{5}$$

Assume that the cost paid by the selfish smart vehicle/parcel P^C_{mi} is given as follows :

$$y_{mi}(u_{mi}(s)) = \frac{c_{mi}}{2}u_{mi}(s)^2 \tag{6}$$

and the cost paid by the malicious smart vehicle/parcel P^D_{nj} is defined as follows :

$$y_{nj}(u_{nj}(s)) = \frac{c_{nj}}{2}u_{nj}(s)^2 \tag{7}$$

The selfish smart vehicles/parcels are rational, a selfish smart vehicle P^C_{mi} seeks to maximize the instantaneous probability $x(s)$ and maximize the smart parcels $f_{mi}(t)$ that have been scheduled to transport and have successfully arrived at the centers of destination; and minimize resource consumption incurred by their actions. Assume that the game is a perfect information game. A selfish smart vehicle/parcel P^C_{mi} chooses the optimal amount of network resource $u_{mi}(s)$ to invest in information security contingent upon the state of game in order to maximize the individual utility function U_{mi}.

The utility function for the selfish smart vehicles/parcels is defined as follows :

$$U_{mi} = \int_{t_0}^{T}\left[q_{mi}x(t) + f_{mi}(t)g_{mi} - \frac{c_{mi}}{2}u_{mi}(s)^2\right]e^{(-r(s-t_0))}\mathrm{d}s + e^{(-r(T-t_0))}S_{mi}x(T) \tag{8}$$

where $r > 0$ denotes discount factor, values received t time is discounted by the factor r, and $S_{mi}x(T)$ denotes the marginal utility of the smart vehicle/parcel P^C_{mi} at time T.

The malicious smart vehicles/parcels, unlike the selfish smart vehicles/parcels, are motivated by the reward for disrupting or jamming the transmissions of other center. Similarly, for a malicious smart vehicle/parcel P^D_{nj}, we can model the malicious smart vehicles/parcels' utility function as follows

$$U_{nj} = \int_{t_0}^{T}\left[q_{nj}(1 - x(s)) + (1 - x(s))h_{nj}(i, s)c - \frac{c_{nj}}{2}u_{nj}(s)^2\right] e^{(-r(s-t_0))}\mathrm{d}s + e^{(-r(T-t_0))}S_{nj}[1 - x(T)] \tag{9}$$

When $x(s) = 1$, the selfish smart vehicles/parcels's instantaneous probability that discovers the malicious smart vehicles/parcels is 100 %, according to our model, the increase of $x(s)$ means increase the intensity of the defensive countermeasures to against attacks, which may cost their large amounts of resources, so the selfish smart vehicles/parcels have to decrease their intensity of the defensive countermeasures to save their resources. Similarly, assume that $x(s) = 0$, the probability that discovers the malicious smart vehicles/parcels is 0, the selfish smart vehicles/parcels don't take any defensive countermeasure to against attacks, which may threat the network security. So the selfish smart vehicles/parcels have to increase their intensity of the defensive countermeasures to against attacks. For simplicity, the dynamics of the probability $x(s)$ that discovers malicious smart vehicles/parcels is governed by :

$$\frac{dx(s)}{ds} = \sum_{i=1}^{m} u_{mi}(s)[1 - x(s)]^{1/2} - \sum_{j=1}^{n} u_{nj}(s)x(s)^{1/2}$$
$$x(0) = x_0 \tag{10}$$

4 Nash equilibrium solutions

In the game theory, the Nash Equilibrium is an important concept. The players will meet an agreement if NE exists.

In this section, we study the node's behavior and characterize the equilibria structure of the selfish smart vehicles/parcels and the malicious smart vehicles/parcels. In fact, none of the selfish smart vehicles/parcels and the malicious smart vehicles/parcels make a pre-agreement, so they seek a Nash equilibrium solution, $u_i(s)$.

4.1 Nash Equilibrium Solution for the Selfish smart vehicles/parcels

Let $u^{(t_0)*}_{mi}(t, x) = [u^{(t_0)*}_{m1}(t, x), u^{(t_0)*}_{m2}(t, x), \ldots, u^{(t_0)*}_{mm}(t, x), u^{(t_0)*}_{n1}(t, x), u^{(t_0)*}_{n2}(t, x), \ldots, u^{(t_0)*}_{nn}(t, x)]$, for $t \in [t_0, T]$, denotes a set of strategies that provides a Nash equilibrium solution to the non-cooperative game $\Gamma(x_0, T - t_0)$ and $W^{(t_0)t}(t, x) : [t_0, T] \times \mathbb{R}^n \to \mathbb{R}$ denotes the value function of the selfish smart vehicles/parcels P^C_{mi}.

For the selfish smart vehicles/parcels P^C_{mi}, $i \in \{1, 2, \ldots, m\}$ a feedback Nash equilibrium solution to the equations (8), (10) satisfies the following conditions :

$$-V^{(t_0)mi}_t(t, x) = \max_{U_{mi}}\{[q_{mi}x(t) + f_{mi}(t)g_{mi} - \frac{c_{mi}}{2}u_{mi}(t)^2]e^{-r(t-t0)} + V^{mi}_x(t, x)[u_{mi}(t)[1 - x(t)]^{1/2} + \sum_{k\neq i=1}^{m} u^*_{mk}(t, x)[1 - x(t)]^{1/2} - \sum_{j=1}^{n} u^*_{nj}(t, x)x(t)^{1/2}]\}$$

$$V^{mi}(T, x) = e^{-r(T-t_0)}S_{mi}x, \tag{11}$$

The Nash equilibrium solution for the selfish smart vehicles/parcels by performing the indicated maximization in equation (11) yields

$$u^{(t_0)*}_{mi}(t, x) = \frac{V^{mi}_x(t, x)}{c_{mi}}(1 - x)^{1/2}e^{r(t-t_0)} \quad \text{for } i \in \{1, 2, \ldots, m\}. \tag{12}$$

4.2 Nash Equilibrium solution for the malicious smart vehicles/parcels

For malicious smart vehicles/parcels P^D_{nj}, $j \in \{1, 2, \ldots, n\}$ a feedback Nash equilibrium solution to the equations (9), (10) has to satisfy the following conditions

$$-V_t^{(t_0)nj}(t,x) = \max_{U_{nj}}\{[q_{nj}(1-x(t)) + (1-x(t))h_{nj}(i,t)c - \frac{c_{nj}}{2}u_{nj}(t)^2]e^{-r(t-t0)} + V_x^{nj}(t,x)[-u_{nj}(t)[x(t)]^{1/2} + \sum_{i=1}^{m} u_{mi}^*(t,x)[1-x(t)]^{1/2} - \sum_{l\neq j=1}^{n} u_{nl}^*(t,x)x(t)^{1/2}]\}$$

$$V^{nj}(T,x) = e^{-r(T-t_0)}S_{nj}(1-x), \quad (13)$$

Similarly, the Nash equilibrium solution for the malicious smart vehicles/parcels by performing the indicated maximization in equation (13) yields,

$$u_{nj}^{(t_0)*}(t,x) = \frac{-V_x^{nj}(t,x)}{c_{nj}}x^{1/2}e^{r(t-t_0)} \quad \text{for } j \in \{1,2,\ldots,n\}. \quad (14)$$

Substituting $u_{mi}^{(t_0)*}(t,x)$ and $u_{nj}^{(t_0)*}(t,x)$ into (11) and (13), respectively, and solving ((11) and (13), we obtain the value functions

$$V^{mi}(t,x) = e^{-r(t-t_0)}[A_{mi}(t)x + B_{mi}(t),], \text{for } i \in \{1,2\ldots,m\}, \quad (15)$$

$$V^{nj}(t,x) = e^{-r(t-t_0)}[A_{nj}(t)(1-x) + B_{nj}(t),], \quad (16)$$
for $j \in \{1,2,\ldots,n\}$.

where $A_{mi}(t)$, $B_{mi}(t)$, $A_{nj}(t)$, and $B_{nj}(t)$ satisfy

$$\frac{A_{mi}(t)}{dt} = rA_{mi}(t) - q_{mi} + \frac{A_{mi}^2(t)}{2c_{mi}} + \sum_{k\neq i=1}^{m}\frac{A_{mi}(t)A_{mk}(t)}{c_{mk}} \quad (17)$$

$$+ \sum_{j=1}^{n}\frac{A_{mi}(t)A_{nj}(t)}{c_{nj}} \quad (18)$$

$$A_{mi}(T) = S_{mi} \quad (19)$$

$$\frac{B_{mi}(t)}{dt} = rB_{mi}(t) - f_{mi}g_{mi} - \frac{A_{mi}^2(t)}{2c_{mi}} - \sum_{k\neq i=1}^{m}\frac{A_{mi}(t)A_{mk}(t)}{c_{mk}} \quad (20)$$

$$B_{mi}(T) = 0 \quad (21)$$

$$\frac{A_{nj}(t)}{dt} = rA_{nj}(t) - q_{nj} - h_{nj}c + \frac{A_{nj}^2(t)}{2c_{nj}} + \sum_{l\neq j=1}^{n}\frac{A_{nj}(t)A_{nl}(t)}{c_{nl}}$$

$$+ \sum_{i=1}^{m}\frac{A_{mi}(t)A_{nj}(t)}{c_{mi}} \quad (22)$$

(23)

$$A_{nj}(T) = S_{nj} \quad (24)$$

$$\frac{B_{nj}(t)}{dt} = rB_{nj}(t) - q_{nj} - h_{nj}c + \sum_{i=1}^{m}\frac{A_{mi}(t)A_{nj}(t)}{c_{mi}} \quad (25)$$

For the symmetric case, (17) becomes

$$B_{nj}(T) = 0. \quad (26)$$

$$\frac{A_{mi}(t)}{dt} = rA_{mi}(t) + (m-1/2)\frac{A_{mi}^2(t)}{c_{mi}} + n\frac{A_{mi}(t)A_{nj}(t)}{c_{nj}} - q_{mi} - h_{nj}c. \quad (27)$$

Similarly, we have that

$$\frac{A_{nj}(t)}{dt} = rA_{nj}(t) - q_{nj} - h_{nj}c + \left(n-\frac{1}{2}\right)\frac{A_{nj}^2(t)}{c_{nj}} + m\frac{A_{mi}(t)A_{nj}(t)}{c_{mi}}. \quad (28)$$

Upon substituting the relevant partial derivatives of $V^{mi}(t,x)$ and and $V^{nj}(t,x)$ from (15) and (16) into (12) and (14) yields the feedback Nash equilibrium strategies

$$u_{mi}^{(t_0)*}(t,x) = \frac{A_{mi}(t)}{c_{mi}}(1-x)^{1/2} \quad (29)$$

for the selfish smart vehicle/parcel P_{mi}^C, $i \in \{1,2,\ldots,m\}$

and

$$u_{nj}^{(t_0)*}(t,x) = \frac{A_{nj}(t)}{c_{nj}}x^{1/2} \quad (30)$$

for the selfish smart vehicle/parcel P_{nj}^D, $j \in \{1,2,\ldots,n\}$.

Substituting the game equilibrium strategies above into (10) yields the optimal state trajectory :

$$\frac{dx(s)}{ds} = -\left(\sum_{i=1}^{m}\frac{A_{mi}(t)}{c_{mi}} + \sum_{j=1}^{n}\frac{A_{nj}(t)}{c_{nj}}\right)x(s) + \sum_{i=1}^{m}\frac{A_{mi}(t)}{c_{mi}} \quad (31)$$

$$x(0) = x_0 \quad (32)$$

$$x^*(t) = \bar{\omega}(t_0,t)\left(\int_{t_0}^{t}\left[\frac{A_{mi}(t)}{c_{mi}}\right]\bar{\omega}^{-1}(t_0,t)ds + x_0\right), \quad (33)$$

for $t \in [t_0, T]$,

where

$$\bar{\omega}(t_0,t) = e^{-\int_{t_0}^{t}H(\tau)d\tau} \quad (34)$$

and

$$H(s) = -\left(\sum_{i=1}^{m}\frac{A_{mi}(t)}{c_{mi}} + \sum_{j=1}^{n}\frac{A_{nj}(t)}{c_{nj}}\right) \quad (35)$$

The Nash equilibrium strategies, $u_{mi}^{(t_0)*}(t,x)$ and $u_{nj}^{(t_0)*}(t,x)$, which are strategy profile for the non-cooperative game with the property that no nodes can improve its gains by altering its strategy unilaterally while the other nodes keep their strategies unchanged [26,27].

5 Conclusion

In this paper, we have studied and analyzed the security between vehicles and parcels in smart cities by using a non-cooperative game. We have first introduced a non-cooperative game model of the interactions between vehicles and parcels in smart cities. Then, we have analyzed the set of Nash equilibrium of the game and discuss its implications for each strategy of our game and presented some necessary conditions for existence of Nash equilibrium.

References

[1] K. Abdul Khaliq, A. Qayyum and J. Pannek. "Novel Routing Framework for VANET Considering Challenges for Safety Application in City Logistics". In *Vehicular Ad-Hoc Networks for Smart Cities*, Volume 548, pp. 53-67, 2017.

[2] A. Roumboutsos, S.Kapros. "A game theory approach to urban public transport integration policy." *Transport Policy, ELSEVIER*, Volume 19, pp. 209–215, 2008.

[3] Abou El Majd, B., Desideri, J. A., and Habbal, A. (2010). Aerodynamic and structural optimization of a business-jet wingshape by a Nash game and an adapted split of variables. Mécanique and Industries, 11(3-4), 209-214.

[4] E. A. Abdellaoui, K. Nassiri, M. El Moudden. "Framework for Analyzing of Inter-cluster Communication in the DRHT by Using Game Theory", in *The Mediterranean Symposium on Smart City Applications (SCAMS 2017), ACM*, October 25-27, 2017 Tangier, Morocco.

[5] Chunli, L . "Intelligent transportation based on the Internet of things." in *IEEE 2nd International Conference on Consumer Electronics, Communications and Networks, IEEE*, 2012.

[6] R. El-Azouzi, F. Pellegrini, V. Kamble. "Evolutionary forwarding games in Delay Tolerant Networks."In proceedings of the *8th International Symposium o Modeling and Optimization in Mobile, Ad Hoc and Wireless Networks (WiOpt), IEEE* pp. 81–89, 2010.

[7] Dimitrakopoulos, G., Demestichas, P., Koutra, V." Intelligent management functionality for improving transportation efficiency by means of the car pooling concept" . In journal the *IEEE Trans. Intell. Transp. Syst.* pp. 424–436, 2012.

[8] Elhenawy, M., A. Elbery, A., A. Hassan, A." An intersection game-theory-based traffic control algorithm in a connected vehicle environment." In: *IEEE 18th International Conference on Intelligent Transportation Systems, IEEE*,2015.

[9] Pan, F., Xi, B., Wang, L.: "Environmental regulation strategy analysis of local government based on evolutionary game theory." in *2014 International Conference on Management Science and Engineering (ICMSE)*, 2014.

[10] Wei, L., Wei, G.: "Decision-making on reverse logistics for manufacturers: an evolutionary game theory perspective". In: *International Conference on Logistics,Informatics and Service Sciences (LISS)*, 2015.

Persistent Homology applied to location problems

Driss Bennis[1,a], Fouad Gharib[1,b] and Ghita Lebbar[1,c]

[1]*Faculty of Sciences, Research Center CeReMAR, Mohammed V University in Rabat, Morocco*

Abstract. Different approaches to solve location problems in transport and logistics have been developed in the literature. This article introduces a new approach using the concept of persistent homology which has been proved to be an efficient method in topological data analysis; and has been served as an alternative new tool in many and various research areas such as image processing, material science and biological systems. Precisely, inspired by the notions of the first homology groups and the persistent homology which mainly describe the behaviour of connectivity relation between elements during a filtration of specific topological spaces; we develop a new method and approach for the treatment of facility location–network design problems.

1 Introduction

The concept of persistence is first introduced by Edelsbrunner, Letscher, and Zomorodian, in [6]; then refined by Carlsson and Zomorodian in [21]. It was used to provide a rigorous response to the following problem: For a parameterized family of spaces, those topological features which persist over a significant parameter range are to be considered as signal with short-lived features as noise (see [7]). Since then, persistent homology has gathered an enormous attention as a fundamental tool in Topological Data Analysis (TDA). It has proved utility and significance to solve different problems in several domains such as biology, image processing and sensor networks (see, for instance, the papers [1,13,16,19]). For a general background on persistent homology, worth to mention the book by Zomorodian [20], survey by Edelsbrunner, Harer [5] and the survey by Ghrist [7]. The mentioned authors offer an introduction to persistent homology which can be useful for mathematical researchers based on topology.

The main interest of our work is related to location decision using persistent homology. Location decision is a critical element in transport planning. In general, decision makers are asked to select a site (location) among several existing sites (cities, district, airports, ...) on a given network in order to satisfy a number of customers while considering some constraints. Several researchers have developed different mathematical models for location decision which are mainly based on the classical optimisation methods. The present paper provides a new treatment of location problems using the concept of persistent homology. Therefore, for a practical understanding, the second section presents some basic notions and background related to persistent homology. Section 3 highlights the problem related to transport design. In section 4, we will focus on our contribution, based on a location decision issue and the adopted approach using persistent homology to solve the problem.

2 Mathematical background

In this section, we give a short presentation on some basic notions concerning simplicial complexes, simplicial homology and persistent homology (more details can be found in [5,7,20]).

2.1 Simplicial complexes

An abstract simplicial complex or simply a simplicial complex is a set K together with a collection S of subsets of K which satisfies the two following conditions:

- S contains all singletons $\{v\}$ with $v \in K$.
- S is closed under subsets: if $\tau \subseteq \sigma$ and $\sigma \in S$, then $\tau \in S$.

An element σ of S of cardinal a positive integer n will be called an (n-1)-simplex or simply a simplex. So, 0-simplices are the singletons which are usually called points; and 1-simplices consist of two elements of K called edges. There are also the 2-simplices called triangles and so on. In fact, any simplicial complex can be represented in a real affine space using points, edges,

[a]*driss.bennis@um5.ac.ma*; [b]*fouad.gharib1994@gmail.com*; [c]*ghitaa.lebbar@gmail.com*

triangles and so on. Thus, using the induced topology on this geometric representation, one can talk about the connected components of a simplicial complex. Namely, this notion is of great interest in simplicial homology theory. Next section shows that the number of connected components of a simplicial complex is the dimension of some kind of vector spaces. Several examples of simplicial complex can be found in the literature. The most useful ones are geometric simplicial complexes, complexes constructed from graphs, complexes constructed from a covering of a geometric space (Nerve of a covering).

2.2 Simplicial homology

Simplicial homology is a powerful tool for description and characterization of some topological features such as connectivity and the existence of holes. Given K as a simplicial complex. The $\mathbb{Z}/2\mathbb{Z}$ vector space generated by the p-dimensional simplices of K is denoted $C_p(K)$. It consists of all p-chains, which are formal sums

$$c=\sum_j \gamma_j \sigma_j \quad (1)$$

where the γ_j are 0 or 1 (in $\mathbb{Z}/2\mathbb{Z}$) and the σ_j are p-simplices in K. For a positive integer p, consider the linear map ∂: $C_p(K) \rightarrow C_{p-1}(K)$, called boundary map, defined on n-simplices as follows: For every p-simplex σ, $\partial(\sigma)$ is the formal sum of the (p−1)-dimensional faces (i.e., subsets of σ of cardinal p). Thus, the boundary of the chain c (i.e., $\partial(c)$) is obtained by extending ∂ linearly, and so

$$\partial(c) =\sum_j \gamma_j \partial(\sigma_j), \quad (2)$$

where we understand that the addition between coefficients is modulo 2, *i.e.*

$$1 + 1 = 0. \quad (3)$$

It is not difficult to check that:

$$\partial \circ \partial = \partial^2 = 0 \quad (4)$$

The p-chains that have boundary 0 are called p-cycles. They form a subspace Z_p of C_p. The p-chains which are the boundary of (p + 1)-chains are called p-boundaries and form a subspace B_p of C_p. The fact that:

$$\partial^2 = 0 \quad (5)$$

implies that $B_p \subseteq Z_p$. The quotient group

$$H_p(K) := Z_p/B_p \quad (6)$$

is the p-th simplicial homology group of K with $\mathbb{Z}/2\mathbb{Z}$ coefficients. The dimension of $H_p(K)$ is called the p-th Betti number or simply the p-Betti number of K and it is denoted $\beta_p(K)$ or just β_p.

Overall, the homology groups describe spaces through their Betti numbers β_p. Precisely, β_0 refers to the number of connected components of the given simplicial complex, β_1 refers to the number of holes of the simplicial complex.

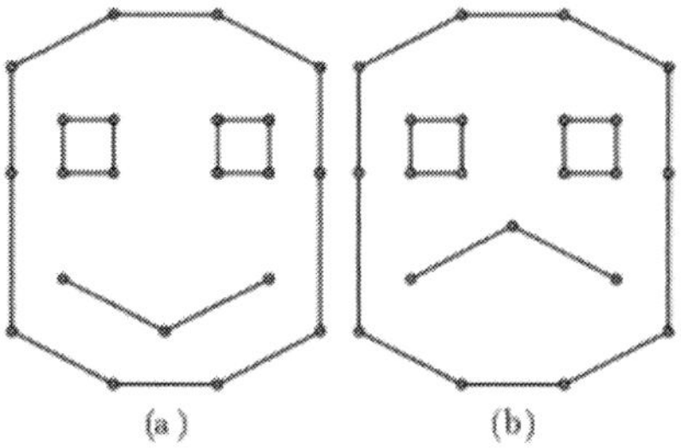

Figure 1: Two simplicial complexes with similar Betti numbers: β_0=4 and β_1=3.

2.3. Persistent homology

Persistent homology is considered as a powerful tool in topological shape analysis which describes the changes in homology that occur to an object with respect to a given parameter.

For a simplicial complex K, a filtration F of K is a finite sequence of complexes:

$$F: \emptyset=K_0 \subseteq K_1 \subseteq \cdots \subseteq K_m =K \quad (7)$$

The idea of persistent homology consists of, instead of examining the homology $H_i(K;F)$ of each individual term K_i of the filtration F, the iterated inclusions are examined:

$$h: H_i(K;F) \rightarrow H_j(K;F) \text{ for all } i < j. \quad (8)$$

These maps reveal which features persist. For a clear understanding, let us give a simple example visualised in the above figures:

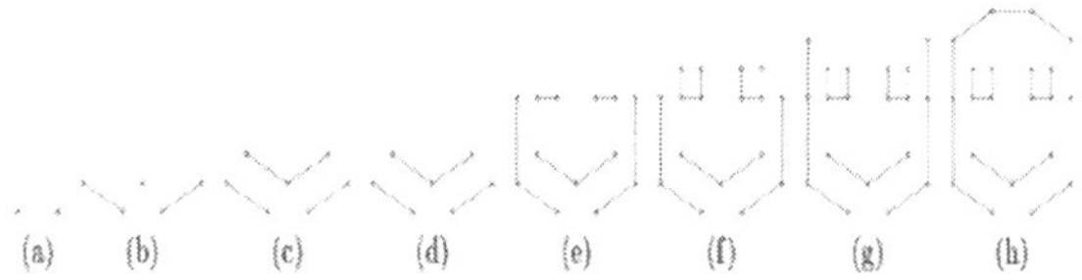

Figure2: Filtration F_1 of Happy face P_1

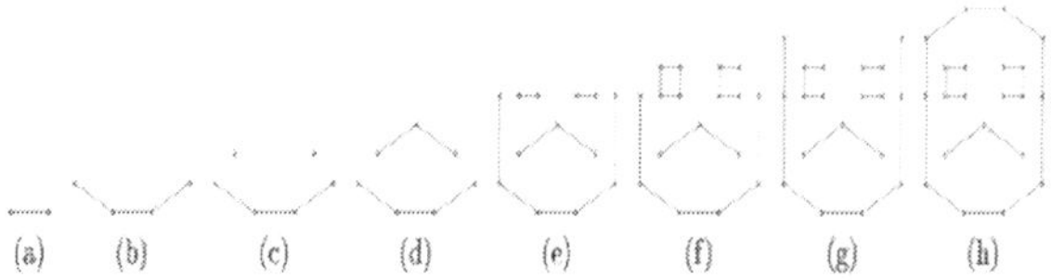

Figure 3: Filtration F_2 of Sad face P_2

Although both pictures P_1 and P_2 are different, they have identical betti numbers. Persistent homology reveals the changes in homologies in the given filtration and prove the difference between both objects. Usually, the parameter intervals arising from the basis for $H_*(K;F)$ are visualised in the form of a barcode. Precisely, a barcode is a graphical representation of $H_k(K; F)$ as a collection of horizontal line segments in a plane whose horizontal axis corresponds to the parameter and whose vertical axis represents an arbitrary) ordering of homology generators (i.e., generators of $H_*(K;F)$). For example, the barcode corresponding to the filtration F_1 above is of the form:

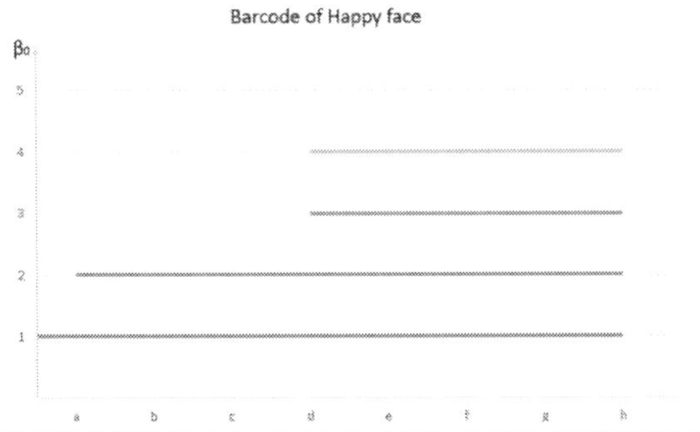

Figure 4: Barcode of happy face

And the barcode corresponding to the filtration F_2 is of the form:

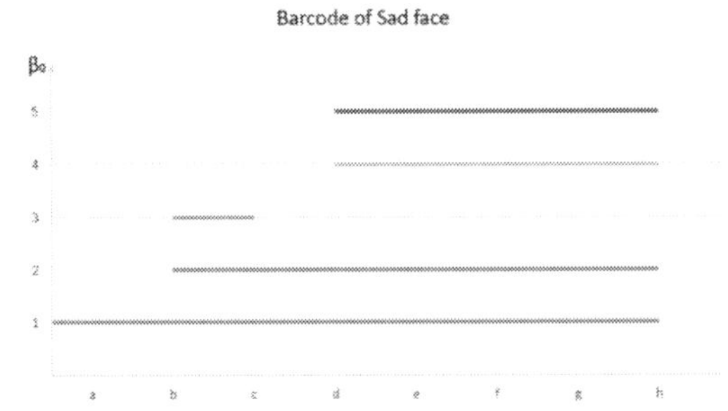

Figure 5: Barcode of sad face

The two barcodes illustrate clearly the difference.

3 Location decision

Location decisions have been analysed since 1909 by Weber [18] who formulated a theory to locate an industry with a minimum transportation cost of raw and final products.

Mainly, the acquisition of a new facility within a network is a costly, time-sensitive target. Hence, experts on transport design focus on solving the following issues:

- The optimal choice of nodes to locate facilities
- Minimize the travel cost
- Reduce travel time
- Optimise networks: Infrastructure (roads, bridges, tunnels...), telecommunication (cables) water pipes, electricity cables.

Hence, it can be understood that the choice of location is not arbitrary but requires strategic and expert planners which are challenged by several factors. As an aid for making decisions, the domain of facility location provides tools and methods for finding optimal locations using mathematical models with respect to quantifiable factors. O'Kelly [11] is one of the researchers who proposed a model for hub facility location based on the optimisation method.

4 Persistent Homology as a tool for facility location

This paper reports a new formulation for facility location by developing algorithms inspired from persistent homology. We treat the problem given in [17] through our method which can be applied in different contexts.

The starting point will be a given network consisting of n districts annotated by (A_1, A_2, A_3,..., A_n) (nodes) in an area of a city with information about distances $d_{i,j}$ between districts A_i and A_j , and numbers of population n_i in each node A_i. A chain of supermarkets decided to open a facility within the districts and they need to examine the most optimum location to serve the whole area.

4.1 Method

The proposed method will follow a procedure which will be explained as follows:

First, it is necessary to fix some nodes as candidates where the facility can be located. The choices are based on transport design criteria which are adopted by planning experts. Then, for a fixed candidate A_i, ponderate each other node A_j by the following coefficient which reflect the "importance" of the node A_j with respect to A_i (see below for the explanation):

$$X_j^i = \frac{1}{n_i}\frac{d_{i,j}}{n_j} \quad (9)$$

In fact, fixing two candidates A_m and A_n , and

consider two nodes A_e and A_f . We discuss the following particular situations:

- If we suppose that our network consists only of three nodes A_m , A_n and A_e. Then, A_m is the optimal solution if $X_e^m \leq X_e^n$. Assume, for example, $d_{e,m}$=$d_{e,n}$, the inequality $\frac{1}{n_m} \leq \frac{1}{n_n}$ can be explained by the phrase : the bigger is n_m, the more attractive is A_m. On the other hand, if n_m= n_n, $d_{e,m} \leq d_{e,n}$ explain the chose of A_m as the optimal solution.
- Now suppose that our network consists only of four nodes A_m , A_n , A_e and A_f. Then, A_m will be the optimal solution if:

$$\frac{X_e^m + X_f^m}{2} \leq \frac{X_e^n + X_f^n}{2}$$

For example, if n_m= n_n and n_e= n_f , the means above can be reduced to the means of distances. So clearly the optimal solution corresponds to the smallest mean. Also, the ratios $\frac{1}{n_e}$ and $\frac{1}{n_f}$ are there since the number of clients has to affect the number in some sense.

Basing on the above observations we construct a complex $K_{m,r}$ associated to a candidate A_m and for a fixed strictly positive real r as follows (we focus only 0-simplices and 1-simplices):

- The 0-simplices are all the points A_j which are different from A_m.

- Two nodes A_i and A_j is related by an edge (i.e., {A_i, A_j} is a 1-simplex) if $\frac{X_i^m + X_j^m}{2} \leq r$.

Clearly, for $r < r'$, $K_{m,r} \subseteq K_{m,r'}$. This means that we can construct a filtration $K_{m,r1} \subseteq \cdots \subseteq K_{m,rs}$, where r1,..., rs is an increasing sequence of strictly positive real numbers. We choose such a sequence in the way that at the last real all complexes (for all candidates) have one connected component (i.e., all nodes are connected). In fact, passing from a complex to the next one in a filtration we obtain more edges (the nodes will be "more" connected) and then less connected component (the provided sequence of 0-Betti numbers decreases). Then, we compare between the sequences of 0-Betti numbers. The one which reaches 1 at the earliest time is the most attractive between all the other ones and then can be adopted as a solution following this approach. This can be seen easily when comparing the provided barcodes.

4.2 Algorithms

There are softwares which deal with the computation of persistent homology in various context. In our context we made our own programs and so all calculation is carried out using the C++ programming by developing the following algorithms. Indeed, to determine the 0-Betti number of a complex associated to one of the candidates, we follow a procedure of three steps as follows:

Fixe a candidate A_i and a real number r.

We need at first an algorithm to calculate all X_j^i.

Algorithm 1

For j=1 , j≠i to n do
$X_j^i = \frac{1}{n_i} \frac{dij}{n_j}$;
End for

Then, we construct the simplicial complex $K_{i,r}$ associated to r using the following algorithm.

Algorithm 2

$\mathscr{C}$= {{A_j}/ j ∈ {1,..., n} and j ≠ i}
For j=1, j ≠ i to n do
For j'=1, j' ≠ i to n do
If ($\frac{1}{C_{jj'}^i}$ < r) $\mathscr{C} = \mathscr{C} \cup$ {{A_j, $A_{j'}$}};
where $C_{jj'}^i = \frac{X_j^i + X_{j'}^i}{2}$
End for
End for

Finally, we determine and calculate the number of connected components of $K_{i,r}$ using the algorithm beneath.

Algorithm 3

Initialize a matrix $M=(m_{ij})_{1\leq j\leq n, 1\leq i\leq n}$ and $m_{11}=A_1$;
i =0;
While (n≠0)
j=0 , k_1=0 ;
While ($m_{ik_1} \neq \emptyset$)
k_2=0, A=m_{ik_1} ;
While (k_2≠n , and $A_{k_2} \notin$ M)
if ({A_{k_2},A} is 1-simplex then
j++ ;
$m_{ij} = A_{k_2}$;
end if

12. k_2++ ;
13. end While
14. k_1++ ;
15. end While
16. s = the number of boxes filled in the ith line of the Matrix M;
17. n = n-s ;
18. End While
19. β_0 = The number of lines containing at least one point;
20. Return β_0 ;

4.2 Example

Suppose that we have a network consisting of 8 districts annotated by (A_1, A_2, A_3,..., A_8) (eight nodes; see [17]):

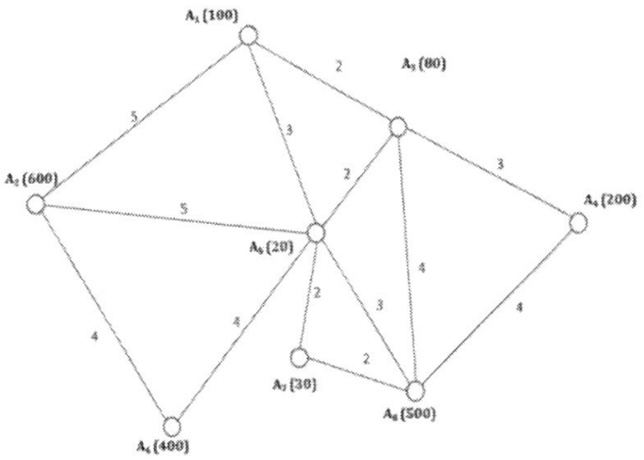

Figure 6: Network of 8 districts

We Fix three candidates A_2 , A_5 and A_8 . Then, we construct the associated barcode for each candidate. Below, the obtained complexes in a filtration associated to A_2. We used the following sequence of real numbers: r_1=1/5; r_2=1/4; r_3=2/7; r_4=1.

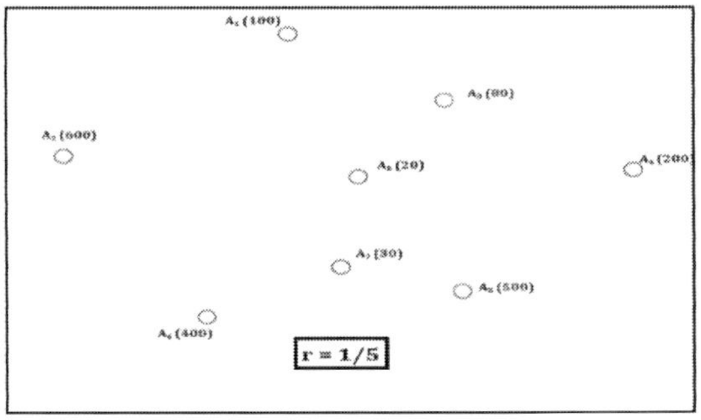

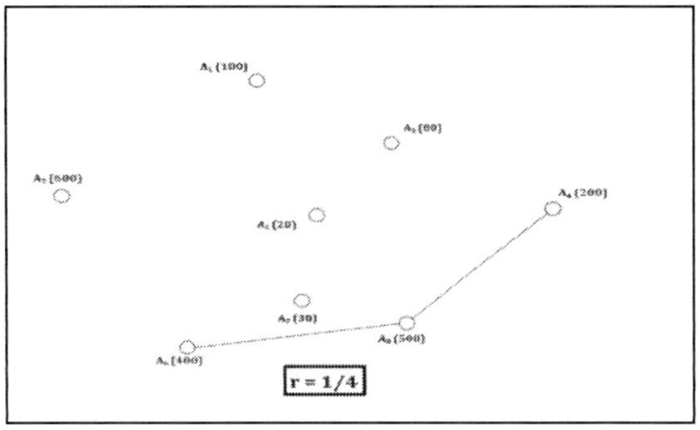

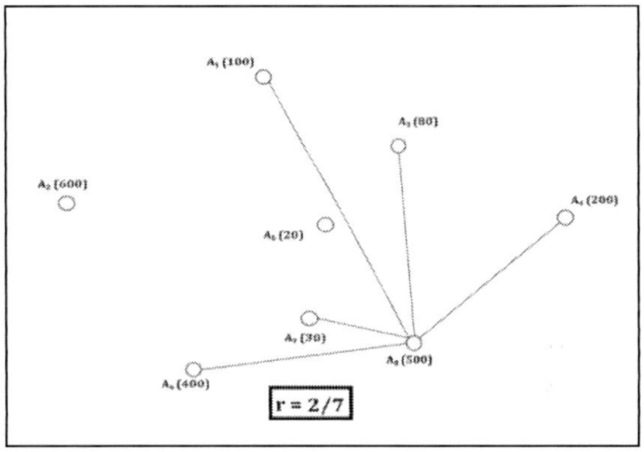

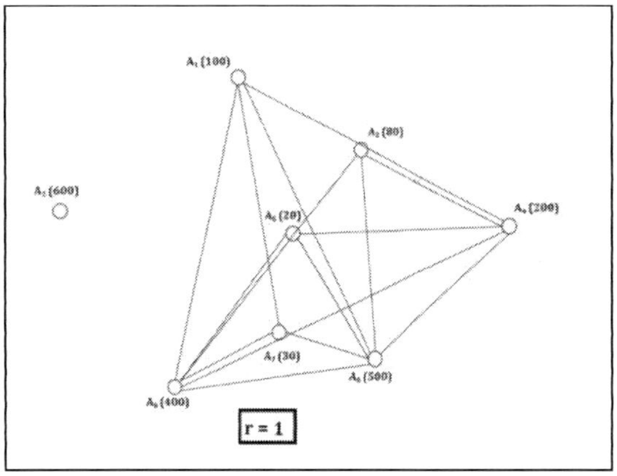

Then we obtain the following barcodes:

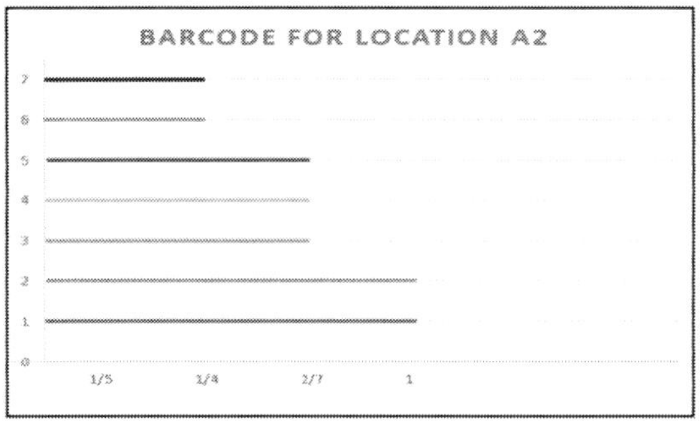

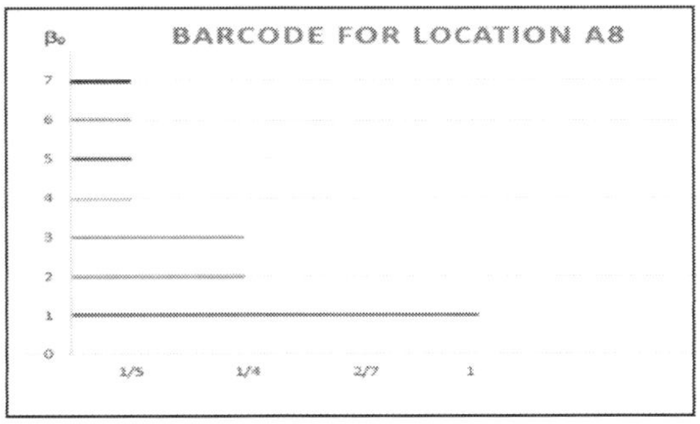

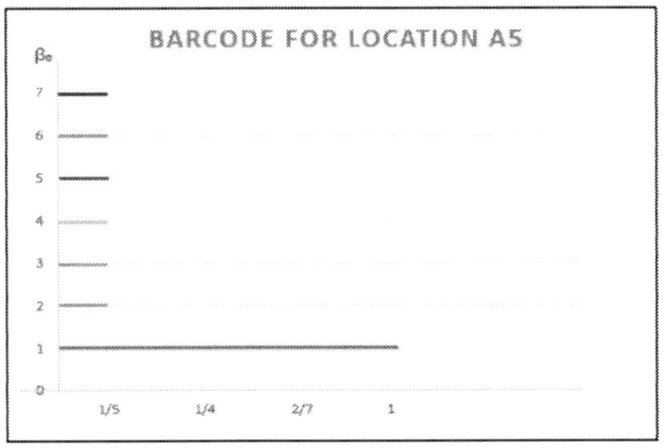

Finally, one can observe clearly that the A_5 is the desired solution.

5 Conclusion

We have applied the proposed method to an example of eight districts; which has provided a successful location decision. Thus, this article is the first significant step made to solve location problems based on the concept of persistent homology. It will certainly attract the attention of several researchers. We expect further research to develop the solution for more complex situations with multiple parameters and for different types of location problems.

References

1. P. Bendich, J. S. Marron, E.Miller, A. Pieloch and S. Skwerer. "Persistent homology analysis of brain artery trees." Ann. Appl. Stat. **10** (2016): 198-218.
2. C. Cocking, *Solutions to facility location–network design problems*; Doctoral thesis at University of Heidelberg (2008).
3. Z. Drezner, G.O. Wesolowsky. "Network design: selection and design of links and facility location." Transp. Res. **37** (2003): 241–256.
4. H. Edelsbrunner, J. Harer, *Computational Topology - An introduction*; American Mathematical Society **26** (2009).
5. H. Edelsbrunner, J. Harer. "Persistent homology - a survey." *Surveys on Discrete and Computational Geometry. Twenty Years Later*, Contemporary Mathematics **453** (2008): 257–282.
6. H. Edelsbrunner, D. Letscher, A. Zomorodian. "Topological persistence and simplification." Discrete Comput. Geom. **28** (2002): 511-533.
7. R. Ghrist. "Barcodes: the persistent topology of data." Bull. Amer. Math. Soc. (N.S.), **45** (2008): 61–75.
8. A.A. Kuehn, M.J. Hamburger. "A heuristic program for locating warehouses." Manag. Sci. **9** (1963).
9. T.L. Magnanti, R.T. Wong. "Network design and transportation planning: models and algorithms." Transport. Sci. **18** (1984) : 643–666.
10. M.T. Melo, S. Nickel, F. Saldanha-da-Gama. "Facility location and supply chain management – A review." Eur. J. Oper. Res. **196** (2009) 92-106.
11. M.E. O'Kelly. "The location of interacting hub lacilitics." Transport Sci. **20** (1986): 92-106.
12. S.H. Owen, M.S. Daskin. "Strategic facility location: a review." Eur. J. Oper. Res. **111** (1998): 423–447.
13. J.A. Perea, J. Harer. "Sliding windows and persistence: An application of topological methods to signal analysis." Found. Comput. Math. (2013): 1–40.
14. R. Rahmaniani, A. Ghaderi. "A combined facility location and network design problem with multi-type of capacitated links." Appl Math. Model. **37** (2013): 6400–6414.
15. C. S. ReVelle, A. H. Eiselt. "Location analysis: A synthesis and survey." Eur. J. Oper. Res. **165** (2005): 1–19.
16. G. Singh, F. Memoli, G. Carlsson. "Topological methods for the analysis of high dimensional data sets and 3D object recognition." Eurographics Symposium on PointBased Graphics, eds Botsch M, Pajarola R (Eurographics Association, Geneva), (2007) 91-100.
17. J. P. Škrinjar, N.Brnjac, D. Žgaljić. "Location problems in transport network." J. Mar. Res. **28** (2014): 94-98.
18. *C. J. Friedrich, Alfred Weber's Theory of the Location of Industries*. University of Chicago Press (1929).
19. K. Xia, Z. Li, L. Mu. « Multiscale persistent functions for biomolecular structure characterization." Perprint, available at arXiv:1612.08311 (2016).
20. A.J. Zomorodian, *Topology for computing*, Cambridge University Press (2005).
21. A.J. Zomorodian, G. Carlsson. "Computing Persistent Homology." Discrete Comput. Geom. **33** (2005): 249–274.

Support Vector Machines for Improving Vehicle Localization in Urban Canyons

Guermah Bassma[1,*], *El Ghazi* Hassan[1,**], and *Sadiki* Tayeb[2,***]

[1]STRS Laboratory, National Institute of Posts and Telecommunication, Rabat, Morocco
[2]Université Internationale de Rabat (UIR), Faculty of Computing and Logistics, TICLab, Sala El Jadida, Morocco

Abstract. Since the middle ages, the need to identify the vehicles position in their local environment has always been a necessity and a challenge. Today, GNSS-based positioning systems have penetrated several field, such as land transport, emergency systems and civil aviation requiring high positioning accuracy. However, the performances of GNSS-based systems can be degraded in harsh environment due to non-line-of-sight (NLOS), Multipath and masking effects. In this paper, for improving vehicle localization in urban canyons, we address a very challenging problem related to GNSS signal reception state detection (LOS, NLOS or Multipath). A SVM-based system for GNSS Multipath detection using the fusion of information provided by two GNSS antennas is proposed. In this work, we aim to explore the potential of machine learning, and more precisely, Support Vector Machines (SVM) to identify GNSS signals reception state. The SVM-based system developed in this work has used the C/N0 of signals provided by RHCP and LHCP antennas, and satellite elevation as classification criteria. The training data set is constructed by several experimental studies done in real environments, Calais, France . Furthermore, four SVM kernel functions are tested, namely, Linear, Gaussian, Cubic and Quadratic.
A GNSS signal reception state detection by applying the proposed SVM-based classifier is demonstrated on real GPS signals, and the efficiency of the system is shown. We obtain empirically an accuracy of signal detection about 93%.

1 Introduction

In harsh environment, the performances of GNSS-based positioning systems can be strongly affected, due to Multipath effects. For improving navigation performance, several solutions are proposed in the literature to filter, mitigate, eliminate, detect and modeling Multipath effects [1–8].
In this paper, the motivating problem is to detect the GNSS signals reception state. In [9] the authors are used the C/N0 ratio to detect the Multipath signals. However, the performance of C/N0 ratio used as indicator for Multipath detection still limited and imprecise. Other investigations have shown that fuzzy logic approach can be applied to classify GPS data based on signal degradation conditions [10–13]. In [10], a GPS-code-based measurements fuzzy processing is applied to detect the optimal observables, using the GDOP (Geometrical Dilution Of Precision) parameter and the signal-to-noise ratio (SNR). Others works have applied fuzzy logic to detect the signal quality using C/N0 and the dilution-of-precision (DOP) [12–14].
In [15], the fuzzy logic approach is applied to classify the data quality using the DOP value and the difference between the measured and excepted C/N0 as input fuzzy variables.

*e-mail: bassmaguermah@gmail.com
**e-mail: elghazi@inpt.ac.ma
***e-mail: tayeb.sadiki@uir.ac.ma

In other study [16], the authors have used the data fusion provided by dual-polarisation antenna for NLOS and Multipath detection basing on C/N0. Other approach is proposed by Monask Socharoentum in [17], it's based on Nav2Nav (Navigation-to-Navigation) technique in which pseudo-range corrections and machine learning work together to detect GNSS Multipath signals.
In this study, we have exploited the potential of the RHCP and LHCP to process the GNSS signal. we have proposed a new classifier of GNSS signals, based on a very robust technique of Machine learning called SVM [18][19]. The main idea of this system is to train the data set of real observations constructed from experimental studies to learn a model able to be used for predicting the GNSS signal reception state in real scenario.
A signal reception state detection by applying the proposed classifier system is demonstrated on real GPS signals, and the efficiency of the classifier is shown.

2 Background

2.1 Support Vector Machines (SVM)

In machine learning, support vector machine, also called support vector network, is a supervised learning technique that performs classification by finding the hyperplane that maximizes the margin between the two classes [20].
The main steps of SVM algorithm is presented as follows (Algorithm (1)):

Algorithm 1 SVM Algorithm

- Define an optimal hyperplane: **maximize margin**.
- Extend the above definition for non-linearly separable problems: **have a penalty term for misclassifications**.
- Map data to high dimensional space where it is easier to classify with linear decision surfaces: **reformulate problem so that data is mapped implicitly to this space**.

2.1.1 Linear SVM

To classify our data set, several hyperplane are present. SVM works to select the separating hyperplane that maximizes the margin (See figure 1).

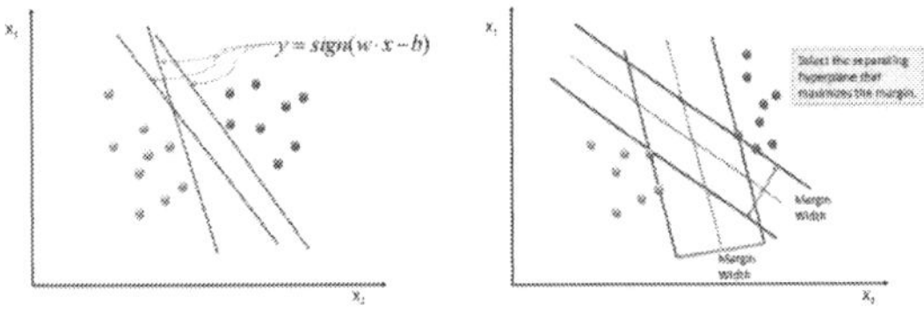

Figure 1. SVM Training.

The vectors that define the hyperplane are called the support vectors. Furthermore, to define an optimal hyperplane we need to maximize the width of the margin (w), which can be presented as :

$$max \frac{2}{\|w\|} \tag{1}$$

Where :

$$(w\ x + b) \geqslant 1,\ \forall\ x\ of\ class\ 1$$
$$(w\ x + b) \leqslant -1,\ \forall\ x\ of\ class\ 2$$

The optimization problem can be defined by the following system :

$$\begin{cases} min \frac{1}{2}\|w\|^2 \\ y_i(w\ x_i + b) \geqslant 1,\ \forall\ x_i \end{cases} \tag{2}$$

We find w and b by solving the system 2 using Quadratic Programming [21]. If the data is linearly separable, there is a unique global minimum value. However, if the data is not linearly separable, the system becomes non-solvable.
Furthermore, the SVM technique consists to search a hyperplane that completely separates the vectors into two non-overlapping classes. However, some data can be not perfectly separable. In this case, SVM aims to find the hyperplane that maximizes the margin and minimizes the misclassifications.
The SVM algorithm aims to maintain the slack variable ξ_i to zero while maximizing margin. We note that the SVM algorithm does not minimize the number of misclassifications but the sum of distances from the margin hyperplanes.
In this case, the optimization problem (system 2) can be reformulated as follows :

$$\begin{cases} min \frac{1}{2}\|w\|^2 + C\sum_i \xi_i \\ y_i(w\ x_i + b) \geqslant 1 - \xi_i,\ \forall x_i \\ \xi_i \geqslant 0 \end{cases} \tag{3}$$

Where C trades-off margin width and misclassifications.

2.1.2 Non-Linear SVM

To separate two groups, the simplest way is with a straight line (1 dimension), flat plane (2 dimensions) or an N-dimensional hyperplane. However, in some case, a nonlinear region can separate the data more efficiently. So, where a linear hyperplane cannot be used to separate the two groups, the SVM uses two solutions, namely, "kernel function" or "Mapping Data to a High-Dimensional Space", to make data possible to perform the linear separation.

Mapping Data to a High-Dimensional Space :

To map data into new space, we must find function $\phi(x)$. The SVM formulation becomes :

$$\begin{cases} min \frac{1}{2}\|w\|^2 + C\sum_i \xi_i \\ y_i(w\ \phi(x) + b) \geqslant 1 - \xi_i,\ \forall x_i \\ \xi_i \geqslant 0 \end{cases} \tag{4}$$

Figure 2 illustrates the linear classifiers in high-dimensional spaces.

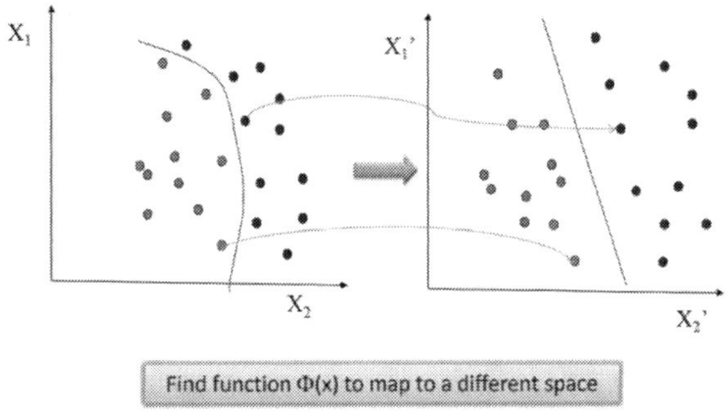

Figure 2. Linear Classifiers in High-Dimensional Spaces.

The data appear as $\phi(x)$ and weights w are now weights in the new space. Furthermore, If $\phi(x)$ is very high dimensional, the explicit mapping is expensive.

The kernel trick solution to simplify calculations:

As long as mapping data to a high-dimensional space is so expensive, as an alternative solution, it come the kernel trick solution that uses the kernel function, defined as follows :

$$K(x_i, x_j) = \phi(x_i)^T.\phi(x_j) \tag{5}$$

With kernel trick solution, the computation of the separating hyperplane don't requires the knowledge of ϕ, but only of K.
Thus, it is possible to use any kernel function to redescribe in a larger dimension space. a presentation of the used kernel functions in this paper has been given in following section. We note that each kernel function has its advantages and disadvantages. It is therefore up to the user to choose the kernel, and the parameters of this kernel, which corresponds to his problem.
To summarize, the four great ideas behind the kernel trick solution are :

- The data described in the entry space E are projected in a redescription space E '.
- Linear regularities are sought in this space E '.
- The Search algorithms don't need to know the coordinates of the projections of the data in E ', but only their scalar products.
- These scalar products can be efficiently calculated through the use of kernel functions.

2.2 Proposed SVM-based Classifier Description

In this paper, we propose a new GNSS LOS/Multipath signal classifier based on the SVM and the fusion of information provided by the RHCP and LHCP antennas.
We present two signal reception state, namely, LOS and Multipath reception. The LOS signal transmitted by satellites is right-hand circular polarized (RHCP)[22]. When the signal is reflected, with an odd reflection number, the received signal become left-hand circular polarized (LHCP) [23].
The schematization of the proposed SVM-based classifier system is given in figure 3.

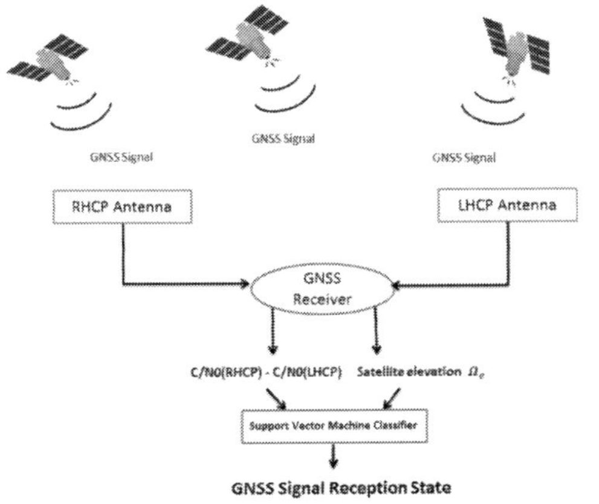

Figure 3. Global schematic set-up of the proposed SVM classifier system.

The SVM-based classifier has two inputs, namely, C/N0-R-L and satellite elevation, provided by RHCP and LHCP antennas, and one output is the "SRS" which indicate the signal reception state (Multipath or LOS signal).

3 GNSS LOS/Multipath Signal Classification

Our proposed SVM classifier is a system based on the SVM defined as a supervised technique of machine learning. At the heart of a SVM-based system exists a Training data set. The proposed SVM classifier aims to learn a model from the training data set, where the resulted model will be used to detect the GNSS signal reception state.
To construct our training data set, we have defined a decision rules and the boundary values that characterizes each class (LOS or Multipath) from several experimental studies done in real environments. The decision rules are :

- If the direct signal is blocked and only reflected signals are being received, then the signal received by the RHCP antenna will be multiplied by an attenuation coefficients, which their value are less than 1.
- If the LHCP antenna can not track the incoming signal, the received signal can be considered as a direct signal.
- If the RHCP antenna can not track the incoming signal, the signal can be considered as a reflected signal.
- The C/N0-R-L based thresholds for Multipath signals are inspired from Ziyi Jiang and Paul D. Groves study [16].
- When the signal is at high elevation, the received signal will be Weakly attenuate.
- N reflections has been considered, where N is an odd number.
- The delay dependent carrier phase between the RHCP and LHCP signal is null.

3.1 SVM Classifier Design

The architecture of the proposed SVM-based classifier system for GNSS Multipath detection is shown in figure 4. The main inputs of our SVM-based classifier is the training data set. Therefore, to construct our data set, we have used the results of the experimental studies and the decision rules. Consequently, a data set about 100 observations (LOS and Multipath) is constructed (See figure 5).

3.2 Hardware and Data Acquisition

To test the efficiency of the proposed SVM-based classifier, real GPS signals data are used.
The experimental data are acquired in real environment (See figure 6), Calais, France, in collaboration with LISIC Laboratory at the University of the Littoral Opal Coast (ULCO).

3.2.1 Training Step

The Training step aims to learn a model from the input data set. In this work, four kernel functions are used : Linear, Gaussian, Quadratic and Cubic.

Model 1 : Linear SVM

The Linear kernel function is the simplest kernel function. It is given by the inner product $< x_i, x_j >$ plus an optional constant c.
Therefore, the Kernel algorithms using a linear kernel are often equivalent to their non-kernel counterparts.
The linear kernel function can be written as follows :

$$k(x_i, x_j) = x_i^T \; x_j + c, \; (c \; is \; a \; constant) \qquad (6)$$

By applying our training data set, the resulting trained model is presented as follows :

 https://doi.org/10.1051/matecconf/201820000004

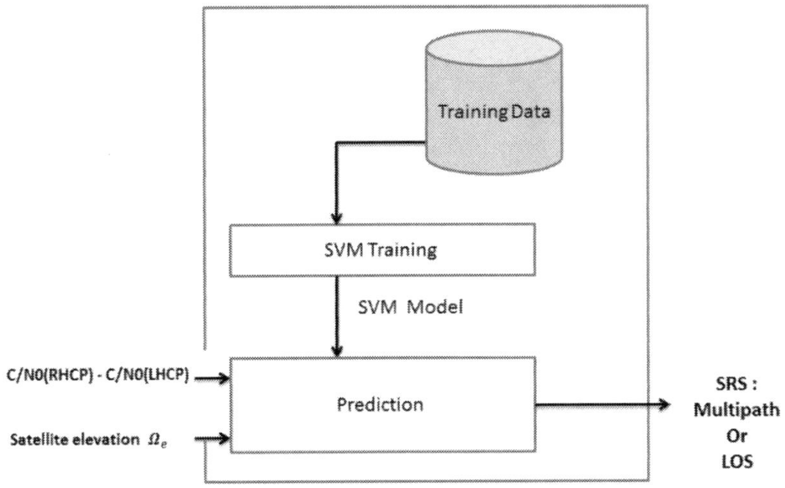

Figure 4. Conceptual structure of our SVM classifier system

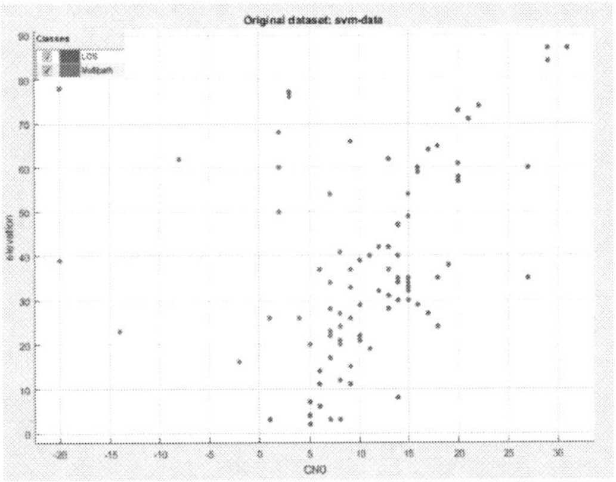

Figure 5. Training Data Set.

Figure 6. The data collection point at the park of the University Littoral Opal Coast, Calais, France

Resulting Trained model	
Accuracy	85%
Prediction speed	8000 obs/sec
Training time	0.31936 sec
Classifier characteristics	
Preset	Linear SVM
Kernel function	Linear
Kernel scale	automatic
Box constraint level	1
Multiclass method	One-vs-One
Standardize data	true

Table 1. SVM using Linear kernel.

Furthermore, the model confusion matrix of the resulting SVM model of linear kernel function is presented in figure 7.

From figure 7, we show that we have 70 LOS observations, 68 of them are correctly learned and 2 are incorrectly learned (considered as Multipath). for the Multipath observations, we have 30, 13 of them are incorrectly learned and 17 are correct.

From the obtained results, the error rate for signals classification using Linear SVM model is 15%, hence the necessity to test another kernel function.

Model 2 : Gaussian SVM

The Gaussian kernel function of SVM is an example of radial basis function kernel. It can be written as :

$$k(x_i, x_j) = exp(-\frac{\|x_i - x_j\|^2}{2\sigma^2}) \tag{7}$$

Alternatively, it could also be expressed by :

$$k(x_i, x_j) = exp(-\sigma\|x_i - x_j\|^2) \tag{8}$$

Where σ must be carefully tuned, because its plays a major role in the performance of the kernel.

So, by applying our training data set, the resulting trained model is presented as follows :

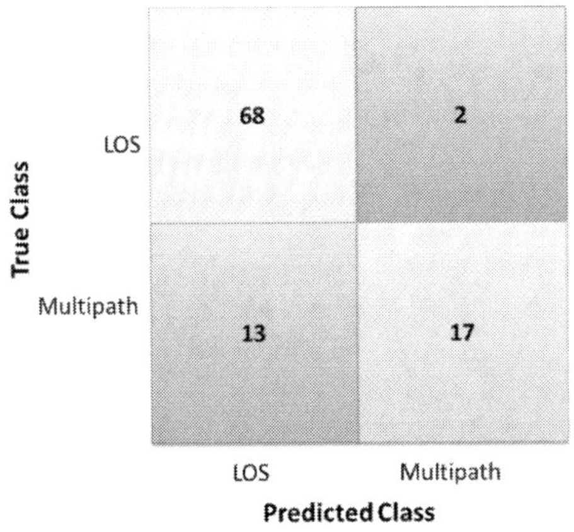

Figure 7. Linear SVM model confusion Matrix.

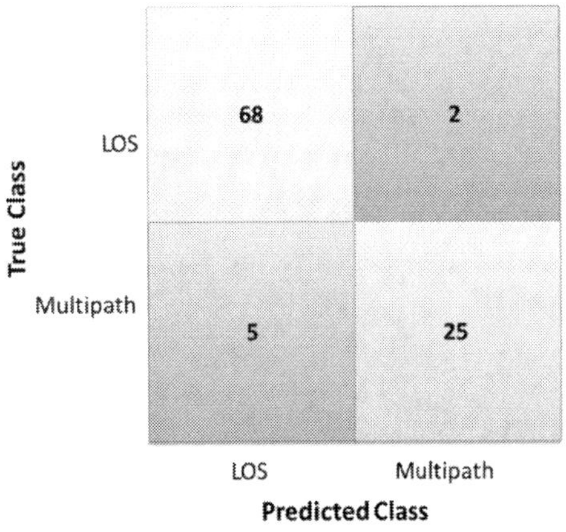

Figure 8. Gaussian SVM model confusion Matrix.

Resulting Trained model	
Accuracy	93%
Prediction speed	7900 obs/sec
Training time	0.13598 sec
Classifier characteristics	
Preset	Medium Gaussian SVM
Kernel function	Gaussian
Kernel scale	1.4
Box constraint level	1
Multiclass method	One-vs-One
Standardize data	true

Table 2. SVM using Gaussian kernel.

Resulting Trained model	
Accuracy	89%
Prediction speed	8000 obs/sec
Training time	0.183 sec
Classifier characteristics	
Preset	Cubic SVM
Kernel function	Cubic
Kernel scale	Automatic
Box constraint level	1
Multiclass method	One-vs-One
Standardize data	true

Table 3. SVM using Cubic kernel.

The model confusion matrix of the resulting SVM model of Gaussian kernel function is presented in figure 8.We show that we have 70 LOS observations, 68 of them are correctly learned and 2 are incorrectly learned (considered as Multipath). For the Multipath observations, we have 30, 25 of them are correctly learned and 5 are incorrect.

From the obtained results, the error rate for signals classification using Gaussian SVM model is 7%. We note that Gaussian kernel gives best results compared to Linear model. Furthermore, 7% of error is a no-negligible rate, hence the necessity to test another kernel function.

Model 3 : Cubic SVM

The Cubic polynomial kernel can be presented as :

$$k(x_i, x_j) = (x_i^T\ x_j + 1)^3 \qquad (9)$$

So, by applying our training data set, the resulting trained model is presented as follows :

The model confusion matrix of the resulting SVM model of Cubic kernel function is presented in figure 9. We show that we have 70 LOS observations, 64 of them are correctly trained and 6 are incorrectly learned (considered as Multipath).

For the Multipath observations, we have 30, 5 of them are incorrectly trained and 25 are correctly trained.

From the obtained results, the error rate for signals classification using SVM Cubic Kernel is 11%, hence the necessity to test another kernel function.

Model 4 : Quadratic SVM

The Quadratic kernel function is less computationally intensive than the Gaussian kernel function and can be used as an alternative solution when using the Gaussian kernel becomes too expensive. The Quadratic function can be written as follows:

$$k(x_i, x_j) = 1 - \frac{\|x_i - x_j\|^2}{\|x_i - x_j\|^2 + c} \qquad (10)$$

By applying our training data set, the resulting trained model is presented as follows :

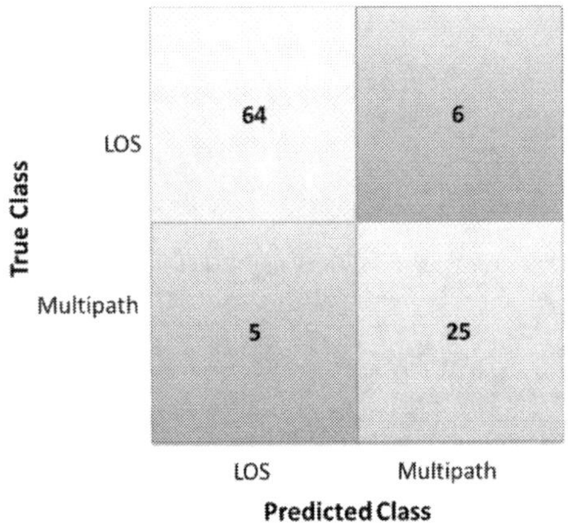

Figure 9. Cubic SVM model confusion Matrix.

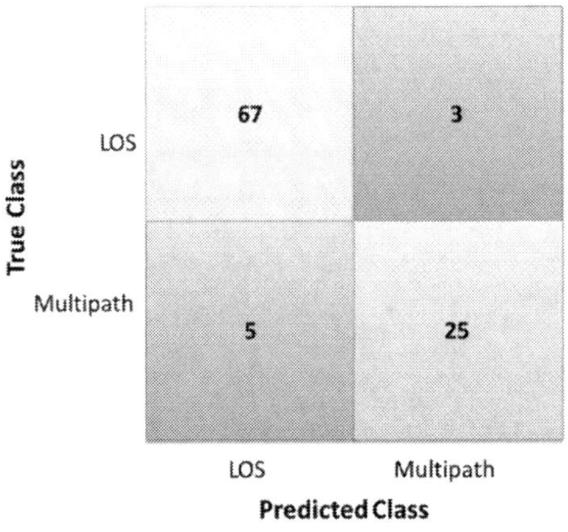

Figure 10. Quadratic SVM model confusion Matrix.

Resulting Trained model	
Accuracy	92%
Prediction speed	7400 obs/sec
Training time	0.14757 sec
Classifier characteristics	
Preset	Quadratic SVM
Kernel function	Rational Quadratic
Kernel scale	Automatic
Box constraint level	1
Multiclass method	One-vs-One
Standardize data	true

Table 4. SVM using rational Quadratic kernel.

Furthermore, the model confusion matrix of the resulting SVM model of Quadratic kernel function is presented in figure 10.

From figure 10, we show that we have 70 LOS observations, 67 of them are correctly learned and 3 are incorrectly trained. For the Multipath observations, we have 30, 25 of them are correctly learned and 5 are incorrect.
From the obtained results, the error rate for GNSS signals classification using SVM Quadratic Kernel is 8%. Consequently, we can conclude that the Quadratic kernel function has given best classification accuracy compared to Cubic and Linear kernel.

SVM Trained model Discussion

From the resulting SVM trained model using the Gaussian, Cubic, Quadratic and Linear kernels, we have shown that the Gaussian kernel has given the best signal classification accuracy.
Therefore, for the LOS observations, the trained models using Linear and Gaussian kernels have given the same results (68 observations are correctly trained from 70 LOS observations). Furthermore, the Gaussian, Cubic and Quadratic kernels have given the same results of Multipath observations (25 Multipath observations are correctly trained from 30 Multipath observations).

3.2.2 Prediction Step

In this step, we used a new data set composed by new observations(each line is a couple of C/N0-R-L and satellite elevation) and we apply the trained models resulting in the "Training step".
The table 5 shows the results of the prediction step by applying the proposed SVM-based classifier system. From table 5, we can confirm that the SVM-based model using Gaussian kernel function has given the best results over Cubic, Quadratic and Linear kernels. Furthermore, our objective is to classify the incoming GNSS signals from satellites. As a results, the efficiency of the proposed SVM-based classifier system to detect the signal reception state has been clearly demonstrated. We have obtained the same the expected results.

4 Conclusions

In this paper, we have proposed a new technique to classify GNSS signal reception state using the fusion of information provided by two antennas (RHCP and LHCP) and a machine learning technique.
The GNSS signals classification is characterized by two measures, satellite elevation and C/N0-R-L which is the difference between the C/N0 from RHCP and LHCP antennas. As long as the classification measures are uncertain and imprecise, a SVM-based approach has been applied. The SVM-based classifier uses a data set of two inputs (C/N0-R-L and satellite elevation). The trained data set is constructed by observations from real GPS signals. The classification Learner toolbox of Matlab has used to design our SVM-based classifier, with training and prediction steps.
Using real GPS signals, the efficiency of the SVM-based classifier for GNSS signals classification has been shown.

Inputs of SVM-based classifier system		SVM Prediction using different kernels			
C/N0-R-L	Satellite elevationΩ_e	Gaussian	Cubic	Quadratic	Linear
11.99	33.18	LOS	LOS	LOS	LOS
14.23	33.94	LOS	LOS	LOS	LOS
51	60.31	LOS	LOS	LOS	LOS
12.63	31.79	LOS	LOS	LOS	LOS
18.05	65.94	LOS	Multipath	Multipath	LOS
20.62	58.39	LOS	Multipath	LOS	LOS
18	35.88	LOS	LOS	LOS	LOS
17.29	87.68	LOS	Multipath	Multipath	LOS
13.44	37.82	Multipath	Multipath	Multipath	Multipath
18.49	70.25	LOS	LOS	LOS	Multipath
15.77	79.01	LOS	LOS	LOS	Multipath
6.69	17.31	Multipath	Multipath	Multipath	Multipath
7.27	34.22	Multipath	Multipath	Multipath	Multipath

Table 5. Signal reception state detection using the proposed SVM-based classifier system.

Acknowledgments

This work was financially supported by the National Center of Scientific and Technical Research (CNRST), Morocco.
We thank Mr Serge Reboul, Mr Georges Stienne and Mr Jean-Bernard Choquel from ULCO university (LISIC laboratory) for their collaboration in designing the proposed GNSS classifier based on RHCP and LHCP antennas, and their assistance in data acquisition.

References

[1] A. Rabaoui, N. Viandier, E. Duflos, J. Marais, P. Vanheeghe, *Dirichlet process mixtures for density estimation in dynamic nonlinear modeling: Application to GPS positioning in urban canyons* (IEEE, 2012), Vol. 60, pp. 1638–1655

[2] L. Wang, P.D. Groves, M.K. Ziebart, *Urban positioning on a smartphone: real-time shadow matching using GNSS and 3D city models* (The Institute of Navigation, 2013)

[3] L. Wang, P.D. Groves, M.K. Ziebart, *GNSS shadow matchin: Improving urban positioning accuracy using a 3D city model with optimized visibility scoring scheme* (Wiley Online Library, 2013), Vol. 60, pp. 195–207

[4] L. Wang, P.D. Groves, M.K. Ziebart, *Smartphone shadow matching for better cross-street gnss positioning in urban environments* (Cambridge Univ Press, 2015), Vol. 68, pp. 411–433

[5] Y. Gu, L.T. Hsu, S. Kamijo, *GNSS/On-Board Inertial Sensor Integration with the Aid of 3D Building Map for Lane-Level Vehicle Self-Localization in Urban Canyon* (IEEE, 2015)

[6] B. Guermah, T. Sadiki, H. El Ghazi, *Fuzzy logic approach for GNSS signal classification using RHCP and LHCP antennas*, in *Ubiquitous Computing, Electronics and Mobile Communication Conference (UEMCON), 2017 IEEE 8th Annual* (IEEE, 2017), pp. 203–208

[7] B. Guermah, T. Sadiki, H. El Ghazi, S. Reboul, E. Ahouzi, *A Map-Matching Based Approach to Compute and Modelize NLOS and Multipath Errors for GNSS Positioning in Hard Areas* (2017), Vol. 13, pp. 256–269

[8] A. Pirsiavash, A. Broumandan, G. Lachapelle, *Characterization of signal quality monitoring techniques for multipath detection in GNSS applications* (Multidisciplinary Digital Publishing Institute, 2017), Vol. 17, p. 1579

[9] G. Bassma, S. Tayeb et al., *GNSS Positioning Enhancement Based on NLOS Multipath Biases Estimation Using Gaussian Mixture Noise*, in *International Journal of Mobile Computing and Multimedia Communications (IJMCMC)* (IGI Global, 2018), Vol. 9, pp. 21–39

[10] N. Crocetto, S. Ponte, G. Puglianoc, L. Savino, *Fuzzy-Logic Based Methodologies for Mobile Mapping: Enhancing Positioning Accuracy of GPS/GNSS Measurements* (2007), Vol. 36

[11] C.J. Lin, Y.Y. Chen, F.R. Hang, *Fuzzy processing on GPS data to improve the position accuracy*, in *Fuzzy Systems Symposium, 1996. Soft Computing in Intelligent Systems and Information Processing., Proceedings of the 1996 Asian* (IEEE, 1996), pp. 557–562

[12] A. Ghalehnoe, K. Mohammadi, M. Mosavi, *Improve determining the location of a moving body on map by low cost GPS receiver*, in *Research and Development, 2002. SCOReD 2002. Student Conference on* (IEEE, 2002), pp. 242–244

[13] J. Wang, Y. Gao, *Evaluating the accuracy of GPS positions under severe signal-degradation using adaptive-network-based fuzzy inference systems (ANFIS)*, in *50th CASI Annu. General meeting and Conf., Canadian Aeronautics and Space Inst., Montréal, Quebec, Canada* (2003)

[14] C. Cheng, J.Y. Tourneret, Q. Pan, V. Calmettes, *Detecting, estimating and correcting multipath biases affecting GNSS signals using a marginalized likelihood ratio-based method* (Elsevier, 2016), Vol. 118, pp. 221–234

[15] J.H. Wang, Y. Gao, *Identification of GPS positioning solutions deteriorated by signal degradations using a fuzzy inference system* (Springer, 2004), Vol. 8, pp. 245–250

[16] Z. Jiang, P.D. Groves, *NLOS GPS signal detection using a dual-polarisation antenna* (Springer, 2014), Vol. 18, pp. 15–26

[17] M. Socharoentum, H.A. Karimi, Y. Deng, *A machine learning approach to detect non-line-of-sight GNSS signals in Nav2Nav* (2016)

[18] L.T. Hsu, *GNSS multipath detection using a machine learning approach*, in *Intelligent Transportation Systems (ITSC), 2017 IEEE 20th International Conference on* (IEEE, 2017), pp. 1–6

[19] Z.M. Miao, L.W. Zhao, W.W. Yuan, F.L. Jin, *Application of One-Class Classification in NLOS Identification of UWB Positioning*, in *Information System and Artificial Intelligence (ISAI), 2016 International Conference on* (IEEE, 2016), pp. 318–322

[20] S. Tong, D. Koller, *Support vector machine active learning with applications to text classification* (2001), Vol. 2, pp. 45–66

[21] M. Frank, P. Wolfe, *An algorithm for quadratic programming* (Wiley Online Library, 1956), Vol. 3, pp. 95–110

[22] K. Zhang, B. Li, X. Zhu, H. Chen, G. Sun, *NLOS Signal Detection Based on Single Orthogonal Dual-Polarized GNSS Antenna* (Hindawi Publishing Corporation, 2017), Vol. 2017

[23] M. Brenneman, J. Morton, F. van Grass, C. Yang, *Mitigation of GPS multipath using polarization and spatial diversities* (2008)

A Novel Hybrid Approach for Optimizing the Localization of Wireless Sensor Networks

Halima LAKHBAB[1,*]

[1]LIMSAD Laboratory, Faculty of Sciences Aïn Chock - Casablanca, Hassan II University, Morocco

Abstract. Wireless sensor networks are used for monitoring the environment and controlling the physical environment. Information gathered by the sensors is only useful if the positions of the sensors are known. One of the solutions for this problem is Global Positioning System (GPS). However, this approach is prohibitively costly; both in terms of hardware and power requirements. Localization is defined as finding the physical coordinates of a group of nodes. Localization is classified as an unconstrained optimization problem. In this work, we propose a new algorithm to tackle the problem of localization; the algorithm is based on a hybridization of Particle Swarm Optimization (PSO) and Simulated Annealing (SA). Simulation results are given to illustrate the robustness and efficiency of the presented algorithm.

1 Introduction

Research on ad hoc wireless sensor networks receives a growing interest over the last few years. Sensor networks usually consist of a large number of sensors which are densely deployed in a geographical area. Such networks facilitate monitoring and controlling of physical environments from remote locations with better accuracy.
Information gathered by the sensors is only useful if the positions of the sensors are known. One of the solutions for this problem is Global Positioning System (GPS). However, this approach is prohibitively costly, both in terms of hardware and power requirements. The Sensor Networks Localization Problem (SNLP) can be formally presented as follows: Assuming that the accurate positions of some nodes are known, how to use them and partial pair-wise distance measurements to estimate the positions of all sensor nodes in the sensor network? Particle swarm optimization algorithm is a population-based algorithm where several artificial particles search for good solutions. PSO was originally introduced by Kennedy and Eberhart in 1995 [1]. In this approach, each particle, of the population, called swarm, adjusts its trajectory toward its own previous best position, and toward the previous best position attained by any member of its topological neighborhood. Thus, the particles profit from the discoveries and previous experience of all other companions during the search for promising regions of the landscape. Simulated Annealing [2, 3] is a probabilistic technique for finding the global minimum of a cost function that may possess several local minima. SA it is commonly said to be the oldest among the metaheuristics and surely one of the first algorithms which had an explicit strategy to avoid local optima. SA exploits an analogy between the way in which a metal cools and freezes into a minimum energy crystalline structure (the annealing process) and the search for a minimum in a more general system; the SA algorithm employs a random search which not only accepts changes that decrease the objective function, but also some changes that increase it. The probability of accepting a trial point is decreased during the search.

Several techniques have been proposed over the last few years in the literature to approximate the sensor network localization problem and its variants . Global optimization techniques [5], popular metaheuristic methods, such as simulated annealing SA [4], evolutionary algorithm [6] or particle swarm optimization [7–9]. Semidefinite Programming relaxations is also proposed [10, 11]. The semidefinite programming approach is further extended to a gradient search technique in [12]. In[13], we have proposed a combination of PSO method and a gradient based method, to approach the solution of the SNLP.

In the present work, the best characteristics of the PSO and of SA are combined to develop a novel hybrid algorithm, for tackling the localization problem. In the following section mathematical modeling of SNLP is presented. Our hybrid approach is outlined in section 3. In section 4, numerical results are presented. The paper is concluded in section 5.

2 Mathematical Modeling of the SNLP

The sensor network localization problem with some distance information can be studied in the framework of graph theory. Given a graph $G(V, E)$ and a set of non-negative weights $\{d_{ij} : (i, j) \in E\}$ on its edges, the goal is to place the vertices of G in $\mathbb{R}^p$, $(p = 2, 3)$ such that the Euclidean distance between every pair of adjacent vertices (i, j) equals to the prescribed weight d_{ij}. The vertices of

*e-mail: halimalakhbab@yahoo.fr

G correspond to the sensors, the edges correspond to the communication links, and the weights correspond to the distances between sensors.

To solve this problem, we assume only a few nodes are equipped with GPS hardware. These nodes are called anchor nodes and they know their positions without communicating with other nodes.

A two-dimensional sensor network example, and its graph realization are presented in figures (1) and (2).

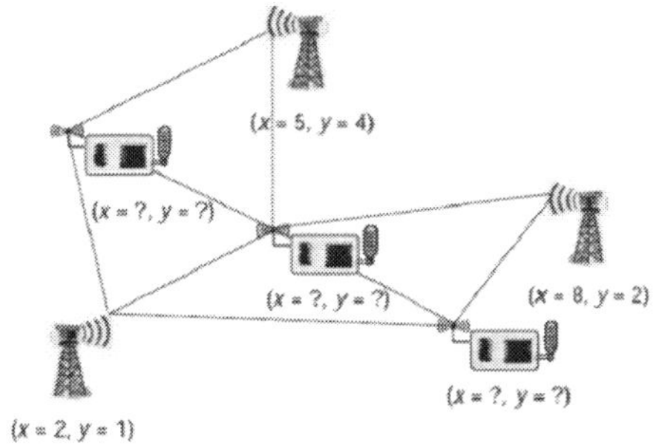

Figure 1. A two-dimensional sensor network example

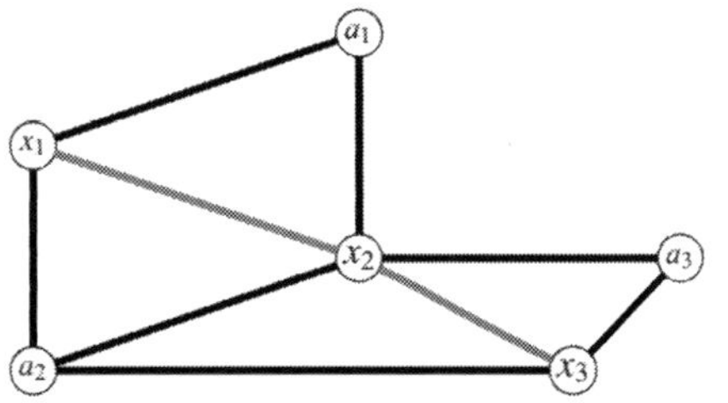

Figure 2. A graph (realization) corresponding to the sensor network in "Figure 1"

In the SNLP, we have $m \geq 0$ anchor points $a_1, ..., a_m \in \mathbb{R}^p$ whose locations are known, and $n > 0$ sensor points $x_1, ..., x_n \in \mathbb{R}^p$ whose locations we wish to determine. Furthermore, we are given the Euclidean distance values $\bar{d}_{kj}$ between a_k and x_j for some $(k, j) \in E$ and d_{ij} between x_i and x_j for some $(i, j) \in E$ with $i < j$. Specifically, let $N_a = \{(k, j) \in E \ : \ \bar{d}_{kj}$ is specified$\}$ and $N_x = \{(i, j) \in E \ : i < j, \ d_{ij}$ is specified$\}$.

The goal of SNLP is to find the coordinates of $x_1, ..., x_n \in \mathbb{R}^p$ such that:

$$\begin{aligned} \|a_k - x_j\| &= \bar{d}_{kj} \quad \forall (k, j) \in N_a \\ \|x_i - x_j\| &= d_{ij} \quad \forall (i, j) \in N_x \end{aligned} \tag{1}$$

The SNLP defined by (1) can be naturally formulated as a nonlinear global minimization problem, where the objective function is given by

$$f(x_1, ..., x_n) = \sum_{(i,j)\in N_x} (\|x_i - x_j\| - d_{ij})^2 + \sum_{(j,k)\in N_a} (\|x_j - a_k\| - \bar{d}_{jk})^2 \tag{2}$$

Here, all constraints are substituted into the objective function.

This function is everywhere infinitely differentiable and has a large number of local minimizers. Also, it is easy to see that 0 is the global minimum of f.

3 The HyPSOSA Method

3.1 Particle Swarm Optimization (PSO) method

PSO is a powerful and promising optimization method which has a wide range of applications in many areas, especially in Wireless Sensor Networks Localization Problem [7–9]. Like the other evolutionary computation techniques, PSO is a population-based search algorithm inspired by social behavior of bird flocking. PSO is initialized with a population (or a swarm) of random solutions, called particles where each particle has its own velocity. Particles spread in the multi dimensional search space, representing a possible solution, with velocities which are dynamically adjusted according to their historical behaviors. Therefore, the particles have a tendency to fly toward a better region of search space over the course of search process. The basic PSO algorithm consists of three steps, namely, generating particle's positions and velocities, velocity update, and finally, position update. A particle changes its position from one move (iteration) to another based on velocity updates. First, the positions, x_k^i, and velocities, v_k^i, of the initial swarm of particles are randomly generated using upper and lower bounds on the design variables values, in the second step, velocities and positions of all particles are updated to persuade them to achieve better objective or fitness values which are functions of the particles current positions in the design space at time k. The fitness function value of a particle determines which particle has the best global value in the current swarm, p_k^g, and also determines the best position of each particle over time, p_k^i , i.e. in current and all previous moves. The velocity update formula is given by

$$v_{k+1}^i = \omega v_k^i + c_1 r_1 (p_k^i - x_k^i) + c_2 r_2 (p_k^g - x_k^i) \tag{3}$$

Here, and c_1 and c_2 are constants, and r_1 and r_2 are random numbers uniformly distributed in [0, 1].
At the final step, PSO update the position by using (4).

$$x(k+1)^i = x_k^i + v_k^i \tag{4}$$

3.2 Simulated Annealing method

The simulated annealing (SA) is a stochastic strategy for selecting the minimal configuration of states of a system. It was first put forward by Metropolis and successfully applied to the optimization problems by Kirkpatrick [2, 3]. In the basic SA algorithm, a random value is considered to be the initial solution in the search space. Then the new solution value of the objective function is calculated and compared with the current solution. A move towards the new solution is made from the current solution if it provides a better result than the previous iteration. The probability of accepting a new solution is given as follows:

$$P = e^{(-\Delta f)/T} \tag{5}$$

Where Δf is the energy difference of the two states and T is a control parameter called temperature, which controls the magnitude of the perturbations of the objectif function f.
To avoid getting trapped at a local minimum point, the rate of reduction of T should be slow. Thus, at the start of SA random and impractical moves may be accepted, but in the end only improving ones are likely to be allowed, which can help the procedure not to converge at a local minimum. A stop criterion is used to determine the points to stop the temperature cooling and finish the optimization.

3.3 The proposed hybrid Approach

In the present method, the best characteristics of the PSO, that has strong global-search ability, are combined with the good local search characteristics of the SA, to develop a novel hybrid algorithm; the proposed algorithm is called **HyPSOSA**. Other applications of hybrid PSO and SA algorithm can also be found in [14–18].

In every iteration of PSO, a random number is generated and is compared with r_k

$$r_k = 1 - e^{-\gamma k}$$

Where γ is a constant and k is the current number of iteration.
If the random number is less than r_k, a local search strategy by SA, around the worst current particle in the swarm, is detonated. We note x_k^{SA}, the solution given by SA.
If $f(x_k^{SA}) < f(p_k^g)$, then we take $p_k^g = x_k^{SA}$.

4 Numerical Results

Simulation of the SNLP and its performance evaluation is done in Scilab 5.5.2.
PSO parameters are set as follows:

- Population size = 50
- Maximum number of iterations = 50
- Acceleration constants $c_1 = c_2 = 1.7$

The linearly decreasing method is adopted for the inertia factor, which is

$$w_k = w_{max} - \frac{(w_{max} - w_{min}) \times k}{Nmax}$$

where $Nmax$ is the maximum number of iterations, w_{max} and w_{min} are maximum and minimum weights, respectively, and k is the current iteration number.
We implement the SA algorithm with the following parameters: the initial temperature is set to 50 and the maximum number of iteration is 20. We set the constant γ to 0.05.
Figure (4) gives the node deployments in the square area, the anchor node rate is 10% from the 100 nodes. In this figure, we also illustrate the communications links between nodes[1].

The figure (5) shows the localizations given by HyPSOSA method for the previous example. The diamonds and the circles indicate the true positions of the anchors and the sensors, respectively.

Next, we consider three problems: S14A6, S27A3 and S90A10 that consist of 14 sensors and 6 anchors, 27 sensors and 3 anchors, 90 sensors and 10 anchors, respectively. In the table (1) we report the smallest value in the objectif function f_{best} reached by the algorithm for 10 runs and the average of minimums f_{mean}.

[1] $R = 0.3$: node communication radius

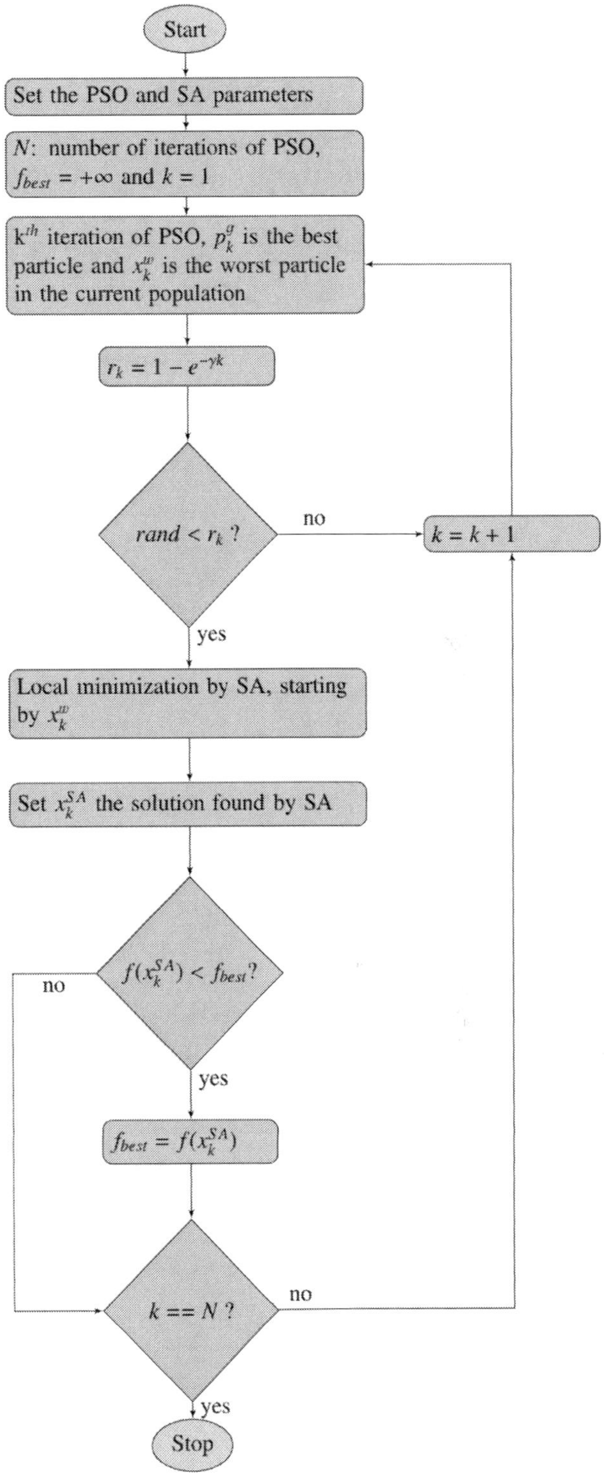

Figure 3. Flow chart of the HyPSOSA algorithm

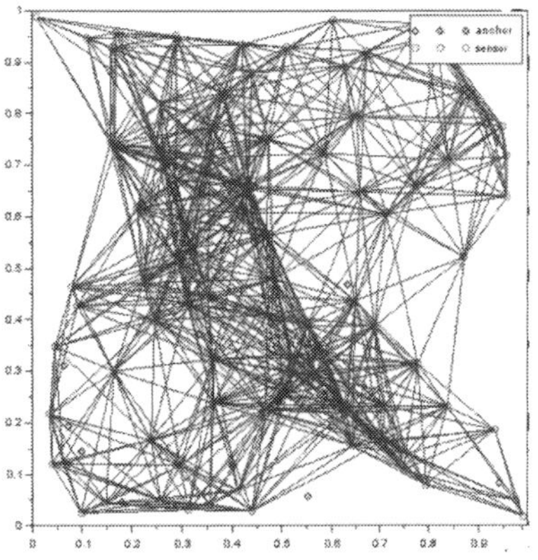

Figure 4. Neighbor relationship diagram

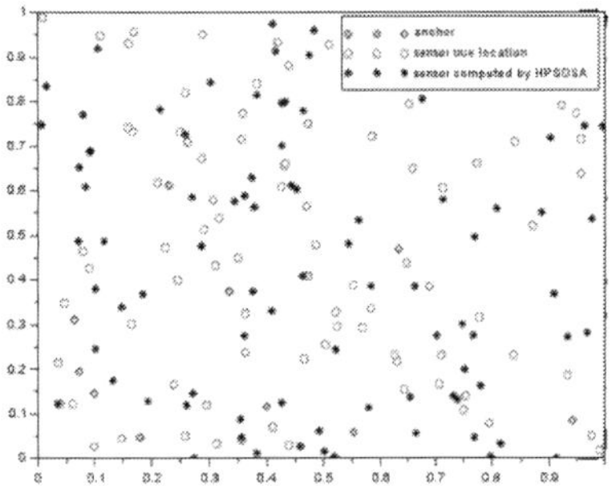

Figure 5. Example S90A10. True location and HyPSOSA localization

Table 1. Results of PSO and HyPSOSA for 3 examples

	f_{best}		f_{mean}	
	PSO	HyPSOSA	PSO	HyPSOSA
S14A6	0.0000017	$2,894.10^{-8}$	0.0003665	0.0000076
S27A3	$3,427.10^{-10}$	$4,387.10^{-11}$	0.0000017	$5,748.10^{-8}$
S90A10	$1,613.10^{-9}$	$4,035.10^{-10}$	0,0000287	0,0000011

Note that the average number of function evaluations by PSO is 2550 for the tree example, and by HyPSOSA is 3182, 2761 and 3196 for S14A6, S27A3 and S90A10 respectively.

We conclude that the minimums and the average of minimum given by HyPSOSA, for the tree examples, are better than those given by PSO.

5 Conclusion

In the literature many methods and systems are proposed and described to estimate the location of nodes in Sensor Network Localization Problem. In this paper we presented a new hybrid approach for tackling this problem by combining a Particles Swarm Optimization and Simulated Annealing. Experiment results show the efficiency of the proposed method.

Note that HyPSOSA can potentially be used in other minimization problem.

References

[1] J. Kennedy, R. Eberhart, Particle Swarm Optimization, Proc. IEEE Int. Conf. Neural Networks, 1942-1948 (1995)

[2] N. Metropolis, A.W. Rosenbluth, M.N. Rosenbluth, A.H. Teller, E. Teller, Equations of state calculations by fast computing machines, J. of Chemical Physics **21**(6), 1087-1091 (1953)

[3] S. Kirpatrick, C.D. Gelatt, M.P. Vecchi, Optimization by simulated annealing, Sc. **220**, 671-680 (1983)

[4] A. A. Kannan, G. Mao, B. Vucetic, Simulated annealing based localization in wireless sensor network, Proc. IEEE Conf. LCN'05, 513-514 (2005)

[5] J. Moré, Z. Wu, Global continuation for distance geometry problems, SIAM J. Opt. **7**, 814-836 (1997)

[6] G.F. Nan, M.Q. Li, J. Li, Estimation of node localization with a real-coded genetic algorithm in WSNs, Proc. Int. Conf. on Machine Learning and Cybernetics **2**, 873-878 (2007)

[7] R. C. Abreu, J. E. C. Arroyo, A Particle Swarm Optimization Algorithm for topology Control in Wireless Sensor Networks, Inter. Conf. Chilean (2011)

[8] J.J. Gnana Chandran, S. P. Victor, An Energy Efficient Localization Technique Using Particle Swarm Optimization in Mobile Wireless Sensor Networks, American J. of Sc. Res., 33-48 (2010)

[9] R. V. Kulkarni, G. K. Venayagamoorthy, A. Miller, C. H. Dagli, Network-centric Localization in MANETs based on Particle Swarm Optimization IEEE Swarm Intelligence Symp., 1-6(2008)

[10] J. Aspnes, W. Whiteley, Y. R. Yang, IEEE Trans. mobile computing, A theory of network localization, **5**(12), 1663-1678 (2006)

[11] P. Biswas, Y. Ye, Proc. 3rd Int. Symp. on Information Processing in Sensor Networks, 46-54 (2004)

[12] T.-C. Liang, T.-C. Wang, Y. Ye,A gradient search method to round the semidefinite programming relaxation solution for ad hoc wireless sensor network localization, Technical report, Stanford University (2004)

[13] H. Lakhbab, S. El Bernoussi, Int. J. of Math. Analysis **6**, (2012)

[14] F.Javidrad, M.Nazari, A new hybrid particle swarm and simulated annealing stochastic optimization method, Appl. Soft. Comp. **60**, 634-654 (2017)

[15] M. Basu, P. Deb, G. Garai, Hybrid of Particle Swarm Optimization and Simulated Annealing for Multidimensional Function Optimization, Inter. J. Information Technology **20**(1),34-45 (2014)

[16] M. Bahrepour, E. Mahdipour, R. Cheloi, M. Yaghoobi, Super-sapso: a new sa-based pso algorithm, Appl. of Soft Comp. **58**, 423-430 (2009)

[17] G. Yang, D. Chen, G. Zhou, A new hybrid algorithm of particle swarm optimization, Lecture Notes in Comp. Sc. **4115**, 50-60 (2006)

[18] N. Sadati, M. Zamani, H. Mahdavian, Hybrid particle swarm-based simulated annealing optimization techniques Proc. IEEE Industrial Electronics Conf. 644-648 (2006)

ABM-GIS simulation for urban freight distribution of perishable food

Hanane EL Raoui[1,2], Mustapha Oudani[2] and Ahmed EL Hilali Alaoui[1]

[1]Modeling and scientific computing laboratory, Sidi Mohamed Ben Abdellah University, Morocco
[2]Université Internationale de Rabat, FIL, TIC lab, Morocco

Abstract. Freight transport is essential to modern urban civilization. No urban area could exist without a powerful freight transport system. However, the distribution of perishable foods in urban areas is seen as a source of problems, due to traffic congestion, time pressures, and environmental impact. In this paper, an Agent-Based Model integrated with Geographic Information Systems (ABM-GIS) is designed for a time-dependent vehicle routing problem with time windows. This simulation model consists of determining the quickest routes to transport fresh products, estimating Vehicle kilometer traveled VKT and vehicle hour traveled VHT where speeds and travel times depend on the time of the day. Based on a case study, analyses of changes on traffic condition were conducted to get an insight into the impact of these changes on cost, service quality represented by the respect of time windows, and carbon emissions. The results reveal that traffic jams and restrictive time windows lead to additional cost, cause delays, and increase co2 emission. As for a short-term planning, time-dependent scheduling algorithm was proposed and assessed while extending time windows. Results have proved the potential saving in cost, travel time, and carbon emission.

1 Introduction

Freight transport is fundamental to modern urban civilization. No urban area could exist without an efficient freight transport system. Considering the demand for high-quality fresh food, transportation requirements for fresh food delivery have been continually increasing in urban areas [1].

The delivery of these goods is perceived as a source of problems. This is owing to specific characteristics of perishable foods, traffic congestion, the increasing requirement of customers in terms of delivery time, and environmental impact. Traffic growth presents a new challenge for carriers in vehicle routing and scheduling to deliver products. Besides, it brings environmental problems due to the increase in carbon emission. Therefore, establishing the fastest routes, optimal departure from the distribution center to deliver these time-sensitive products is a major problem encountered by carriers.

This study focused on a time-dependent vehicle routing problem with time windows for distributing perishable foods in urban areas. Vehicle routing problem (VRP) had a spatial dimension which is practically neglected in research studies. Therefore, handling geographic data is requisite for efficient routes based on real distances. The most promising solution for so is GIS. In fact, we propose an Agent-Based simulation Model integrated with the Geographic Information System (ABM-GIS) to use real-case while performing distances between customers. In this paper, we aim at producing the fastest routes to deliver perishable foods as fast as possible and estimating VHT and VKT which are valuable for transportation, through the simulation model. We aim also to study the impact of congestion on commercial vehicle tours in an urban area. And as for short-term planning, for daily operations, we propose a time-dependent scheduling approach to optimize departure times from the distribution center.

The remainder of this paper is organized as follows: section 2 provides a background on integrating ABM and GIS, section 3 describes the simulation model, section 4 presents the estimation of VKT and VHT, section 5 highlights the impact of congestion, section 6 presents experimental design and results, section 7 introduces the proposed scheduling methods as well as the enhancement they provide, and we end up with a conclusion.

2 Agent Based Model and GIS integration

Agent-based modeling is a relatively new method compared to system dynamics and discrete event modeling. This modeling paradigm is developed to simulate complex systems through the study of active entities behavior, known as agents. Agent Based System (ABS) has been adopted to solve complex problems, from various domains, such as logistics optimization, traffic, and urban planning. ABS can be used for different purpose: (1) Understanding observed dynamics, processes, and systems, (2) Designing or engineering of processes or systems, (3) Managing a system or process,(4) Formulating theory and explanatory

[a]Corresponding author: hanane.elraoui@usmba.ac.ma

models,(5) Predicting, (6) Optimizing resources, capabilities, and processes. For further details refer to [2]. **Geographic Information Systems (GIS)** are recognized as one of the new technologies which can be usefully introduced to Decision Support System (DSS) for vehicle routing [3]. GIS is well defined as an information technology which stores, analyzes, and displays both spatial and non-spatial data.
Combining these two techniques provides us with the advantages of both and allows a better modeling of the vehicle routing problem which is also a spatial problem.

3 Simulation model

3.1. Case study description

This paper presents an agent-based model of a food service distributor that provides perishable foods to restaurants, associated with GIS map of Casablanca, one of the most crowded cities in Morocco, where the Distribution Center (DC) and different customers are located.

3.2 Modelling environment

For our modeling environment, we use AnyLogic software, Version 8.2 developed by XJTechnologies.The advantage of this software is its ability to support different simulation methodologies, including discrete event simulation, system dynamics and agent-based simulation, in which we are interested. Anylogic has an integrated GIS space where we can drag different objects (agents) and define their properties.

3.3 Agents

In our model, agents represent the distribution center, customers, and vehicles used for delivering. After generating agents, the setting is defined in order to make agents live in the GIS Space.

Locations of agents (DC, customers) were defined in a database to let Anylogic place it on the GIS map, through sending the address to an open street map server to obtain the corresponding coordinates.

3.4 Vehicle controlling

In the developing model, we use a state chart to provide refrigerated vehicle movement. Anylogic state chart consists of a sequence of states and transitions that enable the user to define the behavior of an object.

Figure 1, 2 and 3 illustrate the state charts used to control vehicles movements during the simulation of different scenarios presented below.

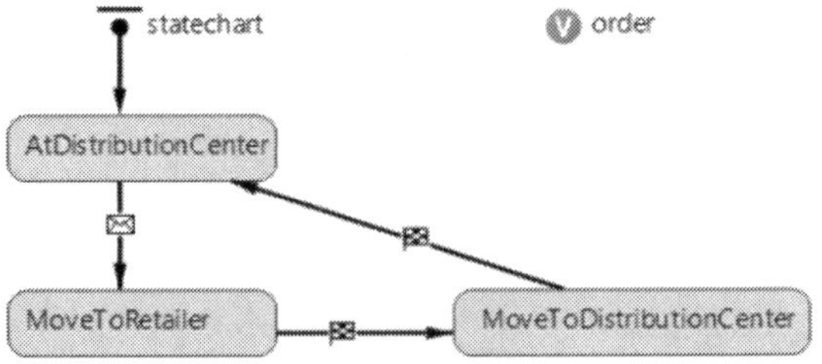

Figure 1. State chart of vehicle in scenario A

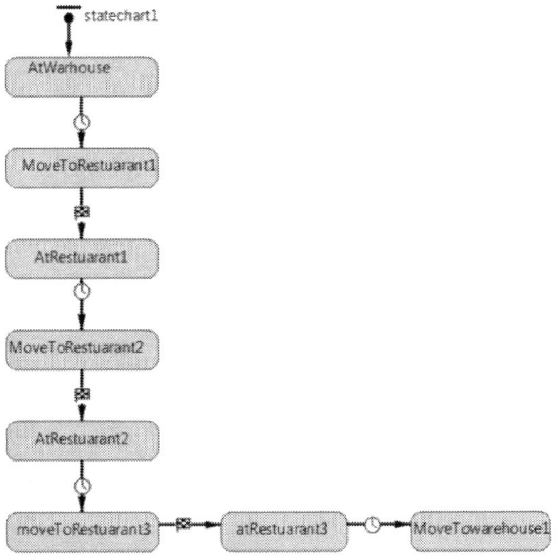

Figure 2. State chart of vehicles in scenarioB -Tour 1

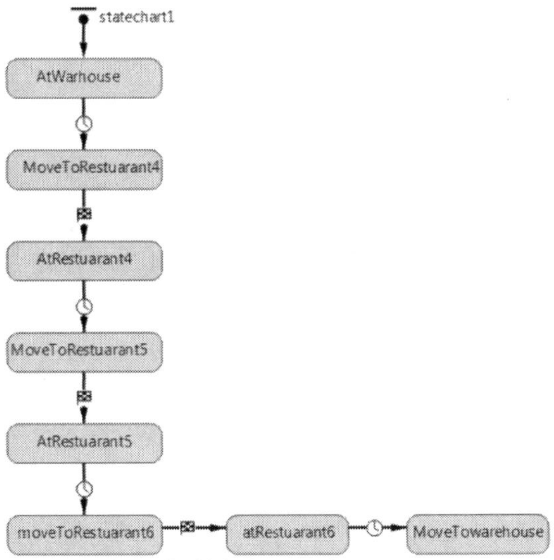

Figure 3. Statechart of vehicles in scenarioB -Tour 2

3.5 scenarios simulated

In our study case, we simulate two scenarios corresponding to two types of tours. Within scenario A, the customer order is equal to the capacity of the truck, otherwise known as a Full TruckLoad (FTL). It's a degenerated type of VRP since the vehicle will visit only one customer before returning to the DC.

In scenario B, the customer's order is relatively small and the delivery is Less Than Truckload (LTL).

3.6 Routes

During the distribution of frozen, refrigerated, and fresh goods it may be important to minimize the travel time, because these products have a short shelf life and their quality decay with the variation of temperature. And since keeping the adequate temperature adds significant costs, selecting efficient fastest routes would consequently reduce the whole transportation costs.

To find a suitable path for distributing perishable foods, several techniques have been applied in the literature. however, the shortfall of most researches is that authors apply mathematical equations instead of using spatial data for efficient routes. For example, [4] undertook the fresh meat distribution problem by applying several algorithms to find optimal sets of routes.

To reach customers in time, real route network should be considered, hence the interest of GIS that was adopted by [5] to determine optimal transportation routes. In [6] authors apply GIS to transportation to find optimal roads among sets of routes. To identify the best routes for waste collection, [7] developed an application of ArcGIS Network Analyst tool. In view of this, we opt for GIS in our study.

In the first scenario, customers get delivered directly. Routes that connect the distribution center to each customer are established by using GIS map route on Anylogic. Since we are dealing with time-sensitive products that should be delivered as fast as possible, we choose the fastest routing method option in the routing properties.

In the second scenario, customers are subdivided into balanced tours determined according to this criterion: (1) tours have the same number of stops and (2) the service areas are the same across tours [8]. In both tours, each node is linked to the nearest one through the fastest path set up by the same technique used in the first scenario.

4 Estimation of VKT and VHT

In transport planning, it's necessary to estimate the vehicle kilometer traveled by a vehicle to satisfy customer demand as well as the travel time VHT. These two elements are important for carriers to get an insight into variable costs. Furthermore, VHT is valuable for scheduling vehicles departure from the DC.

In the literature, there are several works that deal with the approximation of VKT. In a typical work [9] proposes a continuous approximation to estimate VKT for a Capacitated Vehicle Routing Problem (CVRP). In [10] authors Combine simulations with linear regressions to estimate the length of Traveling Salesman Problem (TSP). In [11] they have used the same techniques in addition to neural networks for an accurate approximation. Most of the previous research, authors compute the VHT and VKT based on direct links. However, in reality, vehicles travel on a real road network. Thus, the Euclidean distances used is far from accuracy and cannot hold anymore. This fact prompts us to adopt GIS function to approximate based on a real road network.

After building the AB-GIS model, a simulation run was performed on anylogic software to estimate the VKT and VHT for both scenarios corresponding to FTL and LTL. We assume that roads are not congested, and then the vehicle travel with the speed limit in each arc of the tour. Travel speed limit can be found by open street browser www.openstreetbrowser.orgReferences. And based on statistic outputs about the model execution in the log file, VHT and VKT are measured. Results are shown in figure 4 and figure 5.

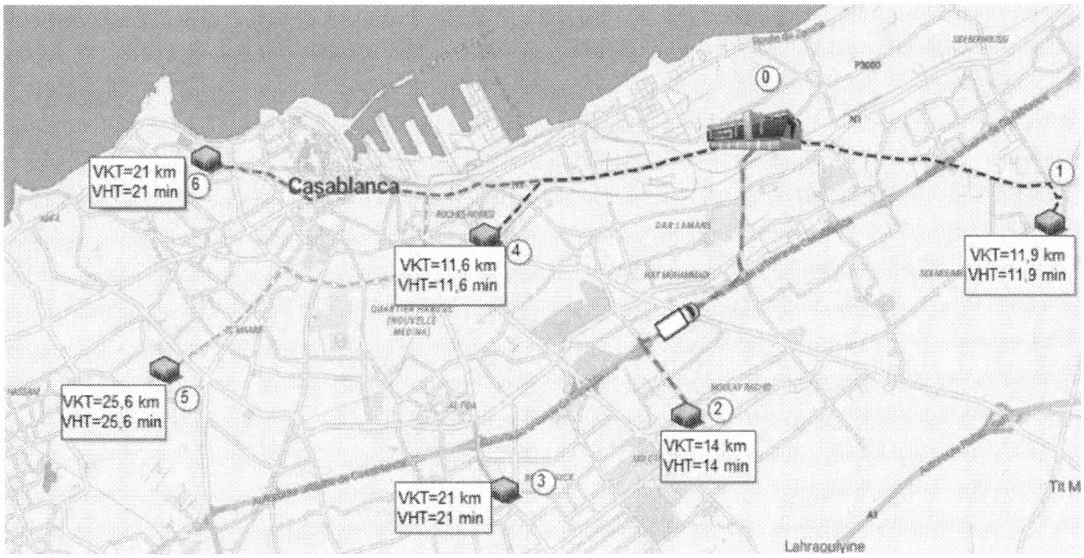

Figure 4. VKT and VHT in scenario A

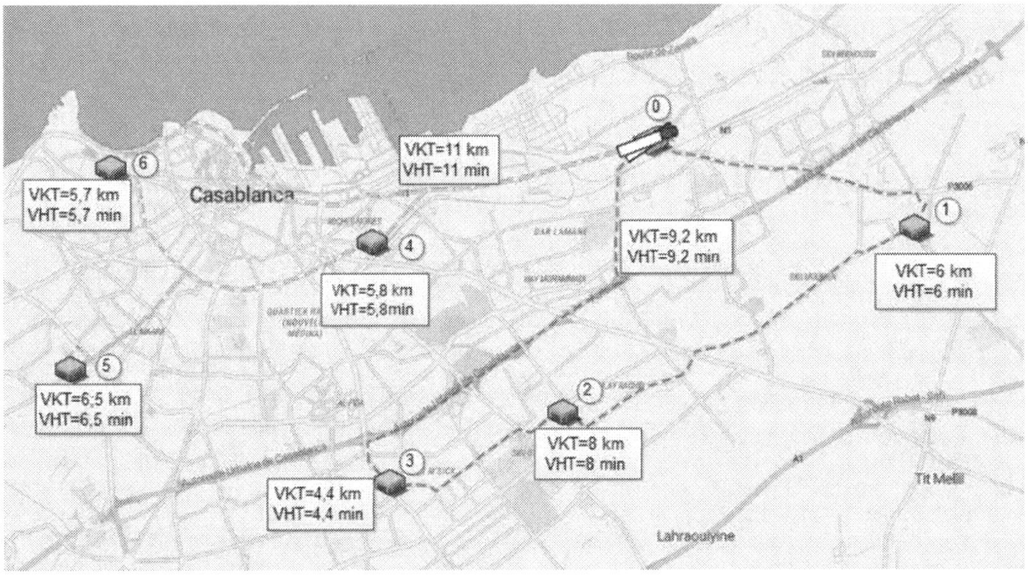

Figure 5.VHT and VKT in scenario B

5 Study of congestion impact

5.1. Impact of congestion on VHT

Nowadays, customers' requirement for fresh food delivery conditions is increasing in urban areas. Customers have become more exigent about the delivery time; they require to be delivered within a specific time window.

In urban areas, the traffic flow is growing steadily, which makes the delivery of fresh food in time seen as a problem, and this is mostly due to traffic congestion.

Traffic jams may be reached during different times of the day, therefore the delivery time will also depend on the time of the day. However, in today's planning systems, the time dependency is not recognized, and the speed is assumed to be constant. The travel time estimation on those bases cannot be accurate, which may lead to additional costs, customer frustration and penalties due to the violation of time windows.

In this paper the variation of speed during the travel time due to the congestion will be considered, thus we are dealing with a time-dependent travel speed. Our study is based on the distribution of speed shown in figure 6.

To each time interval corresponds a congestion level represented by a coefficient C_t computed as:

$$C_t = \frac{S_t}{S_l} \qquad (1)$$

Where S_t the average speed at time t, and S_l the speed limit. We assume that the congestion level is the same on all arcs.

5.1.1 Impact of congestion in scenario A

Before evaluating the impact of road congestion on travel time, we start by getting an insight into the possibility of serving different customers in time under normal conditions. Thus, we calculate the speed required on each arc using the moveToInTime() anylogic function that makes the agent move to a destination in a specified duration. To evaluate how these changes on road condition may affect the VHT, and the service quality represented by the respect of the time window, we test 2 scenarios:

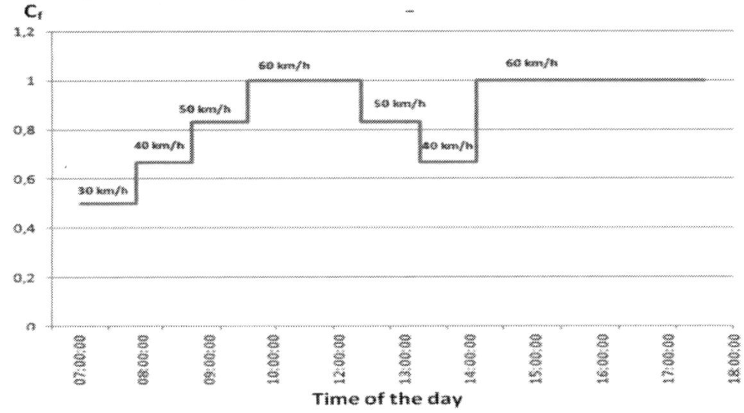

Figure 6. speed distribution

- Scenario 1:
  - Customer Demand is received during a traffic peak (rush hour).
  - The speed required to deliver this customer in time lower than or equal to the average speed during the traffic peak interval.
- Scenario 2:
  - Customer Demand is received during a traffic peak (rush hour).
  - The speed required to deliver this customer in time is higher than the average speed during the traffic peak interval.

For an accurate estimation of the VHT, speed changes within the travel time are taken into account. Two cases are considered, in the first case the departure and arrival time of vehicle are in the same time interval [K], thus no speed changes have occurred. In the second case when the arrival time of a vehicle is estimated to be in the interval [K+1] it means the speed has changed during this period.

1: If D_0 and $AT_n \in [K]$

$$2: VHT_0^n = \frac{VKT_0^n}{V_k}$$

3: Else

$$4: VHT_0^n = VHT_0^{n\,K} + VHT_0^{n\,K+1}$$

$$5: VHT_0^n = (T_{k\sup} - D_0) + (AT_n - T_{k\sup}) \times \frac{V_k}{V_{k+1}}$$

5.1.2 Impact of congestion in scenario B

In this scenario customers have the same time windows form 8 to 9 in the morning, because customers are restaurants and this time demand ensures that they can process and serve fresh food to their customers.

5.2. Impact of congestion on costs

To provide insight into the relationship between cost, and changes in the road network, the distribution costs considered in this study include the transportation costs, damage costs, refrigeration, and penalty costs. Since we are interested in measuring the variation, fixed cost will not be included.

5.2.1 Transportation costs

The transport costs include maintenance and repair costs, tires and depreciation costs, and the major component related to the fuel consumption. Fuel consumption, in our study, is weight and time-dependent, because the travel speed and travel time, which depend on departure time, and the loading weight, are taken into consideration in the procedure.

Fuel consumption calculation:

Let KPL_{ij} be the kilometer per liter for a vehicle traveling from the customer i to customer j. The consumption per unit time corresponding LPH_{ij} is calculated as follow:

$$LPH_{ij} = \frac{V_{ij}}{KPL_{ij}} \tag{2}$$

V_{ij} Is the corresponding traveling speed form i to j.

The transportation costs can be expressed as

$$C_{tran} = \sum_{v=1}^{m} \sum_{i=0}^{n} \sum_{j=0}^{n} (C_{ij}^v x_{ij}^v d_{ij} + C'^v_{ij} x_{ij}^v VHT_{ij}^v) \tag{3}$$

C_{ij}^v Is the sum of variable costs dependent onthe distance traveled (maintenance, tires, depreciation)

x_{ij}^v Is a 0–1 variable, $x_{ij}^v = 1$ if the vehicle v passes the road between customer i and customer j, otherwise0.

d_{ij} Distance traveled from node i to j

VHT_{ij}^v Vehicle hour traveled from i to j

C'^v_{ij} Fuel consumption by unit time of the vehicle v moving from node i to j

5.2.2 The refrigeration costs

Refrigeration is crucial during the transportation of perishable food. Refrigeration costs include the cost caused by energy consumption to keep the adequate temperature during delivery, as well as the cost of additional energy supplied by the refrigeration system during the unloading process.

Refrigeration cost during transportation can be expressed as:

$$C_T = c_e \sum_{v=1}^{m} \sum_{i=0}^{n} \sum_{j=0}^{n} x_{ij}^v T_{ij}^v \tag{4}$$

$$T_{ij}^v = VHT_{ij}^v + t'_j \tag{5}$$

Ce: The refrigeration costs during the transportation process of unit time.

T_{ij}^v :The traveling time of the refrigerated truck v from the customer i to the customer j.

t'_j :The waiting time of refrigerated vehicle in customer j before unloading.

Refrigeration cost during the unloading process:

$$C_u = c'_e \sum_{v=1}^{m} \sum_{j=0}^{n} y_j^v U_j \tag{6}$$

$$C_{ref} = C_T + C_u \tag{7}$$

[a]Corresponding author: hanane.elraoui@usmba.ac.ma

$c_e^{'}$: The refrigeration costs during the unloading process of unit time.

U_j : The unloading time which is needed toserve the customer i.

y_j^v is a 0–1 variable, $y_j^v = 1$ if the vehicle k services for customer i, otherwise $y_j^v = 0$.

5.2.3 Penalty cost

Generally, in the distribution of food, customers require to be delivered within a time window. And if the goods do not reach the destination within the time agreed on by the customer, thus the time window is violated and additional penalty cost will be applied. These costs can be expressed as follow:

$$C_p = \sum_{v=1}^{m} \sum_{j=0}^{n} (\alpha \max[Tw_{\inf} - AT_j^v, 0] + \beta \max[AT_j^v - Tw_{\sup}, 0]) \quad (8)$$

AT_j^v : The arrival time of vehicle v to customer j.

$Tw_{\inf}, Tw_{\sup}$:The lower and upper bound of the time window.

α : The cost of waiting for the unit time if refrigerated truck arrives at a customer in advance.

β : The cost of punishing for the unit time if the refrigerated truck is late to the customer.

5.3. Impact of congestion on co2 emission

The traffic growth in urban areas brings further problems of environmental aspects. In recent years there has been increasing interest in estimating the environmental effects of vehicle routing policies.

In our work, we aim to discover the relationship between traffic congestion, restrictive time windows, and co2 emission. For the purpose, we use the "ASIF" equation [12] to quantify carbon emissions in the simulated scenarios.

6 Experimental design and results

In this section, we use a numerical example to quantify the impact of traffic jams on the VHT, costs, and emissions. Section 6.1 describes the problem and parameter setting; experimental results are analyzed in section 6.2.

6.1 experimental design and parameter settings

The delivery service provider places a premium on service quality, hence all scenario use hard time windows, to guarantee that promised delivery times would be met.

To simplify the problem, we make the following assumptions:

(1) The service provider has a homogenous fleet with one type of refrigerated vehicles.

(2) In the scenario A, the customer order does not exceed the capacity vehicle, and the time spent to serve each customer is 15 min.

(3) In the scenario B, the vehicle can serve 3 customers and spend 10 min at each one.

The distribution center DC start operating at 7.In the first scenario corresponding to direct delivery, orders in the simulation model are generated as an event sent to the distribution center randomly during the day. Since we are interested in evaluating the impact of congestion we took a sample of orders received during the morning, characterized by the higher congestion level. We assume that order preparation requires 30 min and the time windows length is 15min.

In the second scenario, the time window is [8h 9h], since the level of congestion from 7 to 8 is upper, simulation run was performed to estimate the travel time for both tours with an average speed corresponding to this time window figure to determine if this time interval is sufficient for delivery, thus:

If $T_{delivery}$< 1h we start delivery at 8 to avoid congestion

Else$T_{departure}$=8h- $VHT^{i}_{w,K-1}$

$VHT^{i}_{w,K-1}$: vehicle hour traveled to deliver the client i from the warehouse during the time interval K-1.

Parameters of refrigerated vehicles are shown in table 1.

Based on the mileage and the estimated VHT, costs and carbon emissions were measured and compared to those in normal conditions.

Table 1. Vehicle parameters

parameter	*value*
Load capacity (kg)	795
Fuel type	gasoline
maximum speed (km/h)	120
fuel consumption when loaded(km/l)	2,65
fuel consumption when empty(km/l)	6,06

6.2 Results and analysis

In this section, we will illustrate experimental results. The VHT has evidently increased during rush hours. This augmentation did not affect customers in scenario A.1, since the demand can be delivered in agreed time. However, in the scenario A.2, the time is violated which leads to additional penalty costs and impacts customer satisfaction. In this case, the time window agreed with these customers should be reviewed to be less restrictive. Another way is to limit the reception of demands to a specific period less congested, but this solution is not practical since we are dealing with restaurant that needs food in a precise time so they can process and serve fresh food to customers (breakfast, lunch…).

The fuel consumption is time-dependent; hence its heavy increases in the scenario A.2.

As we are delivering perishable foods, refrigeration costs are a very important component that should be considered. Regarding the latter, we can say that congestion has a deeply negative impact since even for a slight increase on the VHT, the costs increased by an average of 30%.

Since the co2 emissions are directly related to the fuel consumed, it remains' relatively flat as the trip length and the travel time increase.

In the direct delivery, we have demonstrated that fuel consumption increases with the VHT, however strangely we have noticed in scenario B that with a slight increase in VHT the fuel consumption is less under congestion than under normal traffic conditions in some arcs of the tour and consequently, the carbon emissions are reduced because the travel time saved is not really important. In these arcs it is not really wise to speed up for saving 1 or 2 min in the travel time.

These findings give us an idea about the optimal speed in each segment of the tour. But since we have another crucial component related to the VHT that should be considered while deciding either the speed is advantageous or not. For this reason, we have analyzed the impact on refrigeration cost, and we have found that congestion has a strong impact. It increases costs with an average of 37% on each tour trip.

7 Scheduling and time windows extension

Through the analysis of different scenarios, we have proved that travel time during rush hours is longer than in other periods, which leads to additional cost, impact the service quality and increases the co2 emission. With strict time windows, businesses have difficulties in optimizing these costs.

These findings imply that carriers should reduce travel time. After showing how the departure time of each vehicle from the distribution center affects costs and

Service, this goal can be achieved by changing route start times to avoid congested times and traveling as fast as is allowed by the traffic conditions, and combined with more flexibility in delivery time windows.

For this purpose, we have developed two algorithms to schedule departure in the direct delivery scenario and the LTL delivery while extending the time window. To discover how these changes in policy may lead to cost saving.

7.1. scheduling algorithm

To plan the departure of the vehicle from the distribution center, we have developed two algorithms for each scenario (direct delivery and LTL). These algorithms are used when the demand is received during a rush hour followed by a time interval with less congestion. The concept consists of programming the departure as late as possible to avoid congestion and travel as fast as possible while respecting the time windows.

7.1.1 Scheduling technique for scenario A

The demand is received during the time interval K, let Tw be the length of time windows. For a delivery from the distribution center '0' to a customer 'n', the departure is programmed as follow:

1: If $\overset{k+1}{VHT_0^n} + (T_{k\sup} - T_{DP}) < TW$

$2: D_0^n = T_{k\sup}$

3: **Else**

$4: \overset{k+1}{VHT_0^n} = TW_{\sup} - T_{k\sup}$

$5: D_0^n = T_{k\sup} - \frac{D_r}{V_k}$

$6: D_r = VKT_0^n - \overset{k+1}{VHT_0^n} \times V_{k+1}$

7.1.2 Scheduling technique for scenario B

The procedure starts by setting the arrival date to the last customer in the route to the upper bound of the time windows. The departure from the upstream customer is then calculated figure 7.

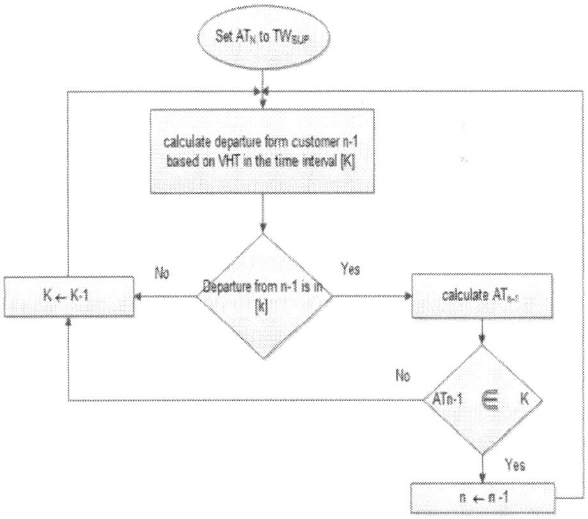

Figure 7. scheduling procedure for scenario B

7.2. Experience and results

To validate the proposed algorithms, they were applied to plan departures for scenario A and B.

In the scenario A, the algorithm is tested to determine the departure from DC to customer 5, which is critical as shown in our previous analyses. The time window is extended by 15 min. In the scenario B, we extend the time windows by 1 hour and we plan departures for 2 vehicles serving the customers in 2 tours.

Experimental results have proved the efficiency of these policy changes in saving costs and reducing emission. Interestingly, for a slight delay of the departure in the scenario A, fuel consumption as well as the co2 emissions have been reduced by 12% and refrigeration cost by 19%.

In the scenario B, policy changes lead to good results, since the travel time was decreased by 20 % even if the improvement in terms of costs is negligible.

8 conclusions

This study aims to solve one of the major problems encountered by carriers, which consist of choosing the optimal routes and program departure to deliver customers intime. Therefore, agent-based simulation combined with GIS is used. Simulation runs were performed to estimate the VKT and VHT. Analyses of network changes (congestion) were conducted to provide insight into the impact of these changes on these 2 values (VHT and VKT) also on cost, service quality, and emissions.

A time-dependent scheduling technique was developed and proved its success in achieving a potential saving, in terms of costs and co2 emission. As a future research, this time-dependent vehicle routing problem will be represented as an optimization model and integrated with ABM-GIS.

Acknowledgment

The first author is supported by the national center for scientific and technical research in Morocco (CNRST) through an excellence scholarship.

References

1. C. I. Hsu & W. T. Chen. Optimizing fleet size and delivery scheduling for multi-temperature food distribution. Applied Mathematical Modeling, **38(3)**, 1077-1091, (2014).
2. N. A. Gómez-Cruz, I. Loaiza Saa, & F. F. Ortega Hurtado. Agent-based simulation in management and organizational studies: a survey. European Journal of Management and Business Economics, **26(3)**, 313-328 , (2017).
3. P. B.Keenan. Spatial decision support systems: a coming of age. Control and Cybernetics, **35(1)**, 9, (2006).
4. C. D.Tarantilis, & C. T. Distribution of fresh meat. Journal of Food Engineering, **51(1)**, 85-91, (2002).
5. Y. W.Chen, C. H., Wang, & S. J. Lin. A multi-objective geographic information system for route selection of nuclear waste transport. Omega, **36(3)**, 363-372, (2008).
6. M.Sharifi,M. Hadidi, E. Vessali, P. Mosstafakhani, K. Taheri, S.Shahoie, &M. Khodamoradpour. Integrating multi-criteria decision analysis for a GIS-based hazardous waste landfill sitting in Kurdistan Province, western Iran. Waste management, **29(10)**, 2740-2758, (2009).
7. A. V. Bhambulkar. Municipal solid waste collection routes optimized with arc gis network analyst. International Journal Of Advanced Engineering Sciences And Technologies Vol, **(11)**, 202-207, (2011).
8. M. A. Figliozzi. Analysis of the efficiency of urban commercial vehicle tours: Data collection, methodology, and policy implications. Transportation Research Part B: Methodological, **41(9)**, 1014-1032, (2007).
9. A. L. Erera. *Design of large-scale logistics systems for uncertain environments*. University of California, Berkeley, (2000).
10. T. W. Chien. Operational estimators for the length of a traveling salesman tour. Computers & operations research, **19(6)**, 469-478, (1992).
11. O.Kwon, B. Golden, and E. Wasil. Estimating the length of the optimal TSP tour: an empirical study using regression and neural networks. Computers & operations research, **22(10)**, 1039-1046, (1995).
12. L. Schipper, M. Cordeiro, & W. S. Ng. Measuring the carbon dioxide impacts of urban transport projects in developing countries. *In Proceedings of the Transportation Research Board Conference*. Washington, DC. USA. (2007, November)

Challenges of cloud computing use: A systematic literature review

Ibtissam M'rhaoaurh[1,*], *Chafik* Okar[2], *Abdelwahed* Namir[1], and *Nadia* Chafiq[3]

[1]Laboratory of Technological Information and Modelisation (LTIM), Faculty of Sciences Ben M'Sik, University Hassan II Casablanca, Morocco
[2]Laboratory of Analysis and Modeling Systems for Decision Support (LAMSAD), University Hassan 1st, EST Berrechid, Morocco
[3]Multidisciplinary Laboratory in Sciences and Information, Communication, and Education Technology (LAPSTICE), Faculty of Sciences Ben M'Sik, University Hassan II Casablanca, Morocco

Abstract. Background: In recent years, cloud computing has grown vastly. Cloud computing represents a new model for IT service delivery and it typically provides over-a-network, on-demand, self-service access, which is dynamically scalable and elastic, using pools of often virtualized resources. However, this new paradigm is facing diverse challenges from many fronts. Methods: We conducted a systematic literature review of potential challenges of cloud computing. Documents that described challenges of cloud computing were collected of routinely. We grouped identified challenges in taxonomy for a focused international dialogue on solutions. Results: Twenty-three potential challenges were identified and classified in three categories: policy and organizational, technical and legal. The first three categories are deeply rooted in well-known challenges of cloud computing. Conclusions: The simultaneous effect of multiple interacting challenges ranging from technical to intangible issues has greatly complicated advances in cloud computing adoption. A systematic framework of challenges of cloud computing will be essential to accelerate the use of this technology for working well in fact and in order to face with respect to mitigating IT-related cloud computing risks. Keywords: Cloud Computing, Challenges, Systematic literature review

1 Background

Recently, Cloud computing has emerged as a buzz word in the ICT industry [1]. Cloud computing has been captivating since 2007 the interest of the Information and Communications Technology community ensuing in a massive amount of industry developments [2]. It has been appreciated as a significant step after Grid computing towards realizing this utility computing, where computing resources are delivered as subscription utility service just such as water and electricity [2]. From business perspective, it is widely viewed as an economic model for renting computing resources [6]. It enables companies to adopt ready-to-use application services through a "pay-as-you go" model that saves cost, resources and time.

Cloud computing has been quickly making spectacular advances in different domains for developing, designing. Besides, a various set of applications like social networking sites, scientific workflow systems, multiplayer gaming portals, and enterprise applications are deploying [7, 27].

According to the National Institute of Standards and Technology (NIST) [3], "*Cloud computing is a model for enabling ubiquitous, convenient, on-demand network access to a shared pool of configurable computing resources (e.g., networks, servers, storage, applications, and services) that can be rapidly provisioned and released with minimal management effort or service provider interaction.*".

This definition is constituted of cloud architectures, security, and deployment strategies. Specially, five essential elements of cloud computing are precisely identified as shown in below [3]:

On-demand self-service: Allows the consumer to be capable of availing real time distant computing resources (such as CPU time, network storage, software use, and so on) according to needs, and without depending upon human intervention.

Broad network access: All the resources must be available and accessible to the user universally and simply through the network (e.g. Internet), whatever the customers use (mobile phones, laptops, PDAs, and workstations).

Resource pooling: For the purpose of serving multiple consumers using either the multi-tenancy or the virtualization model, the supplier's computing resources are pooled together, with diverse physical and virtual resources dynamically attributed and reassigned conforming to consumer need. Economies of scale and

* Corresponding author: mrhaouarhibtissam12@gmail.com

specialization are two important factors, that pool-based computing paradigm rely on, in order to be set up. The effect of a pool-based model is that physical computing resources become 'invisible' to clients, who in general do not have control or knowledge over the exact location, formation, and originalities of these resources (e.g. database, CPU, etc.) . For example, consumers can't indicate the location where their data will be stocked in the Cloud.

Rapid elasticity: One of the characteristics of the cloud computing is the elasticity of the resources. This characteristic allows the users to find quickly new resources so as to be able to answer a rise or a descent in sudden load. It is never obvious to plan the resources

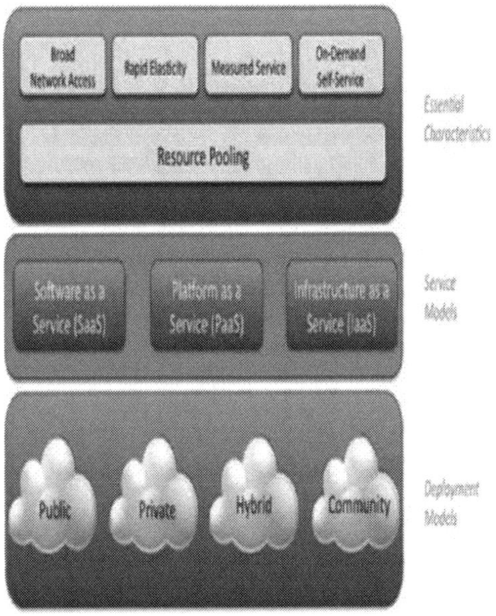

Fig. 1. NIST defined Essential characteristics, Service models and Deployment models

which will be vital for the implementation of any IT service, in particular when this need is constantly evolving. The cloud computing so offers a way to release the computing resources necessary for an evolution or for a peak of use of this service.

Measured Service: With view to measure the usage of these resources for each individual consumer via its metering capabilities, the cloud infrastructure can use appropriate mechanisms despite of the fact that computing resources are pooled and shared by multiple customers (i.e. multi-tenancy)[36].

As per NIST mainly the Service Model consists of Infrastructure as a Service (IaaS), Platform as a Service (PaaS), and Software as a Service (SaaS) (as shown in Fig. 1).

Software-as-a-service provides business processes and applications [9] that enable clients to use cloud services running on cloud infrastructure [10] through either a thin client interface, like a Web browser. For instance, Google offers services such as Web applications with similar functionality to traditional office suites, including Gmail, Google Calendar, and Docs, among others. Customers do not need to control or manage the underlying infrastructure design for the reason that all new installations (software and hardware) are automatically updated by application vendors.

Platform-as-a-service delivers a computing platform as an integrated solution, solution stack or services like hardware, operating systems (OSs), and storage through an Internet connection. PaaS enables customers to develop, test, and deploy IT services over a cloud platform. By abstracting the complexity of software and infrastructure, this concept facilitates accordingly the efficient and quick development of Web applications. This model assists businesses in leasing virtual IT services through which to run existing applications, as well as to develop, test, or deploy a new application [10, 11, 12].

Infrastructure-as-a-service is a virtual delivering of computing resources in the form of servers, hardware, networking, and storage services. Rather than installing and purchasing the required resources in their own data center, Clients lease these resources as needed; they don't need to control the underlying cloud infrastructure [13]. IaaS generates potential benefit through controlling and paying for the amount of resources demanded by customers [14].

Moreover, The Cloud model comprises four deployment models:

Private cloud is a cloud computing model that involves a distinct and secure cloud based environment in which only the specified client can operate. By utilizing an underlying pool of physical computing resource and within a virtualized environment, private clouds will offer computing power as a service. However, under the private cloud model, the cloud (the pool of resource) is only accessible by a unique organization, hence providing that organization with confidentiality and greater control. But data transfer cost [15] from local IT infrastructure to a Private Cloud is still rather considerable.

Community clouds are controlled and shared by multiple organizations and support a particular community that has shared interests, such as mission, policy, security requirements and compliance considerations. It may be maintained by the organizations or a third party and may exist at on or off premise, and the members of the community share access to the data and applications in the community cloud. Community cloud users therefore seek to exploit economies of scale while minimizing the costs related to private clouds and the risks related to public clouds [3].

Public cloud is the most known model of cloud computing to many clients is the public cloud model, under which cloud services are offered in a virtualized environment, built via pooled shared physical resources, and available over a public network like the internet. To some degree they can be determined in contract to private clouds which ring-fence the pool of underlying computing resources, creating a distinct cloud platform which is operated solely within a singular organization.

Public clouds deliver services to multiple customers using the same shared infrastructure [3].

Hybrid cloud is a combination of two or more clouds (private, community, or public) that stay unique entities but are related together by standardized or proprietary technology that allows data and application portability. Concerning applications with lacking rigid security, legal, compliance and service level requirements can be deployed in the public cloud, while maintaining business-critical data and services in a protected private cloud [3].

In spite of the widespread adoption of cloud computing, practitioners and researchers have been actively reporting issues and challenges with this new paradigm. Some of the challenges have the aspects of being crucial like issues with confidentiality and security. Other issues such as suboptimal performance and limited bandwidth are a natural result of pushing the barriers of this new model to achieve more [16]. Unless these barriers are better understood, solutions may remain inefficient. We conducted a systematic literature review of potential challenges of cloud computing and used this evidence to group these barriers in a taxonomy that can be used as a framework to facilitate an international dialogue on solutions and instruments. The objective of our research is to gain an understanding of the type of issues and challenges that have been emerging.

2 Methods

In this section, we present the research steps followed to perform this review. We conducted a systematic review according to Kitchenham and al. guidelines [18] to elaborate the review methodology in detail and identify documents that reported on challenges of cloud computing. The challenges were defined as obstacles that could impede or delay the adoption of cloud computing or that could limit the usage of cloud computing in companies. As per these authors, the research methodology for systematic review should contain different strategies that are employed for searching the most significant research works like search strings and the chosen digital libraries. At last, the existing studies selection is realized through set criteria.

The following search string represents our generic search query used for this SLR.

("cloud computing" OR "cloud" OR "cloud technologies") AND ("issues" OR "challenges" OR "barriers" OR "threats" OR "limits" OR "risks" of cloud computing).

Additional documentation was identified through a variety of databases (such as ACM Digital Library, IEEE Explore, DBLP, Google Scholar, Science Direct, Scopus, Springer, Taylor & Francis and Wiley Online Library) as shown in Table 1.

Table 1. Electronic data sources

Electronic database	Url
ACM Digital Library	http://dl.acm.org/
IEEE Explore	http://ieeexplore.ieee.org/
DBLP	http://dblp.uni-trier.de/
Google Scholar	https://scholar.google.com/
Science Direct	http://www.sciencedirect.com/
Springer	http://www.springer.com/
Scopus	https://www.scopus.com/
Taylor & Francis	http://taylorandfrancis.com/
Wiley Online Library	http://onlinelibrary.wiley.com/

With the regard to make this research up-to date and well-intentioned in the area of cloud computing quick search strategy is used. For this end in order to add recent 2015–2017 publications, we have used the quick search strategy for this research by using the filtering tools in the databases. After using the quick search strategy, we considered the publication from 2008 to 2017 overall for the reason that cloud computing publications started around 2008.

As shown in Fig. 2, the initial search resulted in a total of 800 studies, which were condensed to 200 studies on the basis of their titles, and 100 studies on the base of their abstracts. After that, 100 selected studies were reviewed thoroughly for obtaining a final list of 60 studies on the basis of their content.

Sixty studies were ultimately involved in this review. These studies were primarily read and an initial list of challenge descriptions was extracted. This list was grouped into preliminary categories. Challenge descriptions then classified and generalized within their categories. A final taxonomy and description of barriers were established. For each barrier, we also categorized available evidence to identify knowledge gaps.

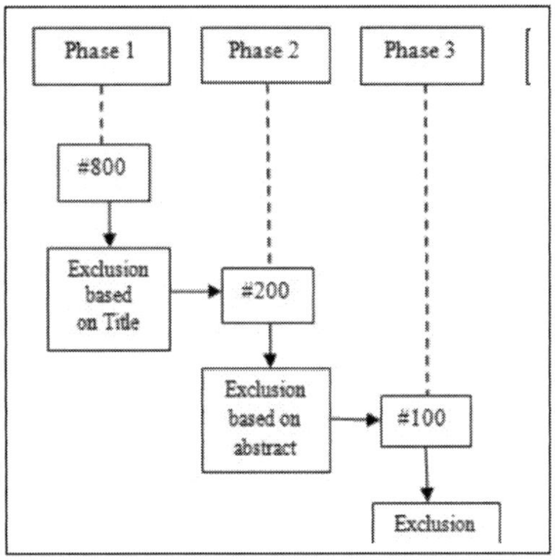

Fig. 2. Study selection process

3 Results

We identified 23 potential issues of cloud computing and classified these in taxonomy of three categories, according to the organization European Union Agency for Network and Information Security (ENISA) [19, 20, 21, 34, 58]: policy and organizational, technical and legal issues (Table 2). These issues and categories represent a landscape of challenges that is highly dynamic, interconnected, and hierarchical.

Policy and organizational issues

These are business-related IT issues that companies may confront when considering cloud computing service providers [68]. Such issues include lock-in, loss of governance, Compliance challenges, supply chain failure [22].

Lock-in vendor lock-in is one of the principal interests declared by IT experts when considering a move from one provider's cloud environment to another [16,20,23]. Lock-in refers to the incapability of a client to move their data and/or programs away from a Cloud Computing risks cloud computing service provider. It is generally the result of proprietary technologies that are incompatible with those of rivals [12, 32, 35, 73].

Loss of governance: When adopting Cloud services, the Cloud Customer necessarily concedes control to the Cloud provider on a number of issues which may impact security [20, 21, 58, 67, 73].

Compliance challenges customers are accountable for the security of their solution, as they can choose between providers that enable to be analysed by 3rd party organizations that control levels of security and providers that don't [10, 45, 60].

Supply chain failure As per ENISA [58], a Cloud provider can deploy parts of its production chain to third parties, or even, as part of its service, use other Cloud Providers. Thus, a potential for cascading failures is produced [20].

Technical issues

These issues for the most part are well understood as part of resilient challenges of cloud computing adoption and continue to form a major obstacle to the availability and use of this technology [78]. They are specified by the failures associated with the technologies and services furnished by the Cloud service vendor [68].

Malicious insiders A malicious insider in the cloud might get hands on an unusual quantity of information and on a widely scale [64] and produce various kinds of damage to a Cloud Customers' assets [9, 12, 19, 20].

Shared technology Cloud service providers offer their services in a scalable way by sharing infrastructure, platforms, and applications. This way, the threat of shared vulnerabilities exists in all delivery models of Cloud computing [34, 39, 40].

Encryption is considered a major risk in cloud computing environments is deficient encryption and key management of data [44-45].

Multi-tenancy is a natural result of trying to achieve economic benefit in Cloud Computing by using virtualization and allowing resource sharing [62-63]. However, it is a technological issue in Cloud computing [27, 33, 45, 61].

Resource and service management One of the important features of cloud computing is the capacity of obtaining and releasing resources on-demand [20]. The purpose of a service supplier in this situation is to allocate and de-allocate resources from the cloud [23]. Nevertheless, it is not obvious how a service provider can achieve this goal [19, 27, 46].

Service level agreement (SLA): it is necessary for customers to have guarantees from suppliers on service offer. Typically, these are offered via Service Level Agreements (SLAs) discussed between the providers and customers [79]. The very first problem is to determine SLA particularizations just like that has a convenient level of granularity expected by a customer from a provider [10, 16, 23].

Denial of service attacks (DOS) are attacks meant to impede users of a cloud service to have the ability to access their applications or their data. The attacker (or attackers, as is the case in distributed denial-of service (DDoS) attacks) attacks typically flood servers, systems or networks with traffic by way of obliging the prevents legitimate users to consume inordinate quantities of finite system resources like memory, processor power, disk space or network bandwidth[20, 21, 23]. Consequently, this produces an intolerable system slowdown and leaves all of the victim cloud service perplexed and furious as to why the service isn't responding [9, 12, 19, 42, 43, 73, 77].

Insecure interfaces and APIs: with the aim of managing and interacting with cloud services, cloud computing providers offer consumers a set of software interfaces or APIs to exploit [23]. Using these interfaces, provisioning, management, orchestration, and monitoring are all executed. The security of these basic APIs impacts the availability and security of general cloud services. Moreover, companies and third parties often depend on these interfaces to provide value-added services to clients. This announces the intricacy of the new layered API; it also raises risk [6, 42, 43, 73].

Data loss or leakage For both consumers and organizations, perpetually losing one's data is terrifying. Due to causes other than malicious attackers, data stored in the cloud can be obviously lost [10-21]. Any accidental removal by the cloud service vendor, or bad, a physical disaster could lead to the permanent leakage of clients' data [42-45].

Integrity The integrity of applications, networks, databases and system software in a shared, globally available cloud environment is threatened by much vulnerability when not adequate and timely patched [4, 9, 16, 28, 54, 67].

Natural disaster such as earthquakes, flooding, and tsunamis can impact the infrastructure of a Cloud vendor [20, 21, 24]. Thus, a Cloud Customers might be affected by natural disasters taking place far away from its own location [33, 34, 41, 50].

Availability means that the data, service, as well as infrastructure are being accessible to authorized clients immediately after a demand has been made. Availability issues may happen at the customer end or the service vendor's end. When a single provider manages a cloud computing service, this way creates a potential environment for a single point of failure [53, 63, 67].

Table 2. Evidence for issues of cloud computing adoption

Category	Issue	Study reference
Policy and organizational	1. Lock-in	[12, 16, 20, 23, 24, 25, 26, 27, 28, 29, 30, 31, 32, 35, 67, 73]
	2. Loss of governance	[20, 21, 58, 67, 73]
	3. Compliance challenges	[10, 19, 21, 33, 41, 44, 45, 47, 58, 60, 67]
	4. Supply chain failure	[20, 58]
Technical	5. Malicious insiders	[9, 12, 19, 20, 23, 24, 26, 27, 33, 34, 37, 38, 39, 40, 41, 42, 43, 73, 76]
	6. Shared technology	[9, 12, 20, 34, 39, 40, 49]
	7. Encryption	[19, 33, 40, 44, 45]
	8. Multi-tenancy	[10, 12, 26, 27, 33, 45,61, 68, 70]
	9. Resource and service management	[19, 20, 23, 24, 27, 46]
	10. Service level agreement (SLA)	[10,16, 23, 24, 27, 28, 33, 45, 47, 48, 49, 50, 51, 74]
	11. Denial of service (DOS)	[9, 12, 19, 20, 21, 25, 27, 29, 37, 38, 41, 42, 43, 73, 77]
	12. Insecure interfaces and APIs	[6, 24, 26, 37, 38, 39, 40, 41, 42, 43, 73]
	13. Data loss or leakage	[10, 19, 20, 21, 26, 27, 34, 37, 40, 42, 45]
	14. Integrity	[4, 9, 16, 22, 23, 24, 25, 28, 33, 34, 38, 41, 44, 45, 50, 53, 54]
	15. Natural disaster	[20, 21, 24, 33, 34, 41, 50]
	16. Availability	[4, 15, 23, 25, 27, 28, 29, 33, 37, 38, 39, 41, 43, 44, 45, 46, 53, 55, 63]
	17. Loss of backups	[4,20, 40, 41, 53, 58, 65, 66, 67, 73]
	18. Data transfer bottlenecks	[15, 23, 24, 36]
	19. Interoperability	[16, 26, 27, 28, 35, 39, 45, 47, 48, 56, 57, 68, 69]
Legal	20. Legal jurisdiction	[16, 19, 20, 21, 23, 24, 33, 34, 36, 67, 73]
	21. Data privacy and protection	[9, 10, 16, 19, 20, 23, 24, 27, 33, 34, 41, 44, 45, 52, 53, 55, 63, 67, 69, 75]
	22. Licensing risk	[15, 19, 20, 29, 58]
	23. Subpoena and e-discovery	[19, 20, 58, 67]

Loss of backups The ability to recover data is salient in business that this is not a guarantee of cloud computing [65]. Rather than retrieval, this model depends on heavy backup, which could result privacy issues, as this is likely to lead to uninformed consent. A critical threat is that of 'data loss or leakage' where original data is deleted and cannot be recovered [40]. Data may also be lost due to dishonest media or data being recorded without a link [66, 67, 73].

Data transfer bottlenecks arises when bandwidth is unable to accommodate huge quantities of system data at designated data transfer rate speeds [23]. Businesses that use cloud computing have to redesign their present technology into new structures of information from being able to care dynamic and great amounts of information, new filing system and storage technologies [15, 24, 36].

Interoperability is the capability of two or more systems work together in order to transfer information and use that transferred data [16-26]. This risk makes difficult for firms to combine their IT systems in the cloud and get productivity gains and cost savings [28, 35, 39].

Legal issues:

These consist of the IT-related issues that are legal in nature, and can also have a negative impact on companies using cloud computing services [16].

Legal jurisdiction Converting to Cloud Computing implicates legal restraints [16, 19, 20]. Considering Cloud Computing providers can be multi-national, it is crucial that such vendors are aware of and stand for by national regulations where they do business [36, 67, 73].

Data privacy and protection: Privacy is one of the longest standing and most essential interests with cloud computing [59]. Furthermore, it is a major issue in cloud computing for the reason that its very nature involves storing unencrypted data on a machine owned and operated by someone other than the original owner of the data [8, 67, 69].

Licensing risk there is also a challenge that firms may pay more than intended to license software on systems hosted by cloud computing service vendors [15, 19, 20, 29, 58].

Subpoena and e-discovery If computer systems are confiscated by law enforcement authorities or through civil suits, the centralization of storage and shared location of physical hardware reveal more risk of undesired data divulgence to cloud computing customers [19-20].

4 Discussions

Using a systematic review of evidence from reviewed literature, we identified 23 real or potential barriers grouped in taxonomy of policy and organizational, technical and legal issues. Checking the criticality of the situation, many of the significant reports by the organizations including CSA, NIST, and European Union Agency for Network and Information Security (ENISA) etc. have been published to focus the effect of aforestated issues. Reports published by these research major organizations are based on their area of specializations, for example, CSA published reports on the security challenges. These reports are highlighting the risks that are applicable to the cloud computing.

As per [58], most of these threats surround privacy, security and service issues. In other words, above presented threats directly or indirectly impact the confidentiality and security of Cloud resources as well as services at various layers.

Nevertheless, privacy and security risk are considered as a major obstacles to cloud computing adoption that are acting as deterrent and retarding its development [80].

According to table 2, the issues that are addressing data protection and privacy, lock-in, multi-tenancy, availability, integrity, malicious insiders and interoperability have been characterized as essential.

Thus, Most technical issues are deeply embedded in much larger challenges of use of cloud computing. Some solutions are being developed as part of addressing some of aforementioned issues [27]. For instance, to overcome the challenges of privacy and protection, two studies [71, 72] present an architecture adopting data location strategies, trusted cloud services, and trusted service vendors. In order to handle the issues of regulatory compliance, a study [71] propose a concept of cloud market. By means of this, users can exchange with the market and demand resources conforming to the applications' needs through the cloud broker. Furthermore, to avoid vendor lock-in, a layered architecture was proposed by [17]. This architecture offers a unified resource model from different cloud environments.

Political and legal issues will require a different approach. Compared to technical challenges, these challenges are less tangible and transparent and will need to be clearly outlined.

Levels of evidence were also different for each issue. The most challenges were very well documented while no empirical evidence was available for other barriers such as supply chain failure, subpoena and e-discovery and data transfer bottlenecks. In-depth formative research is needed to expand the evidence base of these barriers. As knowledge on these challenges will increase, so will opportunities for solutions.

Conclusion

Cloud computing is one of the most advanced digital related technologies, which can bring various business benefits to organizations. Despite several benefits constraints exist with the use of this model impacting on service provision and use of this technology, instead of the conventional in-house technologies, which are physically owned and managed on premises. Thence, a considerable amount of literature on this subject has been published in a nearly short time period. In this research work, the issues of cloud computing adoption are collected and classified in a taxonomy of three categories policy and organizational, technical and legal issues. Thus, these challenges must be addressed by the researchers for making cloud computing work well in reality and in order to further gain the confidence of cloud subscribers.

References

1. R. Buyya, J. Broberg, A. M. Goscinski, "*Cloud Computing: Principles and Paradigms*". ISBN: 978-1-119-14340-6. 664 pages, June (2015).
2. N. Grozev, R. Buyya, "Inter-Cloud architectures and application brokering: taxonomy and survey. Software – practice and experience. Published online 12 December (2012) in Wiley Online Library (wileyonlinelibrary.com). DOI: 10.1002/spe.2168.
3. P . Mell, T. Grance, "The NIST Definition of Cloud Computing, National Institute of Standards and Technology", Information Technology Laboratory, Technical Report Version **15**, (2009).
4. Subashini S., Kavitha V. "A survey on security issues in service delivery models of cloud computing". In: *Journal of Network and Computer Applications* **34**, pp.1-11, (2011).
5. R. Buyya, CS. Yeo, S. Venugopal, J. Broberg, I. Brandic, "Cloud computing and emerging IT platforms: vision, hype, and reality for delivering computing as the 5th utility". *Future Generation Computer Systems,* **25**(6), pp.599–616, (2009).
6. G. Lewis, "Basics about cloud computing". Technical Report September, Software Engineering Institute – Carnegie Mellon, (2010).
7. R. Buyya, S. Pandey, C. Vecchiola, "Cloudbus toolkit for market-oriented cloud computing". In *Cloud Computing*, Jaatun M, Zhao G, Rong C (eds) ed:, **5931**. Springer: Berlin / Heidelberg, pp.24–44, (2009).
8. Y. Hu and G. Bai, "A systematic literature review of Cloud Computing in ehealth". In: *Health Informatics-An International Journal (HIIJ)* **3**(4), November (2014).
9. C. Modi et al., "A survey on security issues and solutions at different layers of cloud computing".

The Journal of Supercomputing, **63**(2):561–592, (2013)

10. S. Hamouda, "Security and privacy in cloud computing". In *Cloud Computing Technologies, Applications and Management (ICCCTAM), IEEE 2012 International Conference on*, (2012)
11. N. Fernando, SW. Loke, W. Rahayu, "Mobile cloud computing: a survey". *Future Generation Computer Systems* (2013), **29**(1), pp. 84–106.
12. S. Iqbal, M.L.M. Kiah, NB. Anuar, B. Daghighi, Wahab AWA., S. Khan, "Service delivery models of cloud computing: security issues and open challenges". Published online 30 August 2016 in Wiley Online Library (wileyonlinelibrary.com). DOI: 10.1002/sec.1585.
13. H. Qi et al. Sierpinski triangle based data center architecture in cloud computing. The *Journal of Supercomputing* (2014); **69**(2) , pp.887–907.
14. A.S. Ibrahim et al. Cloudsec: a security monitoring appliance for virtual machines in the iaas cloud model. in Network and System Security (NSS), IEEE 2011 5th International Conference on, (2011)
15. M. Armbrust, A. Fox, R. Griffith, A. Joseph, R. Katz, A. Konwinski, G. Lee, D. Patterson, A. Rabkin, and I. Stoica, "Above the clouds: A Berkeley view of cloud computing," EECS Department, University of California, Berkeley, Tech. Rep. UCB/EECS-2009-28, 2009.
16. Y. Ghanam, J. Ferreira, F. Maurer. Emerging Issues & Challenges in Cloud Computing— A Hybrid Approach. *Journal of Software Engineering and Applications*, **5**, pp. 923-937, (2012).
17. D. Petcu, G. Macariu, S. Panica, and C. Crăciun, "Portable cloud applications—from theory to practice". In: *Future Generation Computer Systems*, **29**, pp. 1417–1430, (2013).
18. B. Kitchenham, O.P. Brereton, D. Budgen, M. Turner, J. Bailey, S. Linkman, "Systematic literature reviews in software engineering—a systematic literature review". Inform.Softw.Technol. **51**(1), pp. 7–15, (2009).
19. C.L. Liu, W.H. Chen and D.K. Tung, "Identification of Critical Security Issues for Cloud Computing". Applied Mechanics and Materials **145**, pp. 272-276, (2012).
20. C. Daniele, and H. Giles, "Cloud Computing Benefit, risk and recommendations for information security", edited by ENISA, E.U (2009), in press.
21. N. Brender, I. Markov, "Risk perception and risk management in cloud computing: Results from a case study of Swiss companies". In: *International Information Management* **33**, no (5), pp.726-733, (2013).
22. G. Grispos, T. Storer, and W.B.Glisson, "Calm Before the Storm: The Challenges of Cloud Computing in Digital Forensics". In: *International Journal of Digital Crime and Forensics*, **4**(2) pp.28-48, (2012).
23. B.B. Rad, T. Diaby, M. E. Rana, "Cloud Computing Adoption: A Short Review of Issues and Challenges". *ICEEG 2017: Proceedings of the 2017 International Conference on E-commerce, E-Business and E-Government*, (2017).
24. A. O. Akande, N. A. April, J.P. Van Belle. "Management issues with cloud computing". *ICCC '13: Proceedings of the Second International Conference on Innovative Computing and Cloud Computing*, (2013).
25. A. Chandran, C. K. Shyamala, "Data Management Issues in Cloud Integrated Computing: A Big Picture". *International Conference on Advanced Computing and Communication Systems (ICACCS - 2017)*, Coimbatore, INDIA, (2017).
26. P. Harsh, F. Dudouet, R.G. Cascella, Y. Jegou, and C. Morin, "Using Open Standards for Interoperability Issues, Solutions, and Challenges facing Cloud Computing". *6th International DMTF workshop on Systems and Virtualization Management (SVM 2012) / CNSM (2012).*
27. M.A. Chauhan, M.A. Babar and B. Benatallah, "Architecting cloud-enabled systems: a systematic survey of challenges and solutions". In: *Software: Practice & Experience*, (2016).
28. A. Cardoso, P. Simões, "Cloud computing: Concepts, technologies and challenges". *ViNOrg 2011, CCIS 248*, Springer-Verlag Berlin Heidelberg, pp.127-136, (2012).
29. V. Krishna Reddy, B. Thirumal Rao, Dr. L.S.S. Reddy, P. Sai Kiran, "Research Issues in Cloud Computing" *Global Journal of Computer Science and Technology*, **1**(11), (2011).
30. J. Guillén, J. Miranda , Murillo J.M,, and Canal C., "A service-oriented framework for developing cross cloud migratable software," *Journal of Systems and Software*, **86**, pp. 2294–2308, (2013).
31. J. Opara-Martins, R. Sahandi and F. Tian, "Critical analysis of vendor lock-in and its impact on cloud computing migration: a business perspective". In: *Journal of Cloud Computing: Advances, Systems and Applications* **5**(4), (2016).
32. R. Sahandi, A. Alkhalil, J. Opara-Martins, "Cloud Computing from SMEs Perspective: A Survey Based Investigation". *Journal of Information Technology A Publication of the Association of Management* XXIV(1) , pp.1–12, (2013).
33. K. Ullah, and M. N. A. Khan, "Security and Privacy Issues in Cloud Computing Environment: A Survey Paper." *International Journal of Grid and Distributed Computing* **7**(2), pp.89–98, (2014).
34. A. Caldarelli, L. Ferri and M. Maffei, "Expected benefits and perceived risks of cloud computing: an investigation within an Italian setting". *Technology Analysis & Strategic Management*, **29**(2), pp. 167-180, (2017).
35. R. Otuka, D. Preston and E. Pimenidis, "The use and Challenges of Cloud Computing Services in

SMEs in Nigeria". In: *Proceedings of the European Conference on Information Management*, pp.325, (2014).

36. I. Senarathna, W. Yeoh, M. Warren and S. Salzman, "A Conceptual Model for Cloud Computing Adoption by SMEs in Australia". Delivery and Adoption of Cloud Computing Services in Contemporary Organizations, pp.100-128, (2015).
37. S. Chhabra, V.S. Dixit, "Cloud Computing: State of the Art and Security Issues". In: *ACM SIGSOFT Software Engineering Notes*, **40** (2), (2015).
38. R. Barona, E.A. Mary Anita, "A Survey on Data Breach Challenges in Cloud Computing Security: Issues and Threats". In: *International Conference on circuits Power and Computing Technologies [ICCPCT]*, (2017).
39. J. Singh, "Study on Challenges, Opportunities and Predictions in Cloud Computing". In: *International Journal Modern Education and Computer Science*, **3**, pp. 17-27, (2017).
40. CSA (2010). Top ten threats in cloud computing. Cloud Security Alliance, (2010).
41. [41] S.M. Shariati, Abouzarjomehri, M.H. Ahmadzadegan, "Challenges and security issues in cloud computing from two perspectives: Data security and privacy protection". In: *2nd international conference on knowledge-based engineering and innovation*, (2015).
42. CSA (2013). The Notorious Nine: Cloud Computing Top Threats in 2013. Cloud Security Alliance, (2013).
43. G. Ramachandra, M. Iftikhar, F.A. Khan, "A Comprehensive Survey on Security in Cloud Computing". *The 3rd International Workshop on Cyber Security and Digital Investigation (CSDI 2017)*.
44. M. Carroll, A. Merwe, P. Kotzé, "Secure Cloud Computing Benefits, Risks and Controls". In: *Information Security South Africa (ISSA)*, pp. 1-9, (2011).
45. R. Prasad Padhy, M. Ranjan Patra, S. Chandra Satapathy, "Cloud Computing: Security Issues and Research Challenges". In: *International Journal of Computer Science and Information Technology & Security*, (2011).
46. Q. Zhang, L. Cheng and R. Boutaba, "Cloud computing: state-of-the-art and research challenges". In: *Journal of Internet Services and Applications*, **1**(1), pp.7-18, (2010).
47. S. O. Kuyoro, F. Ibikunle and O. Awodele, "Cloud Computing Security Issues and Challenges". In: *MIPRO* (2010).
48. T. Dillon, C. Wu and E. Chang, "Cloud Computing: Issues and Challenges". In: *24th IEEE International Conference on Advanced Information Networking and Applications*, (2010).
49. K. Mualla, D. Jenkins, "Evaluating Cloud Computing Challenges for Non-Expert Decision-Makers". In: *International Journal of Digital Information and Wireless Communications (IJDIWC)* **5**(4), pp. 285-296 The Society of Digital Information and Wireless Communications, (2015).
50. S. Kumar and R.H. Goudar, "Cloud Computing-Research Issues, Challenges, Architecture, Platforms and Applications: A survey". In: *International Journal of Future Computer and Communication*, **1**(4), (2012).
51. J.H. Morin, J. Aubert and B. Gateau "Towards Cloud Computing SLA Risk Management: Issues and Challenges". In: *45th Hawaii International Conference on System Sciences*, (2012).
52. K. Popović, Ž. Hocenski, "Cloud Computing security issues and challenges". In: *MIPRO* (2010).
53. O. Ali, J. Soar, J. Yong, "Challenges and Issues that are Perceived to Influence Cloud Computing Adoption in Local Government Councils". In: *Proceedings of the IEEE 21st International Conference on Computer Supported Cooperative Work in Design*, (2017).
54. A. Monjur and A.T. Litchfield "Taxonomy for Identification of Security Issues in Cloud Computing Environments". In: *Journal of Computer Information Systems*, **58**(1), pp.79-88, (2016), DOI: 10.1080/08874417.2016.1192520.
55. M. Zaigham, "Data Location and Security Issues in Cloud Computing". In: *International Conference on Emerging Intelligent Data and Web Technologies*, (2011).
56. M.G. Avram (Olaru), "Advantages and challenges of adopting cloud computing from an enterprise perspective". In: *The 7th International Conference Interdisciplinarity in Engineering (INTER-ENG 2013)*.
57. P. Goyal, "Enterprise Usability of Cloud Computing Environments: Issues and Challenges". In: *2010 Workshops on Enabling Technologies: Infrastructure for Collaborative Enterprises*, (2010).
58. E. Cayirci, A. Garaga, A. Santana de Oliveira and Y. Roudier, "A risk assessment model for selecting cloud service providers" In: *Journal of Cloud Computing: Advances, Systems and Applications*, **5**(14), (2016).
59. B. Gatewood, "Clouds on the information horizon: How to avoid the storm". Information Management (15352897), **43**(4), pp. 32-36, (2009).
60. D. Yimam and E.B. Fernandez, "A survey of compliance issues in cloud computing". In: *Journal of Internet Services and Applications*, **7**(5), (2016).
61. H. AlJahdali, A. Albatli, P. Garraghan, P. Townend, L. Lau, J. Xu, "Multi-Tenancy in Cloud Computing". In: *IEEE 8th International Symposium on Service Oriented System Engineering*, (2014).
62. P. Saripalli, and B. Walters, "QUIRC: a quantitative impact and risk assessment framework for cloud security," *IEEE 2nd International Conference on Cloud Computing* (2010).

63. A. Wayne Jansen," Cloud hooks: security and privacy issues in cloud computing," *Proceedings of the 44th Hawaii International Conference on System Sciences* (2011).
64. A. Duncan, S. Creese, M. Goldsmith, "Insider Attacks in Cloud Computing". In: *IEEE 11th International Conference on Trust, Security and Privacy in Computing and Communications*, (2012).
65. P. Hemant, N. P. Chawande, A. Sonule and H. Wani, "Development of servers in cloud computing to solve issues related to security and backup". In: *Proceedings of the IEEE International Conference on Cloud Computing and Intelligence Systems*, pp. 158- 163, (2011).
66. O. Ali, J. Soar and J. Yong, "An investigation of the challenges and issues influencing the adoption of cloud computing in Australian regional municipal governments". In: *Journal of Information Security and Applications*, **27/28**, pp. 19-34, (2016).
67. N. Gonzalez, C. Miers, F. Redígolo, M. Simplıcio, T. Carvalho, M. Naslund and M. Pourzandi, "A quantitative analysis of current security concerns and solutions for cloud computing". In: *Journal of Cloud Computing: Advances, Systems and Applications*, **1**(11), (2012).
68. R. Latif, H. Abbas, S. Assar, and Q. Ali "Cloud Computing Risk Assessment: A Systematic Literature Review". In: *Future Information Technology*, Lecture Notes in Electrical Engineering, **276**, pp. 285-295, (2014).
69. S. Pearson, A. Benameur, "Privacy, Security and Trust Issues Arising from Cloud Computing". In: *2nd Int Conference on Cloud Computing Technology and Science* (2010).
70. J. Zhu, D. Li, J. Wu, H. Liu, Y. Zhang, J. Zhang, "Towards bandwidth guarantee in multitenancy cloud computing networks". In *Network Protocols (ICNP), 2012 20th IEEE International Conference*, (2012).
71. R. Buyya, S. Pandey, and C. Vecchiola, "Cloudbus toolkit for market-oriented cloud computing," In *Cloud Computing*, ed: Springer, pp. 24–44, (2009).
72. P. Belimpasakis and S. Moloney, "A platform for proving family oriented RESTful services hosted at home". In: *Consumer Electronics, IEEE Transactions on*, **55**, pp. 690–698, (2009).
73. A. Naim, "Cloud Computing: Technology, Security Issues and Solutions". In: *2nd International Conference on Anti-Cyber Crimes (ICACC)*, (2017).
74. A. Shawish, M. Salama, "Cloud Computing:Paradigms and Technologies". F. Xhafa and N. Bessis (eds.), Inter-cooperative Collective Intelligence: Techniques and Applications, Studies in Computational Intelligence 495, DOI: 10.1007/978-3-642-35016-0_2, Springer-Verlag Berlin Heidelberg, (2014).
75. G. Sumit, "Public vs Private vs Hybrid vs Community -Cloud Computing: A Critical Review". I.J. *Computer Network and Information Security*, **3**, pp. 20-29, (2014), DOI:10.5815/ijcnis.2014.03.03
76. A. Duncan, S. Creese, M. Goldsmith, "An overview of insider attacks in cloud computing". In: *Concurrency And Computation: Practice And Experience*, **27**, pp. 2964–2981, (2014) in Wiley Online Library. DOI: 10.1002/cpe.3243.
77. T.K. Damenu, C. Balakrishna, "Cloud Security Risk Management A Critical Review". In: *9th International Conference on Next Generation Mobile Applications, Services and Technologies*, (2015)
78. A. Dutta, G. C. A. Peng and A. Choudhary, "Risks in Enterprise Cloud Computing: The Perspective of it Experts", *Journal of Computer Information Systems*, **53**(4), pp. 39-48, (2013), DOI: 10.1080/08874417.2013.11645649
79. M. Ahmed and A. T. Litchfield, "Taxonomy for Identification of Security Issues in Cloud Computing Environments", *Journal of Computer Information Systems*, **58**(1), pp. 79-88, (2018), DOI: 10.1080/08874417.2016.1192520
80. D. Zissis, D. Lekkas, "Addressing cloud computing security issues". *Future Generation Computer Systems*, **28**, pp. 583–592, (2012)

Design and sizing of a Hybrid PV-Wind-Grid System for Electric Vehicle Charging Platform

Kamal Anoune[1,2*], Mohsine Bouya[1], Abdelali Astito[2], Abdellatif Ben Abdellah[1,3]

[1]*Laboratory of Renewable Energies and Advanced Materials (LERMA), International University of Rabat (UIR), Sala Al Jadida 11000, Morocco. kamal.anoune@gmail.com.*
[2]*Laboratory of Informatics, Systems &Telecommunications (LIST), Faculty of Science and Technology of Tangier (FST) – Abdelmalek-Essaadi University (UAE), Ziaten. BP: 416, Tangier, Morocco.*
[3]*Laboratory of Engineering, Innovation and Management of Industrial Systems (LEIMIS), Faculty of Science and Technology of Tangier (FST), Abdelmalek Essaâdi University (UAE), Ziaten. BP: 416, Tangier, Morocco.*

Abstract. The recent technological advancement in the power electronics converters, control, and battery technology, allowed the enhancement of the use of the electric vehicle (EV) as a means of transportation, EV requires to be charged with an electrical grid which imposes an extra burden on the electrical supply during the peak demand duration. The use of renewable energy sources (RES) such as solar and wind remains as alternative sources to reduce this negative impact. The main interest of this work is integrating the RES in transportation systems through a smart grid system with a proposed design of a hybrid PV-Wind-Grid charging platform for EV.

Keywords: Electric Vehicle (EV); Vehicle-To-Grid (V2G), EV Charging Platform (EVCP), Hybrid PV-Wind-Grid System (HPWGS), Renewable Energy Sources (RES).

1 Introduction

The use of EV in the transportation sector is a promising vision aiming to solve the environmental issue and energy fossil crisis altogether. In this proposal, numerous governments in the worldwide have taken various initiatives to encourage the EV deployment. The environmental and economic are the major points of interest. Distinguished issues because of EV charging on the power grid mainly, a voltage drops, overloading of the grid, and finally harmonics and system losses. In spite of the fact that EV charging has stances challenges to the power grid operation, the EV deployment can bring a new view to the power grid subsequently EV can be considered as portable energy storage and can be powered from renewable energy such as solar and wind.

The renewable energy is used to charge the EV battery, the excess of this energy is injected into the grid, moreover EV battery can be utilized to power the grid using a bidirectional power converter, the energy exchange between the power grid and EV battery can be feasibly operated using the concept of vehicle to grid (V2G). It could provide various services to the hybrid power system, such as regulation of the power grid, rotating reserve, shaving of peak load, reactive power compensation and a realistic hub for renewable energy resources integration [1]. Several studies have analyzed the design, controls strategies and experimental test of the integration of an EV charging station as DC microgrid then studied the possibilities of the Vehicle to Grid (V2G) operation following different DC architecture [2]. Birnie et al. [3] investigated the effect of daytime recharging utilizing photovoltaic powering system situated at workplace parking lots. Nunes et al. [4] explored the potential complementarities of the energy excess during a daytime for renewable energy production for EV charging. Traube et al. [5] presented an experimental system where a bidirectional DC-DC electric vehicle charger is located between the high-voltage DC bus of a PV system and the EV battery. Chandra Mouli et al. [6] investigated and compared different system architecture for an EV-PV charger. Then they reviewed the power converter topologies that integrate EV and PV for bidirectional power capability for V2G operation. Numerous studies are focused to present the importance of the use of a hybrid system as realistic power sources compared to the conventional fossil sources and to maximize the renewable power production [7], Anoune et al. [8] focused to size and integrate a PV-wind system to power process an industrial prototype using a deterministic approach. Providing a correct hybrid system configuration witch ensure powering load with a minimum cost of investment is the main objective of several research works. In this regard, Anoune et al. [9] presented a review work of current sizing and optimization technique used to size a

* Kamal Anoune: kamal.anoune@gmail.com

hybrid PV wind system following different techniques, such as deterministic/stochastic, program tools and algorithms.

The main objectives of this paper is to size and integrate a hybrid PV-Wind-Grid system (HPWGS) to power a local prototype of EV charging platform (EVCP), and even to promote the use of the renewable energy sources (RES) such as a power source with the minimal energy exchange with the grid [7] . Moreover, an appropriate EVCP could provide a great flexibility to exchange the energy flow between RES, EV battery, and the grid. In this regard, a profile load of a slow charging station is adopted, then the campus of the international university of Rabat (UIR) Morocco is chosen as a location under study, next, a dynamic simulation of the meteorological data is presented, and finally, the long-term performance of the optimized system requirement is done through providing a dynamic simulation to illustrate the energy exchange between all components of EVCP of the chosen period.

2 Hybrid PV-Wind-Grid System for EV Charging Platform

An electric vehicle is powered exclusively by electricity through a charging station. The energy stored in its rechargeable batteries is used to ensure the mobility using an electric motor. The EV is more considerate to the environment than traditional cars when the energy produced by RES resources is used for their charging such as the current HPWGS under study.

The utilization of EV required another necessity to the help supports infrastructure to keep up their charging. EV must be equipped with the straightforward charger with an estimated input power between 3.3 kW to 3.7 kW, this equivalent 16 A current with 230 VAC such as power common in public uses. Another possibility is the use of 3 phase AC chargers with input power 22 kW equivalents to 32 A current with 400 VAC. These chargers may be used to charge an EV in slow charging mode. Although EV Chargers with input control up to 100 kW permitted to charge EV in the quick charging mode.

EV can be charged with using EVCP located in a specific zone, two types of charging speed can be identified.

- Fast Charging Station (FCS) is characterized by a quite short time (15 – 30 minutes) for EV charging process and can reach to 50 – 80% of their maximum capacity. Several battery parameters should be identified and measured during the fast charging cycles to prevent the damage of the EV Batteries.

- Slow Charging Station (SCS) may take approximately between 5 to 8 hours. EV batteries can be charged up to 100% of their maximum capacity.

The use of smaller current during slow charging causes less technical problems and less damage to batteries. The SCS can present the real opportunity to be powered by the renewable energy system (RES) such as photovoltaic or wind turbine through an HPWGS with the maximum power reliability and the minimum cost of investment.

3 Sizing the HPWGS

In this work, the slow charging mode in EVCP is adopted. It can be used to charge EV batteries starting from 3.3 kW to 3.7 kW per hour. Referring to Nissan LEAF catalog [10], the 2011 Nissan Leaf has 24 kWh-capacity batteries and required at the least 3.3 kW per hour for its charging process. The EVCP required at the least an input power up to 3.7 kW with 230 V and should ensure between 5 to 8 hours of charge during the packing day in order to reach the EV batteries charge of 100%.

Under the constraints of the maximum power reliability and the minimum system cost of EVCP, a sizing-simulation approach helps to determine the appropriate HPWGS requirement regarding three fundamental parameters:

- Meteorological input mainly the global irradiation, and the wind velocity of the chosen local implantation.
- The electrical demand profile, which explains the behavior of electrical power consumption.
- The specification of the hybrid system that includes the configuration and the technical specification of the desired equipment such as photovoltaic panels, wind turbine, and battery

The optimized configuration of HPWGS helps to power EVCP that used to charge the EV batteries. An appropriate developed program under TRNSYS (transient simulation package) [11] and Matlab are used to simulate the long-term electric energy output from a photovoltaic array and wind turbine. The campus of the International University of Rabat (UIR), Morocco (at latitude 33.980 N, a longitude 6.709 W and an altitude of 75 m above sea level) will host the HPWGS under study.

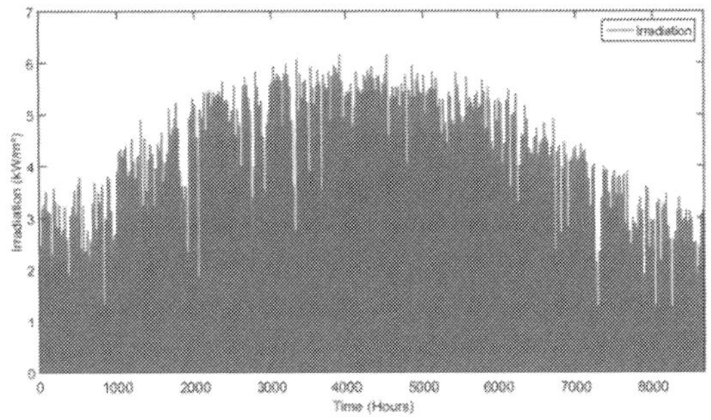

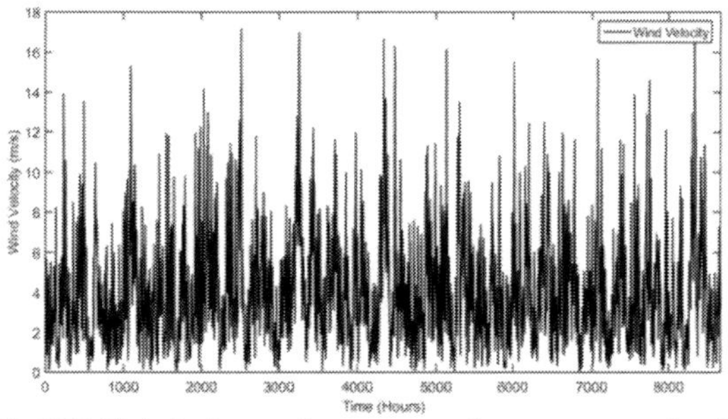

Figure 1: Hourly and monthly weather data for UIR Rabat

In IUR Rabat, the yearly average solar energy per day is 5.47 kWh/m² (See Fig .1). The yearly average wind speed at 10 m hub height is 3.14 m/s. The hourly solar irradiation on a horizontal plane (Wh/m²), as well as and wind speed (m/s), is illustrated respectively in (Fig. 1).

4 Result and discussion

The proposed hybrid system configuration of HPWGS is composed of 10 kWp of PV and 1 kW of the wind turbine. The dynamic simulation results of the HPWGS energy production and exchange are shown in the figure below (Fig.2 & 3). The yearly average of energy production from a PV array is 11287.52 kWh, while the yearly average of energy production by the wind turbine is 1137.19 kWh. The optimized hybrid configuration can ensure 12424.71 kWh/Year, while the electrical load demand of the HPWGS is around 12154.5 kWh/Year.

The energy output of the chosen period (April 01-05) and (January 02-06) is presented respectively in (Fig. 2) and (Fig. 3). The blue curve represents the hybrid energy production of the PV array and WT. The orange curve represents the nominal electrical load demand of an EVCP during a daytime. The excess of energy serves to charge the storage system (green curve). It may happen that the hybrid produced power doesn't cover the load demand at a specific period of the day. In that case, the battery can ensure the needed energy until its discharge (purple curve). When that happens, the system can connect to the grid in order to feed the loads (Fig. 2 & 3).

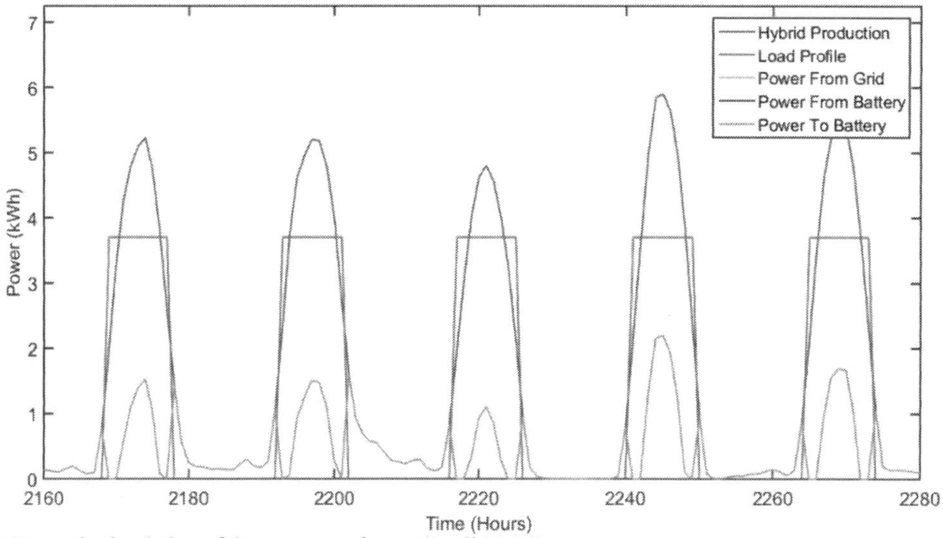

Figure 2: Dynamic simulation of the energy exchange (April 01-05)

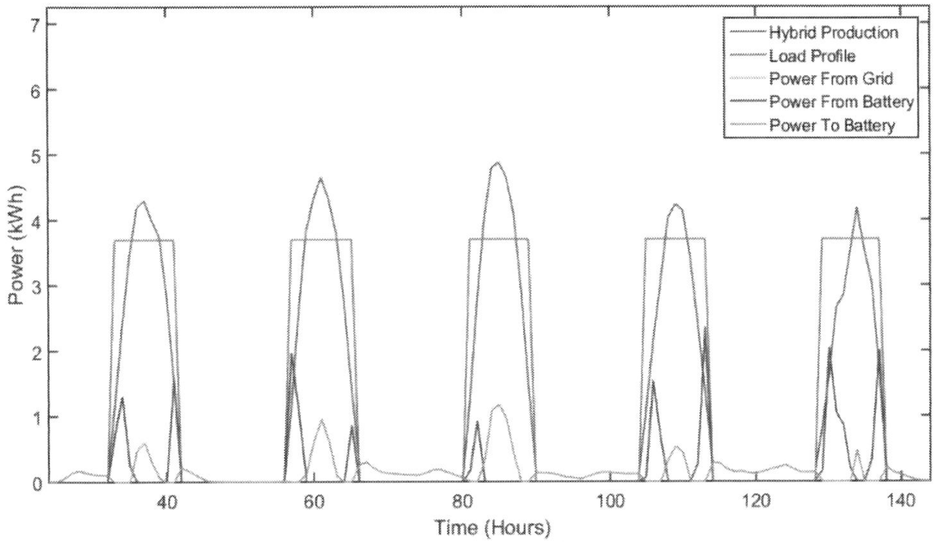

Figure3: Dynamic simulation of the energy exchange (January 02-06)

Conclusions

A modular EVCP using HPWGS are leading to a significant resolution of the low mobility distance caused by the low energy density of EV batteries and insufficient support infrastructure of the EVCP. The integration of Hybrid PV-Wind-Grid System. The Electric Vehicle Charging Platform is considered as a realistic renewable and sustainable solution aiming to power the electrical energy demand during the parking period. The EV charging process can be ensured totally or partially by HPWGS through the energy output of photovoltaic panels (PV) and wind turbine (WT). The excess of energy production after fully charged the EV battery is injected into the grid, the EVCP can be connected to the grid to charge the EV batteries when there is no sufficient energy output from the RES and no sufficient energy stocked at the EVCP batteries.

A simulation study has been conducted for the optimized configuration of HPWGS aiming to evaluate the long-term energy output, then to illustrate the energy exchange between the EVCP batteries, EV batteries and grid using the TRNSYS/Matlab program. The electrical energy will be provided from the renewable energy system installed on the upper surface of the covered parking surface. EVCP will ensure a sustainable gateway for transportation in the future with a direct utilization of the HPWGS.

Acknowledgments

The authors would like to express their appreciation to "IRESEN" by providing financial support to carry out this research under the project "Continued heating of bitumen storage by using high energy solar thermal collectors". We would like to emphasize that. We have not been able to complete this research without the joint support of all employees of the renewable energy laboratory and its hierarchical superior of the UIR (International University of Rabat). We take this opportunity to express our deep gratitude and greeting to M. LAKNIZI Azzeddine for his inspiring and invaluable advice and availability.

References

[1] Yilmaz M, Krein PT. Review of the Impact of Vehicle-to-Grid Technologies on Distribution Systems and Utility Interfaces. IEEE Trans Power Electron 2013;28:5673–89. doi:10.1109/TPEL.2012.2227500.

[2] Capasso C, Veneri O. Experimental study of a DC charging station for full electric and plug in hybrid vehicles. Appl Energy 2015;152:131–42. doi:10.1016/j.apenergy.2015.04.040.

[3] Birnie DP. Solar-to-vehicle (S2V) systems for powering commuters of the future. J Power Sources 2009;186:539–42. doi:10.1016/J.JPOWSOUR.2008.09.118.

[4] Nunes P, Farias T, Brito MC. Day charging electric vehicles with excess solar electricity for a sustainable energy system. Energy 2015;80:263–74. doi:10.1016/j.energy.2014.11.069.

[5] Traube J, Lu F, Maksimovic D. Electric vehicle DC charger integrated within a photovoltaic power system. Conf Proc - IEEE Appl Power Electron Conf Expo - APEC 2012:352–8. doi:10.1109/APEC.2012.6165843.

[6] Chandra Mouli GR, Bauer P, Zeman M. Comparison of System Architecture and Converter Topology for a Solar Powered Electric Vehicle Charging Station. ECCE Asia 2015:1908–15. doi:10.1109/ICPE.2015.7168039.

[7] Anoune K, Bouya M, Ghazouani M, Astito A, Abdellah A Ben. Hybrid renewable energy system to maximize the electrical power

production. 2016 Int. Renew. Sustain. Energy Conf., Marrakech, Morocco: 2016, p. 533–9. doi:10.1109/IRSEC.2016.7983992.

[8] Anoune K, Laknizi A, Bouya M, Astito A, Ben Abdellah A. Sizing a PV-Wind based hybrid system using deterministic approach. Energy Convers Manag 2018;169:137–48. doi:10.1016/j.enconman.2018.05.034.

[9] Anoune K, Bouya M, Astito A, Abdellah A Ben. Sizing methods and optimization techniques for PV-wind based hybrid renewable energy system: A review . Renew Sustain Energy Rev 2018;93:652–73. doi:10.1016/j.rser.2018.05.032.

[10] Nissan Motor. Nissan LEAF specifications. n.d.

[11] TRNSYS. http://www.trnsys.com/ n.d.

A multi-objective approach for the clustering problem in wireless ad hoc networks

Khalil Amine*

MOAD6 Research Team, Mohammadia School of Engineering, Mohammed V University of Rabat, Morocco

Abstract. Ad hoc networks are infrastructure-less computer networks consisting of heterogeneous mobile devices connected by wireless links. The devices are essentially faced to several limitations in terms of embedded energy, processing power, bandwidth, and internal memory. Clustering has been introduced in ad hoc networking to provide hierarchy and stability and to support information circulation reduction, energy management, and scalability. It consists of a process of self-aggregation of nodes into a set of sub-networks called clusters where each cluster would be managed by a master node called cluster centre responsible for gathering, maintaining and updating the cluster topology information as well as ensuring the data routing over the network. The set of cluster centres institutes a temporary network core that enhances the topology maintenance. From a graph perspective, this set represents a dominating set, that is a set of nodes such that each node of the network is either an element or adjacent to at least one element of this set. This paper develops a constrained bi-objective optimisation model for finding a dominating set that provides a trade-off between maximising the energy amount and minimising the cardinality. A bi-objective simulated annealing meta-heuristic for constructing near-optimal solutions is then presented.

1 Introduction

An ad hoc network is a wireless infrastructure-less network containing heterogeneous mobile devices with several limitations in terms of embedded energy, processing power, bandwidth, and internal memory. Ad hoc networks are self-organising networks: nodes are able to establish reactively a temporary network topology tolerant of any increase or decrease in the number of devices, and are able to find multi-hop routes to relay any message, when needed. Typical applications of ad hoc networks include military intervention, emergency operations, remote areas coverage, and vehicular communication networks.

Ad hoc networks are faced to two main challenges, namely the energy management and the dynamic topology. Clustering has been introduced in ad hoc networking to deal with such situation. It consists of a process of self-aggregation of nodes into a set of sub-networks called clusters where each cluster would be managed by a master node called cluster centre responsible for gathering, maintaining and updating the cluster topology information as well as ensuring the data routing. Clustering allows reducing the total network energy consumption by decreasing the amount of 'circulatory' information, and limiting nodes responsibility. It brings temporary network core that enhances the topology stability and increases the network lifetime.

From a graph theoretical perspective, cluster centres represent a dominating set, that is a set of nodes such that each node of the network is either an element or adjacent to at least one element of this set. Selection of cluster centres is the main stage of a clustering process that requires a specific selection criterion. This paper develops a constrained bi-objective optimisation model for finding a dominating set that provides a trade-off between maximizing the energy amount and minimising the cardinality. A bi-objective simulated annealing meta-heuristic for constructing near-optimal solutions is then presented.

*e-mail: khalilamine@research.emi.ac.ma

2 Background and problem statement

Clustering in ad hoc networks consists of a self-aggregation of nodes into a number of sub-networks called clusters. Each cluster is managed by a master node called cluster centre, or more commonly cluster head. Clustering institutes an abstract structure, as illustrated in figure 1, that allows to differentiate between three types of nodes:

- cluster centre, responsible for gathering, maintaining and updating the cluster topology information as well as ensuring data routing;
- ordinary member, a node related to a cluster centre;
- gateway, a router node enabling the access to its related cluster.

This temporary structure enhances the network stability and increases the network lifetime. Furthermore, it allows reducing the amount of circulatory information, thereby reducing the total network energy consumption. From a conceptual perspective, while the communication is managed proactively inside clusters, each cluster can be considered as an independent entity allowing the communication over the network to be managed reactively. That is, information related to ordinary members is locally shared within the associated cluster and would be updated periodically, yet only information related to cluster centres is shared globally when needed.

Selection of cluster centres represents the main stage in a network clustering process that requires a specific selecting criterion. Several clustering criteria have been introduced in the literature, yet few studies [5, 13] have considered a selecting criterion that is related directly to the residual energy in the nodes. This work aims at dealing with energy limitation in ad hoc networks by considering the residual energy as main key of clustering. Accordingly,

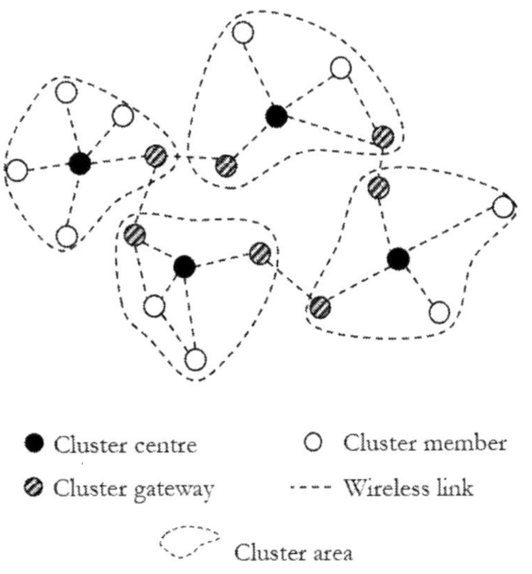

Figure 1: A hypothetical clustered topology

only nodes with high performance would be trusted to be assigned the cluster centre status. Besides, the number of selected cluster centres should be reduced as much as possible in order to preserve the total network energy amount.

From a graph theoretical perspective, clustering would consist of finding a dominating set, that is a set of nodes such that each node of the network is either an element or adjacent to at least one element of this set. Yet, as illustrated in figure 2, a dominating set for a graph is not unique and a network would contain many dominating sets with different cardinality and, then, with different performance. Therefore, considering an ad hoc network as an undirected graph, the considered problem is reduced to find an efficient dominating set that provides a compromise between the energy amount and the cardinality.

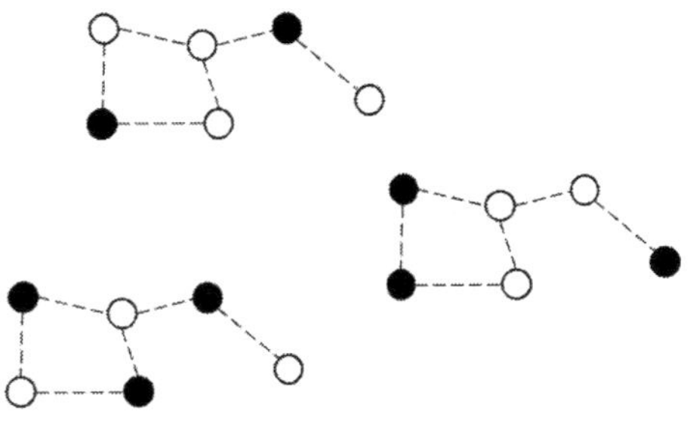

Figure 2: Three dominating sets for the same network

3 Mathematical model

Let $G = (V, E)$ be an undirected graph where V is the set of vertices representing the network nodes and E the set of edges (pairs of nodes) representing the available direct communication links, and let us consider the following notations

$\mathcal{D}$	:	a dominating set in G
$C(v)$	:	a cluster with centre v
e_v	:	residual energy in a node v
ε_v	:	minimal energy required for a node v to manage the communication between each related node
κ	:	maximal number of clusters

Let us consider the variables x and y such that

$$x_{uv} = \begin{cases} 1 & \text{if } u \in C(v) \\ 0 & \text{otherwise} \end{cases}$$

and

$$y_v = \begin{cases} 1 & \text{if } v \text{ is a cluster centre} \\ 0 & \text{otherwise} \end{cases}$$

The proposed mathematical formulation consists of the following model

$$\begin{cases} \max\limits_{v} \sum\limits_{v \in V} e_v y_v \\ \min\limits_{v} \sum\limits_{v \in V} y_v \\ \text{subject to} \\ \quad \sum\limits_{v \in V} x_{uv} = 1 \quad ; \quad \forall u \in V & (1) \\ \quad e_v y_v \geq \varepsilon_v \sum\limits_{u \in V} x_{uv} \quad ; \quad \forall v \in V & (2) \\ \quad \sum\limits_{v \in V} y_v \leq \kappa & (3) \\ \quad x_{uv}, y_v \in \{0, 1\} \quad ; \quad \forall u, v \in V \end{cases}$$

Model constraints explanation is given as follows:

- constraint (1) stipulates that each node is assigned to exactly one cluster centre;
- constraint (2) requires that each cluster centre should have enough energy to manage communications with all its cluster members;
- constraint (3) imposes a largest number κ of constructed clusters.

Furthermore, we consider the following assumptions:

- the network is static or with low mobility;
- nodes have the ability to know their current connectivity degree;
- devices are equipped by an Advanced Configuration and Power Interface (ACPI) module (or similar) allowing managing and monitoring the embedded energy. In other words, nodes are able to access the information about their own residual energy;
- nodes are identified by means of a unique number (label): the Medium Access Control (MAC) address which is given by manufacturers to every wireless interface.

4 Proposed meta-heuristic

Simulated annealing is a local search method introduced by Kirkpatrick et al. [6], initially for global combinatorial mono-objective optimisation problems. It consists at each iteration of generating a candidate solution that will be whether accepted as a result of a comparison with a current solution in hand. A certain transition rule is set according to the objective function (called energy function) that aims at moving regularly to an improving solution. Yet, some uphill solutions would be accepted occasionally in order to guarantee a diversity of choice within the domain exploration and to avoid getting caught at local optima. The number of iterations is controlled by a certain cooling schedule consisting of a temperature parameter and a decreasing function. Simulated annealing has showed significant success for many applications – in spite of the proven asymptotic convergence to the optimal solution, it allows gradual convergence to a near-optimal solution in practice, and it requires low short-term memory. More detailed review can be found in [2, 4, 8, 10].

Multi-objective simulated annealing approach has been introduced in the work of Serafini [7] and have merged in many subsequent works [1, 3, 9, 11, 12, etc.] providing a construction of the near-optimal solutions. In general, multi-objective simulated annealing methods use an archive memory that catches the non-dominated solutions while exploring the feasible domain, and they often consider a variation of a certain composite energy function for solutions evaluation. Like all meta-heuristics, simulated annealing is problem-dependent and the choice of appropriate parameters is rather a challenging task, particularly in multi-objective case.

We propose a bi-objective simulated annealing approach to find dominating sets that each corresponds to a certain compromise between the total energy amount and the number of elements. As illustrated in figure 3, the meta-heuristic consists of two nested loops: the outer loop performs a cardinality reduction by removing a randomly chosen node from the input solution, and the inner loop consists of a swap procedure that generates neighbouring solutions by replacing randomly a node from the current solution by another one from the network.

The meta-heuristic starts from a solution of cardinality κ, and performs a sequence of reductions and swaps to result hopefully in an efficient solution for each cardinality level (cf. figure 4). The number of iterations is controlled by means of an equilibrium and a cooling condition. Some uphill solutions, possibly infeasible, would be accepted occasionally according to a probabilistic parameter that involves both the two objective functions. A 'Pareto archive' is set to store non-dominated solutions. Each efficient solution found is considered as a potential non-dominated solution for the current cardinality level and the Pareto archive is updated in consequence.

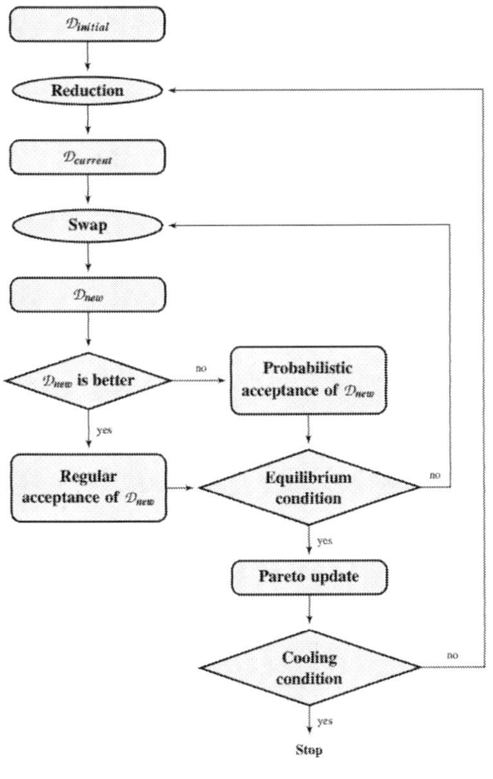

Figure 3: Proposed simulated annealing algorithm flowchart

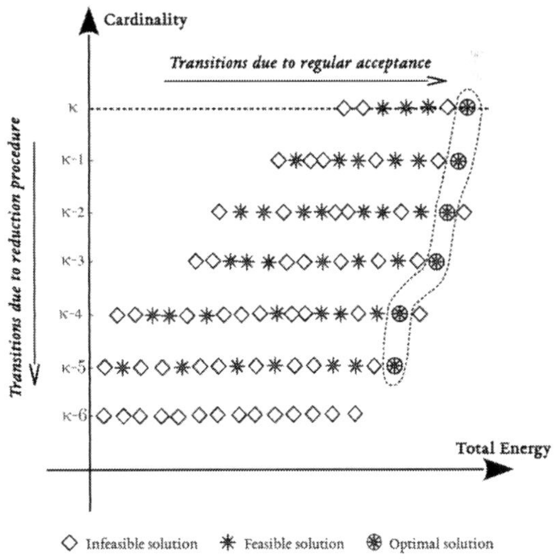

Figure 4: Directions of transitions in the objective space

5 Conclusion

Clustering in ad hoc networks provides hierarchy and stability and supports information circulation reduction, energy management, and scalability. The main clustering stage is the selection of cluster centres that is equivalent from a graph theoretical perspective to finding a dominating set satisfying the considered clustering purpose. A constrained bi-objective optimisation model for finding a dominating set that provides a trade-off between maximizing the energy amount and minimising the cardinality has been introduced. Due to the significant success of simulated annealing and particularly for the requirement of low short-term memory, a bi-objective simulated annealing meta-heuristic for constructing near-optimal solutions has been proposed.

References

[1] M. H. Alrefaei and A. H. Diabat, "A simulated annealing technique for multi-objective simulation optimization," *Applied Mathematics and Computation*, vol. 215, no. 8, pp. 3029–3035, 2009.

[2] Kh. Amine, "Insights into simulated annealing," in *Handbook of Research on Modeling, Analysis, and Application of Nature-Inspired Metaheuristic Algorithms*, S. Dash, B. K. Tripathy, and A. Rahman, Eds. Hershey, PA, USA: IGI Global, 2018, pp. 121–139.

[3] S. Bandyopadhyay, S. Saha, U. Maulik, and K. Deb, "A Simulated Annealing-Based Multiobjective Optimization Algorithm: AMOSA," *IEEE Transactions on Evolutionary Computation*, vol. 12, no. 3, pp. 269–283, 2008.

[4] K. A. Dowsland and J. M. Thompson, "Simulated annealing," in *Handbook of Natural Computing*, G. Rozenberg, Th. Bäck, and J. N. Kok, Eds. Berlin Heidelberg: Springer-Verlag, 2012, pp. 1623–1655.

[5] D. Gavalas, G. Pantziou, Ch. Konstantopoulos, and B. Mamalis, "Clustering of mobile ad hoc networks: An adaptive broadcast period approach," in *Proceedings of the 19th IEEE International Conference on Communications*, ser. ICC'06, vol. 9, Istanbul, Turkey, June 11-15 2006, pp. 4034–4039.

[6] S. Kirkpatrick, C. D. Gelatt, and M. P. Vecchi, "Optimization by simulated annealing," *Science*, vol. 220, no. 4598, pp. 671–680, 1983.

[7] P. Serafini, "Simulated annealing for multi objective optimization problems," in *Multiple Criteria Decision Making: Proceedings of the Tenth International Conference: Expand and Enrich the Domains of Thinking and Application*, G. H. Tzeng, H. F. Wang, U. P. Wen, and P. L. Yu, Eds. New York, NY: Springer-Verlag, 1994, pp. 283–292.

[8] N. Siddique and H. Adeli, "Simulated annealing, its variants and engineering applications," *International Journal on Artificial Intelligence Tools*, vol. 25, no. 6, 2016, 1630001, 25p.

[9] B. Suman, "Simulated annealing-based multiobjective algorithms and their application for system reliability," *Engineering Optimization*, vol. 35, no. 4, pp. 391–416, 2003.

[10] B. Suman and P. Kumar, "A survey of simulated annealing as a tool for single and multiobjective optimization," *Journal of the Operational Research Society*, vol. 57, no. 10, pp. 1143–1160, 2006.

[11] O. Tekinalp and G. Karsli, "A new multiobjective simulated annealing algorithm," *Journal of Global Optimization*, vol. 39, no. 1, pp. 49–77, 2007.

[12] E. Ulungu, J. Teghem, P. Fortemps, and D. Tuyttens, "MOSA method: A tool for solving multiobjective combinatorial optimization problems," *Journal of Multi-Criteria Decision Analysis*, vol. 8, no. 4, pp. 221–236, 1999.

[13] J. Wu, F. Dai, M. Gao, and I. Stojmenovic, "On calculating power-aware connected dominating sets for efficient routing in ad hoc wireless networks," *Journal of Communications and Networks*, vol. 4, no. 1, pp. 59–70, 2002.

https://doi.org/10.1051/matecconf/201820000010

Optimization models for train load and transport planning problems

Lamiaa Dahite[1,*], *Abdeslam* Kadrani[1], and *Rachid* Benmansour[1]

[1] *SI2M, Institut National de Statistique et d'Economie Appliquée , Rabat, Morocco*

Abstract. Optimization of rail transport is complex because of the industry's multiple constraints. The transport of containers is very particular since it is characterized in addition to its specificities as a product to be loaded and transported by a strong instability of the demand. So far research in this area has dealt only with the separate treatment of the train load and transport problem. The present study focuses on optimizing resources facing unstable demand for the combined problem. Mathematical models are proposed to assign customers demands to wagons and for railcars allowance per axis depending on the available park, the locomotive capacity and the train length. An algorithm for the train load problem is also suggested. The models have been tested to measure their efficiency by comparing them to an existing train planning model and to manual assignment adopted in the rail industry. Some test results are finally reported to show how a novel formulation can simplify the resolution of a complex problem.

1 Introduction and literature review

Rail transport is a unique production of services that involves several stakeholders at once to respond to customers orders for different products to be transported. Each product imposes specific constraints. The design of efficient rail supply chain is challenging particularly containers transport and that for several reasons. The orders are divided into forecasted and sudden orders which we know about only the day before. The characteristics of the demand are variable over time, in terms of quantities (number of containers) and technical specifications (nature of containers and their weights). The demand is also very unstable. The delivery times of these products are very low in a highly competitive market. This traffic is also characterized by the multiplicity of stakeholders which makes its management particularly difficult. The rail network and the transport plan impose strict constraints that need to be respected.

The optimization of rail transport can be done by different methods that aim to the most appropriate planning. We can therefore intervene on loading on trains, allocation on the network or strategic planning of needed trains. Some works treated loading on trains particularly intermodal ones. Corry and Kozan [1] propose a mixed-integer programming model to determine the spatial arrangement of containers on a train known as the train planning problem to minimize a weighted total number of wagons required and equipment working time. They then use heuristics local search and simulated annealing algorithm because of the large number of variable. Ambrosino and Siri [2] propose mathematical models to find the optimal assignment of containers to train slots while allowing the possibility of reshuffles in the stacking area and to not sequentially load the train to minimize the unproductive movements. Foti *et al.*[3] present a basic model of the train load planning problem which is then extended in order to be more realistic and to better represent the re-handling operations in the storage area. The results were obtained by boolean optimization. Large problems can be generated even with a low number of containers and wagons. The model gives small formulas to counter this difficulty.

Another stream of research was about allocation in the network. Zhao *et al.*[4] propose a mathematical model for the empty cars allocation problem in railway networks with dynamic demands to minimize the total cost incurred by transferring and storing empty cars in different stages of the demand. They then propose a genetic algorithm to generate approximate optimal solutions. Mutyavavire and Battle [5] establish a linear mathematical model as a decision support system to the empty wagon re-distribution management problem. They focus mainly on network capacity and resources constraints in their formulation. Lam *et al.*[6] propose an approximate dynamic approach for determining optimal strategies to relocate empty containers in the containerized sea-cargo. They use a simulation based approach known as the temporal difference learning for average cost minimization to determinate approximate optimal solutions to the stochastic model (two-ports two-voyages). Azanov *et al.* [7] use graph theory to formulate a model to minimize the number of locomotives used for the transportation of the trains by choosing routes of the trains and locomotives. Mathematical definitions of a railway haul, a train path, a train route, and a locomotive route are suggested to do so. An heuristic search algorithm is then proposed to find an approximate solution of the problem by maximizing the usage of locomo-

* e-mail: lamiaadahite@gmail.com

tives that started earlier compared to others. Newman and Yano [8] address the problem of determining day-of-week schedules for both direct and indirect trains and containers routing decisions in a rail intermodal setting. The integer program proposed seek to minimize operating, handling and yard storage costs while meeting on-time delivery requirements. They also develop a decomposition procedure that yields near-optimal solutions and a method for providing tight bounds of the objective function. The article [9] can be classified in the last category. Jonaitis [9] highlights the importance of planning process of rail transportation. These problems are strategic, tactical and operational. They also can have different targets weather they concern the timetable, the rolling stock and the crew. He focuses on planning the structure and volume of the rolling stock. He proposes two approaches to determine the demand for trains, the first seeks to find quantitative parameters of rolling stock and the second relies on them. The determination of optimal service life of the trains, is important too, considering elements such as repair costs, number of overhauls, current expenses and operational characteristics.

This article deals with the problems of trains loading and allocating railcars to a destination taking into account the constitution of the trains. These problems are treated separately in all the reviewed articles. The models proposed aim to jointly minimize resources and maximize loading on the trains by considering several relevant train constitution and transport constraints. An algorithm suggested provides an optimal solution for the train load problem.

2 Problem definition and formulation

The model proposed to optimize containers rail transport focuses mainly on the train load problem and combines it to take into account the constraints of the axis assignment. This work is based on the reality of the Moroccan rail network and Moroccan containers trains, but a generic model that can be applied to any other network is proposed.

We need first to specify all loading patterns in advance. Our decision variable will then be the number of wagons of each loading pattern. In the Moroccan railways, customer demand is formulated with the number of containers 40 foot (l=2) and 20 foot (l=1) to transport and the tonnage of each container. We have two types of wagons 19.9m (k=1) and 14m (k=2). The weight constraints in this case will always be respected and can therefore be omitted from the model except for the arrangement d= 1 where we can place 3 containers 20 foot given lengths constraints. However, if the three containers are very full, their weights can exceed the maximum weight of the wagon. In this case the wagon 19.9 m will only support 2 containers 20 foot in this disposition. We solve therefore the problem with only four loading patterns since the first one will become similar to the third with a longer wagon. It is interesting to choose in this case this less long loading pattern for the economy of length. Other important assumptions to describe the model can be enumerated. Axis assignment constraints such as the wagons offer, the length of the train to be sent and the maximum capacity offered by the locomotive are considered. The assignment per axis is unique since it considers not only the quantity to be sent in number of wagons, but also the quantity of each loading pattern to be sent, hence the quantity of each type of wagons to send in such a way that customer demand is placed on the trains according to the loading patterns obtained. Figure 1 represents all the five possible configurations of which 4 maximize the loading on trains.

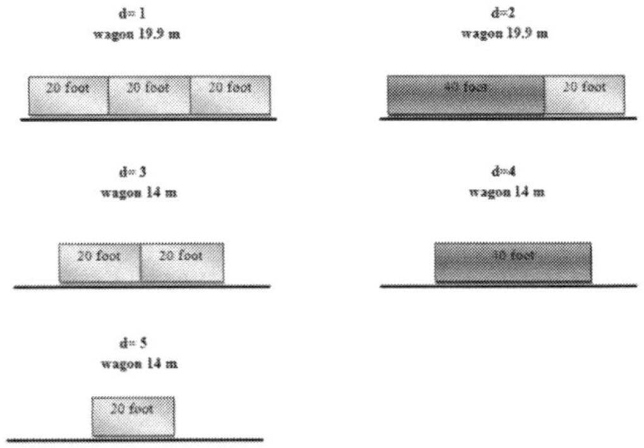

Figure 1: All possible loading patterns in Moroccan railways

2.1 Indices

Let us introduce the following notation:

- c: containers (c=1...C) where C= $max_l(q_l)$
- l: type of containers (l=1...L)
- k: type of wagons (k=1...K)
- a: axis (a=1...A)
- m: locomotive series (m=1...M)
- s: location of the container in the disposition or slot (s=1...$S(d)$)
- d: loading pattern (d=1...N)

Sets of types of containers, types of wagons, axes, locomotives, slots, loading patterns, containers of type l and type of containers in the disposition are defined respectively as follows:

$$\mathcal{L} = \{1,\ldots,L\}, \mathcal{K} = \{1,\ldots,K\}, \mathcal{A} = \{1,\ldots,A\},$$
$$\mathcal{M} = \{1,\ldots,M\}, \mathcal{S}(d) = \{1,\ldots,S(d)\}, \mathcal{N} = \{1,\ldots,N\},$$
$$\mathcal{Q}_l = \{1,\ldots,q_l\}, \mathcal{L}(d) = \{1,\ldots,L(d)\}.$$

2.2 Parameters

The following parameters are used to define the problem throughout the paper:

- n_{dl}: Number of containers type l loaded in the disposition d.
- q_l: Number of containers of type l in the order.

- P_l^v: Empty weight of a container type l.
- T_{cl}: The weight of container c of type l.
- T_k^{max}: Maximum weight a wagon of type k can support.
- e_k: The non useful extremities of a wagon of type k (length that remains unused from the flat railcar after having loaded containers).
- O_k: Available offer of saint wagons type k.
- T_k^{ar}: Tare or empty weight of a wagon of type k.
- L_k: Length of wagons type k.
- $L_{a,t}$: Maximum length of the train allowed on the axis a.
- T_{ijm}^{max}: Maximum weight that a locomotive m can tow on the line section (i,j).
- T_{am}^{max}: Maximum weight that a locomotive m can tow on the axis a where T_{am}^{max}=Min(T_{ijm}^{max}).
- T_a: Total tonnage of the order transported to a. We have:

$$T_a = \sum_{c=1}^{C}\sum_{l=1}^{L} T_{cl}$$

- N_k: Number of patterns/dispositions for wagons type k (N_0 =0).
- N: Total number of possible arrangements of all types of containers and wagons.
- L: Number of type of containers used in the problem.
- K: Number of the type of wagons.
- l_m: Length of the locomotive of series m.
- $S(d)$: Number of containers locations in the disposition or number of slots in the disposition.
- $L(d)$: Number of types of containers in the disposition.

2.3 Variables

In the the next models, the variables considered are:

- x_d: The number of wagons of loading pattern d, to transport for the order of the site.
- y_k: The number of wagons type k to be transported for the site is as the following:

$$y_k = \sum_{d=N_{k-1}+1}^{N_k+N_{k-1}} x_d$$

For model 2, we will need an additional binary variable to ensure compliance with maximum weight of wagons.

- z_{clds}: is equal to 1 if the container c type l is assigned to the location s in the loading pattern d, and 0 otherwise.

x_d and z_{clds} represent the decisions variables. It is interesting to use x_{dk} instead of x_d in the models since this formulation is easily adaptable to others load configurations and industries.

- x_{dk}: The optimal number of wagon type k and loading pattern d. In this case:

$$y_k = \sum_{d=1}^{N_k} x_{dk}$$

2.4 First mathematical model for train load and transport problem

This first model tests the weight constraint of the first disposition since the others are always verified. We set 3 loops to go through the indices of the three containers placed on the first disposition. The containers must naturally not have the same index. We ensure to have that by a condition. Whenever the total weight of the 3 containers do not exceed the maximum weight that the disposition 1 can support, the submodel 1 is executed. If the weight constraints of the first load configuration are not verified, the model eliminates disposition 1 since it becomes equivalent to 3 and the problem is solved for the four remaining patterns (submodel 2). We can also retain the indices of the containers to be placed together for this disposition after testing weight constraints.

The submodel 1 can be formulated as follows:

$$\min \sum_{d=1}^{N} x_d \tag{1a}$$

$$\sum_{d=1}^{N} n_{dl} x_d = q_l, \quad l \in \mathcal{L} \tag{1b}$$

$$\sum_{d=N_{k-1}+1}^{N_k+N_{k-1}} x_d = y_k \leq O_k, \quad k \in \mathcal{K} \tag{1c}$$

$$\sum_{k=1}^{K}\sum_{d=N_{k-1}+1}^{N_k+N_{k-1}} x_d(L_k + e_k) \leq L_{a,t} - l_m, \quad a \in \mathcal{A},\ m \in \mathcal{M} \tag{1d}$$

$$T_a + \sum_{k=1}^{K} T_k^{ar} \sum_{d=N_{k-1}+1}^{N_k+N_{k-1}} x_d \leq T_{a,m}^{max}, \quad a \in \mathcal{A},\ m \in \mathcal{M} \tag{1e}$$

$$\sum_{l=1}^{L} P_l^v q_l + \sum_{k=1}^{K} T_k^{ar} \sum_{d=N_{k-1}+1}^{N_k+N_{k-1}} x_d \leq T_{a,m}^{max}, \quad a \in \mathcal{A},\ m \in \mathcal{M} \tag{1f}$$

$$x_d,\ y_k \in \mathbb{N} \tag{1g}$$

The submodel 2 that considers only four dispositions if weights constraints are not verified for the first disposition is:

$$\min \sum_{d=2}^{N} x_d \tag{2a}$$

$$N_1 = 1 \tag{2b}$$

$$\sum_{d=2}^{N} n_{dl} x_d = q_l, \quad l \in \mathcal{L} \tag{2c}$$

$$\sum_{d=N_{k-1}+2}^{N_k+N_{k-1}+1} x_d = y_k \le O_k, \quad k \in \mathcal{K} \tag{2d}$$

$$\sum_{k=1}^{K} \sum_{d=N_{k-1}+2}^{N_k+N_{k-1}+1} x_d (L_k + e_k) \le L_{a,t} - l_m, \quad a \in \mathcal{A},\ m \in \mathcal{M} \tag{2e}$$

$$T_a + \sum_{k=1}^{K} T_k^{ar} \sum_{d=N_{k-1}+2}^{N_k+N_{k-1}+1} x_d \le T_{a,m}^{max}, \quad a \in \mathcal{A},\ m \in \mathcal{M} \tag{2f}$$

$$\sum_{l=1}^{L} P_l^v q_l + \sum_{k=1}^{K} T_k^{ar} \sum_{d=N_{k-1}+2}^{N_k+N_{k-1}+1} x_d \le T_{a,m}^{max}, \quad a \in \mathcal{A},\ m \in \mathcal{M} \tag{2g}$$

$$x_d,\ y_k \in \mathbb{N} \tag{2h}$$

The objective function (1a) minimizes the total number of wagons used of each loading pattern. The constraints (1b) aim to maximize loading of containers of the customers demand on the train. Constraints (1c) represent the offer constraint. Constraints (1d) guarantee the respect of the train length during the formation. The maximum length permitted depends on the axis (specificities of the design of the network). Constraints (1e) ensure that the tonnage of the train do not exceed the maximum tonnage that the locomotive can support on each axis for full containers. This tonnage varies from one locomotive to another and from one section of the line to the other. Constraints (1f) are dedicated to the case of empty containers transport. So, we will have (1e) or (1f) depending on the case. Constraints (1g) specify the nature of the variables. The objective function (2a) and constraints (2c) to (2h) are similar to the above objective function and constraints but are applied for the second condition. The additional constraint (2b) fixes now the number of loading patterns for wagon k=1 to 1 instead of 2 since one loading pattern that uses this type of wagon is eliminated.

This model generates few numbers of variables but a large number of constraints because of the conditions we test. The second model we present is therefore better.

The first model which executes the appropriate mathematical submodel depending on the case considered is structured as represented in the algorithm 1.

2.5 Second mathematical model for train load and transport problem

This second model tests weight constraints of the first disposition thanks to the binary variable. It also provides us with the indices c type l=1 of the containers to place on each slot of the disposition 1. We do not need the loops

Algorithm 1: Model 1

```
for c1 = 1  to  q1 do
    for c2 = 1  to  q1 do
        for c3 = 1  to  q1 do
            if (c1 ≠ c2  and  c1 ≠ c3  and  c2 ≠ c3) then
                if (T_{c1,1} + T_{c2,1} + T_{c3,1} ≤ T_1^max then
                    Execute  submodel  1
                Execute  submodel  2
```

and the conditions anymore. So we add to constraints (1b) to (1f), the following constraints:

$$\sum_{c=1}^{q_1} \sum_{s=0}^{S(1)} T_{c1} z_{c11s} \le T_1^{max} \tag{3a}$$

$$\sum_{c=1}^{q_1} \sum_{s=0}^{S(1)} z_{c11s} = x_1 n_{11} \tag{3b}$$

$$\sum_{c=1}^{q_1} z_{c11s} \le x_1, \quad s \in \mathcal{S}(1) \tag{3c}$$

$$\sum_{s=0}^{S(1)} z_{c11s} \le 1, \quad c \in Q_1 \tag{3d}$$

$$z_{c110} = 0, \quad c \in Q_1 \tag{3e}$$

$$if \quad T_{c1} = 0 \quad then \quad z_{c11s} = 0 \quad endif, \quad c \in Q_1,\ s \in \mathcal{S}(1) \tag{3f}$$

$$x_d,\ y_k \in \mathbb{N}, \qquad z_{c11s} \quad binary \tag{3g}$$

Constraint (3a) tests the weight of containers to place them on the slots of the disposition 1. Constraint (3b) ensures that the total number of containers assigned to each slot in the disposition is equal to the number of wagons of the loading pattern multiplied by the total number of containers in this loading pattern. Constraints (3c) illustrate that for the same slot and disposition, the number of different containers is at most equal to the number of wagons of the disposition. Constraints (3d) guarantee that one container can be placed only in one slot in a load configuration. Constraints (3e) and (3f) are specific to the formulation of the model and are used to avoid contradictions. Finally (3g) specify the nature of the variables.

This model can be easily generalized to test weight constrains of all load configurations. It generates an acceptable number of variables and constraints. They are here more reduced because we test only one loading pattern.

3 An algorithm for the train load problem

The following algorithm provides an optimal solution for the train load problem considering both weight and length constraints of containers and wagons. We need to start first such as the previous models by determining all possible load configurations considering only length constraints.

https://doi.org/10.1051/matecconf/201820000010

At this step, we know exactly how many containers and from which type to be placed on each type of wagon to maximize loading on the train. We have then to define an order of the dispositions that the algorithm must follow. We order the load configurations from the most optimal one to the least. In our case study for example, to maximize loading on the trains, d =1 is better than d =3, since in the first disposition there is 3 containers for one wagon and in the third disposition only 2 in the wagon. Similarly d =3 is better than d =5, since for the same type of wagon, it contains one container more. Moreover, if we have 2 containers of different types, the second disposition is better than the fourth or the fifth. If there is no particular order for the rest, we just place the remaining dispositions in an arbitrary way. Here the order is d =2, d =1, d =4, d =3 and finally d =5. Consider *sigma* the permutation of the elements of the set of dispositions $\{1, 2, 3, 4, 5\}$ so as to have the predefined order. Let us denote $O(d)$ this predefined order and here this set is O(d) = $\{2, 1, 4, 3, 5\}$ where *sigma*(1)=2, *sigma*(2)=1, *sigma*(3)=4,*sigma*(4)=3 and *sigma*(5)=5.

The algorithm then tests the weight constraints for each disposition from the first in the defined order to the last. If the weight configuration is not verified, it passes to the next disposition. If this condition is verified, it increments the number of wagon of this disposition, change Var_l to one to specify that the containers on that wagon have been assigned and finally it stores the indices of containers to be placed together in this loading pattern and that are already assigned in $indice_d$. $T_l(i_{ls})$ represents the tonnage of container i_{ls} type l. i_{ls} is the index of the container in a location s. In this section we use $L(d)$ presented previously which is the number of types of containers in the disposition. For instance, $L(d)$ =2 for disposition 2 since it carries 2 types of containers 20 foot and 40 foot and $L(d)$=1 for all the remaining dispositions.

This algorithm can easily include train constitution and transport constraints.

Algorithm 2: Solving the train load problem

Set the tables $T_l(q_l)$, $Var_l(q_l)$ with l={1,. . . ,L(d)} and $T_l > 0$, $indice_d(x_d,S(d))$ for d={1,. . . ,N}
Initialize Var_l=0 l={1,. . . ,L}, $indice_d$=0 d={1,. . . ,N}
for $j = 1$ *to* N **do**
  d=*sigma*(j),
  (Enter to the d in the order defined)
  l, l'={1,. . . ,L(d)}, s, s'={1,. . . ,n_{dl}},
  for $l = 1$ *to* $L(D)$ **do**
    for $s = 1$ *to* n_{dl} **do**
      i_{ls}
  for $i_{ls} = 1$ *to* q_l **do**
    (This instruction is structured as nested loops for i_{ls} obtained above, It is applied the number of time the locations are in the disposition, S(d) times)
    if $(T_l(i_{ls}) > 0$ *and* $Var_l(i_{ls}) = 0$ *and* $(i_{ls} \neq i_{l's}$ *or* $i_{ls} \neq i_{ls'})$ *or* $i_{ls} \neq i_{l's'})$ **then**
      (this condition tests all i_{ls} in the nested loops and defined above)
      if
$$\sum_{l=1}^{L(d)} \sum_{s=1}^{n_{dl}} T_l(i_{ls}) \leq T_k^{max}$$
      then
        x_d=x_d+1
        for $l = 1$ *to* $L(d)$ **do**
          for $s = 1$ *to* n_{dl} **do**
            $Var_l(i_{ls})$ =1
            for $s" = 1$ *to* $S(d)$ **do**
              $indice_d(x_d, s")$ =i_{ls}
              -Eliminate the location s" and the container assigned i_{ls} so that to have $(indice_d(x_d, s") \neq indice_d(x_d, s"'))$ for s"'={1,. . . ,S(d)}
              -Assign the next container i_{ls} to the next location by reentering this loop

4 Experiment results

In this section, we first present results of the two mathematical models and the algorithm for the Moroccan railways. A second subsection deals with the comparison between the train planning base model (TPM) and the part of our model that consider only length constrains since it is the case of TPM.

4.1 Comparison with manual planning

We have tested our models and algorithm in the case of Moroccan railways. We compared the operator's manual planning with our results. The table below presents the data of the day 24 May 2014. In the database, we have only the total tonnage of the containers of the same category so we considered that they all have the same weight. However, the models solve even containers of different tonnage. The table 1 shows that our models and algorithm provide an optimal number of wagons to send to the customer compared to manual planning. There is a difference in the dispositions obtained by the models and the algorithm due to the fact that the two mathematical models consider the train's length constraints. The dispositions obtained by them are therefore shorter than the algorithm which solves only the load problem. With the first order of the table, our models propose to send 14 wagons with a total length 266.8 m while the algorithm propose the same number for a length of 272.7 m. In the table 1, note that

x_d^* refers to the optimal solution with the models and x_d^{**} to the optimal solution with the algorithm.

Table 2 illustrates the results for the whole month where 262 could be saved for more orders. This month was chosen because it represents the highest season.

May 2014	# wag. used	# wag. required	Gain
31 days	1470	1208	262

Table 2: Results of the case study for May 2014

They are several indicators that can be used to demonstrate the efficiency of our models and to evaluate the performance of rail transport. Some of them are presented in table 3 and calculated in table 4 to measure improvement.

Indicators	Definition
Routing of the park for a period	# of wagons used in the period /Total wagons in operation during this period
Average tonnage per per wagon for a period	Tonnage in the period/ # of wagons in the period
Average # of containers per wagon for a period	# of containers in the period/ # of wagons in the period

Table 3: Definition of indicators

Indicators May '14	Before	After	Improved
Routing	9	7	22.22 %
Average tonnage per wagon	19.96	24.3	21.74 %
Average number of containers per wagon	1.35	1.64	21.48 %
# wagons used	1470	1208	18 %

Table 4: The improvement of indicators in the case study

4.2 Comparison with train planning model

The objective of the train planning model (TPM) is to decide where each particular container should be loaded on a given train while minimizing the weighted sum of the number of wagons used in the train loading plan. It considers lengths of containers, slots and wagons by using a binary decision variable U_{cwds} which is equal to 1 if container c is assigned to slot s in pattern d of wagon w. Each slot can carry a container of the same length. The mixed integer linear programing model (MILP) of train planning, TPM, generates a large number of variables which makes most authors try to simplify and extend it. For example, TPM generates 40000 binary variables for 50 containers, 40 wagons, an average of 10 patterns each and 2 slots per wagon [1].

In the comparison below, we consider the part in our models that take into account only length constrains of wagons and containers. Let us also choose another example apart the one of the Moroccan industry to show the adaptability of the proposed models to other industries. Consider we have one type of wagon of length 19.9 meters and 3 types of containers, 20 foot, 30 foot and 40 foot. We first need to define all possible loading patterns. We obtain dispositions that can carry two containers. Another specific hypothesis to this example is to limit the number of slots to 2, so we consider that the containers transported are filled to their maximum capacity. In this case the first disposition can only carry two containers 20 foot. Figure 2 presents all the permitted configurations considering length constraints. In this specific example, weight constrains are respected which justifies also not including them.

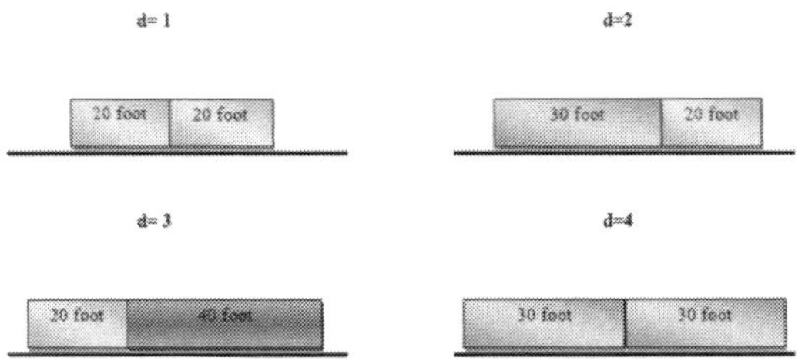

Figure 2: All possible loading patterns of the example

The part in our models which is equivalent to the TPM is the following:

$$\min \sum_{d=1}^{N} x_d \tag{4a}$$

$$\sum_{d=1}^{N} n_{dl} x_d = q_l, \quad l \in L \tag{4b}$$

$$x_d \in \mathbb{N} \tag{4c}$$

The model for this example is as follows:

$$\min x_1 + x_2 + x_3 + x_4 \tag{5a}$$

$$2x_1 + x_2 + x_3 = q_1 \tag{5b}$$

$$x_2 + 2x_4 = q_2 \tag{5c}$$

$$x_3 = q_3 \tag{5d}$$

$$x_1, x_2, x_3, x_4 \in \mathbb{N} \tag{5e}$$

Table 5 illustrates the difference between the two models in terms of number of constraints and variables. The proposed model formulated as an integer linear programing model (ILP) reduces them considerably compared to the basic model existing in the literature which makes it easily extendable to take more elements into account. In the previous section for instance, we integrated weight constraints for a more realistic load problem but also train constitution and transport constraints for a combined resolution.

Computational times for the instances considered are not reported since there are very low for both the models and the algorithm (around one second).

Origin	Destination	T_1	T_2	Total T.	T_1/q_1	q_1	q_2	# wag.used	x_d^*	x_d^{**}	# wag.required	Gain
Casa RN	Marrakech	93.95	15.85	109.8	2.61	36	2	17	12 of d=1, and 2 of d=4	11 of d=1 2 of d=2, and 1 of d=5	14	3
Casa Mita	Marrakech	4.4	8.8	13.2	2.20	2	2	4	2 of d=2	2 of d=2	2	2
Casa RN	Fes	26.89	9.96	36.85	26.89	1	1	2	1 of d=2	1 of d=2	1	1
Casa RN	Casa Mita	582	0	582	26.45	22	0	12	11 of d=3	11 of d=3	11	1
Marrakech	Casa RN/ Mita	1075.94	800.06	1876	22.89	47	40	72	39 of d=2, 4 of d=3 and 1 of d=4	40 of d=2, 3 of d=3 and 1 of d=5	44	28
Tanger Med	Casa Mita	590.44	208.69	799.13	28.12	21	23	34	21 of d=2 and 2 of d=4	21 of d=2 and 2 of d=4	23	11
Fes	Casa RN	24.6	0	24.6	6.15	4	0	3	1 of d=5 and 1 of d=1	1 of d=5 and 1 of d=1	2	1
Total		2398.22	1043.36	3441.58		133	68	144			97	47

Table 1: Comparison between models, algorithm and manual planning the 24 May 2014

	Elements considered/Models	TPM	Proposed model	Gap	Improvement	TPM	Proposed model	Gap	Improvement
Data	Nature of the model	MILP	ILP			MILP	ILP		
	# of containers	8	8 (4 of 20 foot, 3 of 30 foot and 1 of 40 foot)			40	40 (20 of 20 foot, 15 of 30 foot and 5 of 40 foot)		
	# of wagons	4	4			20	20		
	# of slots	2	2			2	2		
	# of patterns	4	4			4	4		
Results	# of variables	256	4	252	98.44 %	6700	4	6694	99.91%
	# of constraints	60	2	58	96.67 %		2		
	Value of the objective function (optimal)	4	4			20	20		
	Patterns used	Container c to place in slot s of loading pattern d of wagon w	3 of d=2 and 1 of d=3			Container c to place in slot s of loading pattern d of wagon w	15 of d=2 and 5 of d=3		

Table 5: Comparison between the proposed model and the TPM

5 Conclusion

The paper deals with the train load and transport planning problem. The objective is to minimize the number of wagons and to maximize loading on trains while satisfying train constitution and transport constraints. We have proposed two mathematical models different in formulation that can be used to solve the combined problem. An algorithm that provides optimal results is also proposed for the train load problem. The models and the algorithm have been tested on real data of railway industry. The part of our model that considers only length constraints has been compared to the base train planning model that takes into account only these elements. Our proposed model simplifies the formulation and therefore reduces considerably the number of constraints and variables. The two models can be extended specially the part of transport per axis to treat multiples destinations in each axis and multiples axes. We also can consider strategic planning over multiple periods.

References

[1] P. Corry, E. Kozan, *Optimised loading patterns for intermodal trains*, O.R. Spectrum **30**, 721-750 (2008).

[2] D. Ambrosino, S. Siri, *Models for train load planning problems in a container terminal*, Advances in Intelligent Systems and Computing **262**, 15-25 (2014).

[3] L. Foti, M. Maratea, S. Sacone, S. Siri, *Solving train load planning problems with boolean optimization*, Proceedings of the 19th RCRA workshop on Experimental Evaluation of Algorithms for Solving Problems with Combinatorial Explosion, (2012).

[4] C. Zhao, L. Yang, S. Li, *Allocating freight empty cars in railway networks with dynamic demands*, Discrete Dynamics in Nature and Society **2014**, Article ID 349341 (2014).

[5] R. P. Mutyavavire, K. Battle, *Optimization of empty rail wagon re-distribution*, SAIIE25 Proceedings, (2013).

[6] S.W. Lam, L.H. Lee, L.C. Tang, *An approximate dynamic programming approach for the empty container allocation problem*, Transport. Res. C-Emer. **15**, 265-277 (2007).

[7] V.M. Azanov, M.V. Buyanov, D.N. Gaynanov, S.V. Ivanov, *Algorithm and software development to allocate locomotives for transportation of freight trains*, Vestnik YuUrGU. Ser. Mat. Model. Progr. **9**, 73-85 (2016).

[8] A.M. Newman, C.A. Yano, *Scheduling direct and indirect trains and containers in an intermodal setting*, Transport. Sci. **34**, 256-270 (2000).

[9] J. Jonaitis, *Planning of the amount of trains needed for transportation by rail* , Transport **22**, 83-89 (2007).

Optimization of the maintenance planning of a multi-component system

Issam Mallouk[1], *Badr* Abou El Majd[1], and *Yves* Sallez[2]

[1]CeReMAR, LMSA Lab, FSR, Mohamed V University, Rabat, Morocco
[2]UPHV, LAMIH, F-59313 Valenciennes, France

Abstract. The vehicle's maintenance costs, uptime and security are the most important goals for owners and transport companies, but these goals are conflictual and the major cause for delays is related to the maintenance policies. The main objective of transporters is to respond properly to their customer's demands. In order to deal with this competitiveness, transport companies are working to improve the management of their fleets by focusing in particular on vehicle maintenance, which impact the vehicles uptime, and generate the most important cost. In addition, a vehicle maintenance policy aims to avoid failures and keep the vehicle up and safe. This objective is reached by ensuring a high reliability; otherwise, an unexpected failure of a component can cause vehicle down and can affect the entire sub-system while generating costs. In this paper, we propose a new maintenance policy based on multi-objective optimization. This problem is solved using the Speed-Constrained Multiobjective Particle Swarm Optimization (SMPSO) for an instance of 18 components and 20 vehicles. First, we give an overview of the existing techniques used for vehicle's maintenance policy, then we present the mathematical model that describes the cost of maintenance and the level of safety. Numerical experiments are presented to demonstrate the efficiency of our approach.

1 Introduction

In the last decade, maintenance of vehicles in the commercial vehicle industry has received a special attention in order to improve transport solutions and competitiveness [3]. Indeed, in terms of costs, scheduling of maintenance activities permit to save assuredly significant charges. Furthermore, preventive maintenance has the main role in order to avoid vehicle's degradation and especially potential accidents. There are several definitions of maintenance in the literature review. We selected here the one given by AFNOR: "The maintenance maintains or restores a system in a specified state so that it is able to provide a specific service" (AFNOR, N. 2001). Thus, a good maintenance scheduling ensure an efficient and robust decision-making, especially when managing a fleet of vehicles. Actually, in a context with rapid evolution of new technologies, the vehicle is become an active and intelligent actor in transport industry [29]. Vehicle uptime is one of the key parameters to improve transport solutions and competitiveness and it is considered as the most important aspect for commercial vehicles owners who are interested mainly to security and cost of ownership. A vehicle is a multi-components complex system composed of subsystems such as "Electrical supply system", "Braking system" and "engine block" [16]. An unexpected failure may cause significant material losses on a certain part or all of the system, or even more dramatically, it may cause human injuries or deaths. The main objective of transporters is to respond properly to their customers demands. Transport companies are working to improve the management of their fleets by focusing in particular on maintenance that impact the vehicles uptime. The aim of preventive maintenance is to avoid failures and keep the system up and running by selecting the right component and decide for change or repair it. The choice is made by following a statistical analysis of the reliability of the components and data reported from the solutions embedded in the vehicle. Nowadays, almost 90% vehicles are equipped with smarts sensors over wireless communication protocols [5]. The maintenance polices benefits of these technologies by using the vehicle's historical data generated and by remotely analysing and provide diagnosis and maintenance prognosis for driver assistance systems which bring us to the level of e-Maintenance [14] [13]. These technological advances also make fault diagnosis and maintenance interventions much more challenging, since these operations require a deep understanding of the entire system [15]. The embedded systems collect the information from the sensors in the format of a raw information. Afterwards, informations are uploaded or synchronized with a remote server over a wireless connection [4]. The authors of [22] [23] have introduced the notion of an 'active product', which although it builds on the concept of product intelligence, it focuses on the ability of a product to trigger actions in its own environment, that is the decision making and execution features. Henceforth, the main challenge is to take benefit from all the collected data based on the analytical capabilities of machine learning algorithms and the performance of cloud computing and artificial intelligence [25].

In order to maximize productivity, carriers have a dual interest: perform maintenance without affecting operational uptime and increase productivity by exploiting fleet capacity as optimal as possible. Several strategies of maintenance scheduling were proposed in literature with the aim to develop a maintenance policy by relying on dynamic programming and on a horizon defined previously [5]. All these techniques try to optimize maintenance operations and fleet management but most of them suffer from a lack of real time information. The idea is to use in real time the current information of the components health in order to update the maintenance schedule. The challenge is to keep the maintenance policy valid for the multi-component case, knowing that the components are dependent on each other [19] and each dependency is considered as constraint for another component.
Therefore, the final model turns into an NP-hard problem [16] with a large number of constraints and dependencies to satisfy.

The paper is organized as follows. Section 1 gives a description of the mathematical model. In section 2 the real instance and the optimization algorithm used to experiment our model are presented. Section 3 presents some numerical results that validate the model and the optimisation process. Finally, section 2 concludes the paper and states some open questions.

2 The Mathematical model

In order to maximize productivity, carriers have a dual interest: Perform maintenance without affecting operational uptime and increase productivity by exploiting fleet capacity as optimal as possible.The main aim is to develop a policy which simultaneously optimize maintenance operations and fleet managment. Essentially this work deals with optimization of costs related to component replacements.

2.1 Notations

We consider a system Sys has n components, each component i has a replacement cost C_i, a replacement time D_i and a level of confidence LC_i degradable according of time and use. Each level of elementary confidence contributes to the overall level of confidence of the system LC_{sys}. The idea is that each inspection should check the system confidence level if it has not yet reached the minimum LC and take advantage of the opportunity to replace the other components in dependence [2] and regaining the level LC_{sys} while minimizing replacement costs and logistical costs associated with downtime.

We summarize in the Table 1 some abbreviations and notations used to explain our mathematical model.

Abbreviations	Notations
C_{rem}	Remplacement cost
n	the number of system components
LC_{sys}	The confidence level of the system
LC	The specified confidence level
Sys	The system
C_i	Unit price of the component
D_i	The replacement period of the componant i

Table 1: Abbreviations and Notations

2.2 Multi-objective optimization

Almost every real-world problem involves simultaneous optimization of several incommensurable and often competing objectives. While in single-objective optimization the optimal solution is usually clearly defined, this does not hold for multi-objective optimization problems. Instead of a single optimum, there is rather a set of alternative trade-offs, generally known as Pareto-optimal solutions. These solutions are optimal in the wider sense that no other solutions in the search space are superior to them when all objectives are considered.

One common approach to multi-objective optimization is the so-called aggregation method, in which the goal is to minimize a single objective, some weighted sum of all objectives. The main advantage of this approach is that any generic optimization algorithm can then be used to minimize this single objective. However, this approach also suffers several drawbacks: it requires some a priori knowledge of the trade- off the decision maker is willing to make between the different objectives. A completely different approach is to use some population based search, and to somehow factorize the efforts by identifying the whole Pareto front at once.

An interesting approach based on game theory was proposed by the authors of [1] [7] [9] to solve multi-objective optimization problems. This approach considers the multi-objective optimization problem as a multi-player co-operative game where each objective function to be optimized is a player in the game. A game is said to be co-operative if the players are able to reach an agreement on strategies. The players are the objective functions, which are ultimately controlled by the decision maker and so can be expected to reach an 'agreement', meaning the game is co-operative. Using the fundamental text on co-operative games [18].

Several multi-objective evolutionary algorithms (MOEAs) have been proposed in literature to solve multi-objective optimization problems [10]. These algorithms are based on different implementation of the Pareto dominance selection and the diversity criterion. In particular, many MOEAs use an archive of solutions, where they maintain the non-dominated solutions ever encountered during the search. Multi-Objective evolutionary algorithms (MOEAs), have demonstrated their ability to do in a flexible and reliable way.

2.3 Mathematical Formulation

The subject of our approach is to find a vector of components to replace or to repair and which is able to give the system a state of good performance. The choice is made by minimizing the replacement cost and by respecting that the confidence level of the system still higher than a specific level (fixed by the decision maker).

In this work, it is assumed that after each maintenance operation where one or several components are replaced, their reliability performances are considered "as good as new". It is also assumed that the reliability performance of the other components is considered unchanged or "as bad as old". Although these assumptions are not necessary they simplify the discussion and do not severely limit the generality of the development. Under these assumptions, the multi-objective optimization problem can be formally stated as follows:

$$\begin{cases} \min\limits_{x} \sum\limits_{i=0}^{n} x_i * [C_i + (D_i * T_{mo})] & (1) \\ \max\limits_{x} \frac{1}{n} \sum\limits_{i=0}^{n} \left(LC_i * (1 - x_i) + (100 * x_i) \right) & (2) \\ st: LC < \frac{1}{n} \sum\limits_{i=0}^{n} \left(LC_i * (1 - x_i) + (100 * x_i) \right) < 100 \\ x \in \{0, 1\}. \end{cases}$$

where x_i is a binary variable which indicates the selection of a maintenance operation on the component.
The first objective function aims to select a vector of components to be replaced while minimizing the replacement cost and by respecting that the confidence level of the system still higher than a specific level. The subject of the second objective function is to keep a certain degree of confidence after the replacement of the components, this allows a good performance after the maintenance action. This degree of confidence LC_{sys} is calculated containing the historical study of the interventions of maintenance, in our case, we summarize this information in a percentage (the sum of perfect elementary confidences = 100%). When a need for maintenance is necessary, a maintenance decision rule to select the components to be replaced or repair during this occasion should be defined.

3 Numerical Results

3.1 Provided Data

This work was conducted in collaboration with *STMF* [1] company as a part of an industrial project. The real data provided by *STMF* is generated manually by an agent who enters the maintenance orders once the maintenance is completed. This helps us to know a complete information after any spare part consumption, but this is limited since real time is missing. Several important informations are available.We give in Table 2 a list of spare parts. These data permits to calculate the actual costs of replacement. These informations change regularly, so we create a dynamic database with the component reference as a unique identifier. Table 3 provides us with information on vehicles, odometer and year of entry into service. Also any other information relating to the vehicle in general. In Table 4, we show the most important database, it includes all the information of change and repair of system components. Also it links the two previous lists. We consider this database as a source of historical data.

Designation	Unit price	Duration	LC
Cushion of air tank	1205	20	80
Wiper blade	61	5	35
Alu junction reduction	480	15	10
Bronze ring	110	5	47
Spanner wrench	75	25	80
Volvo alternator belt	388.64	60	30
Air hose	120	30	30
Front oil sensor	950	20	20
Volvo front mirror	576.66	15	5
Trodume seal	90	10	14
Tank brake pad	761.67	30	40

Table 2: Data Base list of spare parts

Vehicle type	Id	Date of purchase	Registration	Odometer
Tractor	3	01-09-2016	35419-A-7	164250
Tractor	4	01-09-2016	35420-A-7	171231
Tractor	5	01-10-2017	49980-A-7	75092
Tractor	6	02-01-2016	47703-A-7	235648
Tractor	7	01-09-2016	35404-A-7	183540
Tractor	8	15-11-2016	36508-A-7	175354
Tractor	9	01-09-2016	35398-A-7	192463
Tractor	10	02-06-2017	44447-A-7	85632
Tractor	11	01-09-2016	35405-A-7	186321
Tractor	12	02-01-2017	41704-A-7	152304

Table 3: Data Base list of vehicles

Required work	Spare part	Date of maintenance	Price
Cabin winder to replace	Volvo spring cab shock absorber	30-05-2016	940
Cabin carter to be replaced	Front Volvo brake lining	30-05-2016	940
Alternator belt to replace	Volvo alternator belt	03-06-2016	388.64
Fan belt to replace	Volvo fan belt	03-06-2016	529.41
Wrench wrench to replace	Air horn	04-06-2016	75
Oil sensor to replace	Volvo oil sensor	12-06-2016	950
Suspension sensor to replace	Suspension sensor	21-06-2016	1500

Table 4: Data Base list of interventions

3.2 Estimation of the level of confidences

A Monte Carlo simulation was used to estimate the level of confidence of each component according to its

[1] STMF is a Moroccan company specializing in the transport of sensitive materials (petroleum products, gas, refrigeration, equipment, cabling, etc.)

own Weibull parameters. In order to compute the level of confidences, a time unit that varies between 0 and 1,000,000 units with a step of 1 was chosen, this gives us the simulation over 15 months as shown in figure 1. The impact of information uncertainty on the estimation of the costs of maintenance replacement is depicted in figure 2.

The Monte Carlo simulation approach was used in this paper in order to give a numerical solution for real-time system state estimation and prognostics from noisy provided data.

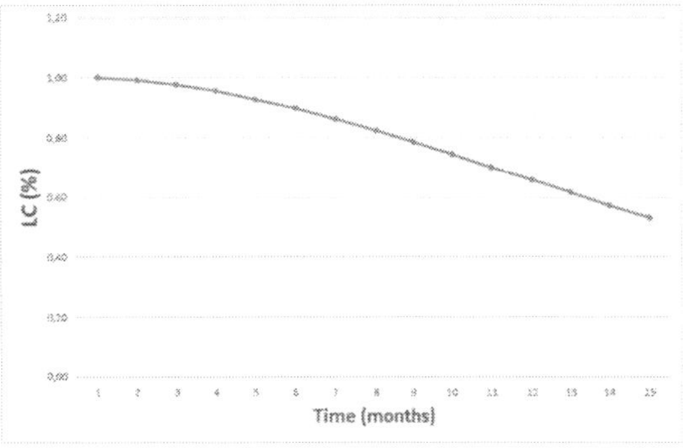

Figure 1: Estimation of the levels of confidences using Monte-Carlo Simulation

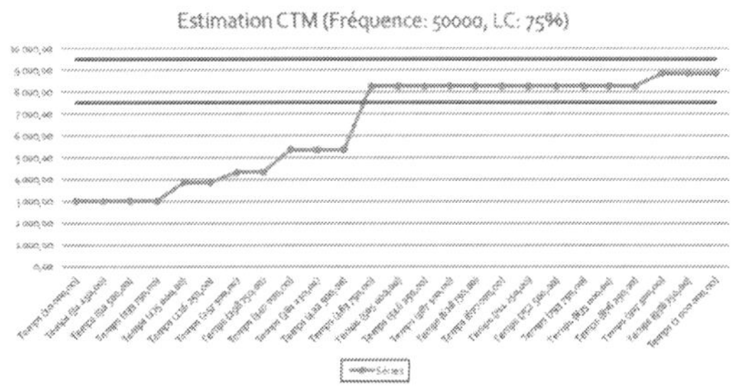

Figure 2: The dependence between time and replacement costs due to Degradation of LC

3.3 Algorithm

This section introduces the algorithms that have been used to validate numerically the theoretical framework described above, namely the Speed-Constrained Multi-Objective Particle Swarm Optimization algorithm (SMPSO).

SMPSO [17] is a particle swarm optimization method (bio-inspired) developed from OMOPSO [20]. The particle swarm optimization (PSO) algorithms are inspired by the flight of birds while trying to find food. In these algorithms, the population is called swarm. The individuals are called particles, which are "flown " through space following the best performing particle at that moment. The position of a particle is given by the current values of its parameters, belonging to an orthogonal representation particle's space. As every particle tries to get closer to the current best particle its parameters are changed. The change takes into account both the current global best and the particle's personal best. Based on this change the particle gets a new position and it needs to be evaluated again. After all the particles are evaluated, the new global best particle is selected, the personal bests are updated and the process is restarted. In Multi-objective PSO algorithms there can be multiple global best particles.

SMPSO has been chosen here for its long record of successes, demonstrating its robustness to find a good approximation of the Pareto front [21] [11].
The flowchart below, Figure 3, explain the process of SMPSO.

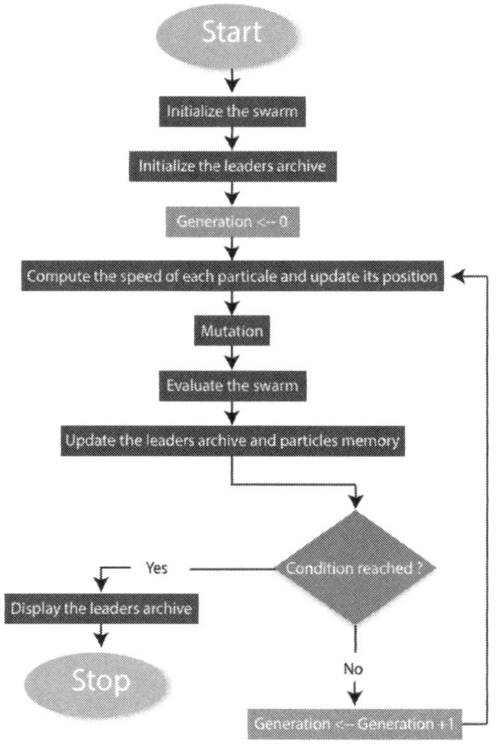

Figure 3: Flowchart of SMPSO algorithm

3.4 Results

Several simulations are done to check the CPU time by considering three subcategory of 50 components. Figure 5 shows the CPU time with respect to the components. However, the execution time remains very low. We consider a list of component given by the table 5. For each component, we generate the level of confidence

(LC) using Monte-Carlo simulation. Figure 4 gives us the Pareto Front obtained. The two optimal solutions of the objective functions appear in the Pareto front, with an execution time of 19s we obtained 15 possible scenarios.

The Pareto front depicted in figure 4 shows the trade-off between minimizing the replacement cost and maximizing confidence level. If the decision maker (DM) wants to minimize the replacement cost no matter the value of confidence level he shall select a solution toward the lower right corner of the graph. Otherwise, if the DM thinks that the confidence level matters, he can choose a compromise in the top lest corner of the graph, resulting in higher confidence level, at the price of a higher cost of replacement.

Id	Designation	Unit price	Duration	LC
1	Cushion of air tank	1205	20	80
2	Wiper blade	61	5	35
3	Alu junction reduction	480	15	10
4	Bronze ring	110	5	47
5	Spanner wrench	75	25	80
6	Volvo alternator belt	388.64	60	30
7	Fan belt	529.41	40	40
8	Air hose	120	30	30
9	Front oil sensor	950	20	20
10	Volvo front mirror	576.66	15	5
11	Trodume seal	90	10	14
12	Tank brake pad	761.67	30	40
13	brake pad	1398	40	20

Table 5: List of component

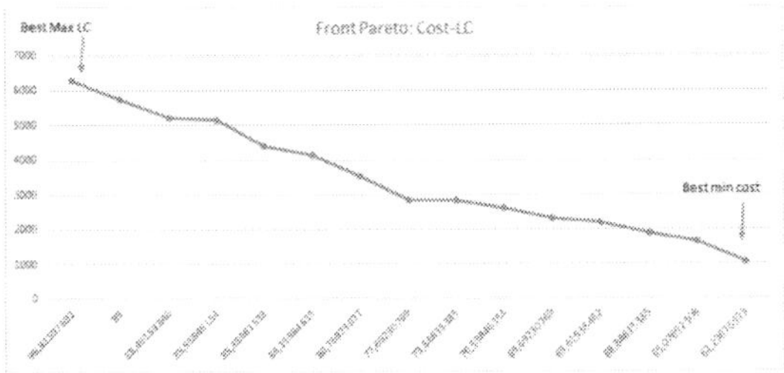

Figure 4: Pareto front Cost-LC

4 Conclusion

In this paper, we have proposed a new approach for vehicle's maintenance. The selection of the components to be replaced will be managed by an evolutionary and dynamic mechanism which loop on the subcategories of the components (Engine, braking, suspension ...) each one has its technical specifications which defines its optimal *LC*. On-going and further work will be focusing on integrating the process of simulation and optimization into a single application and try to generate the state of

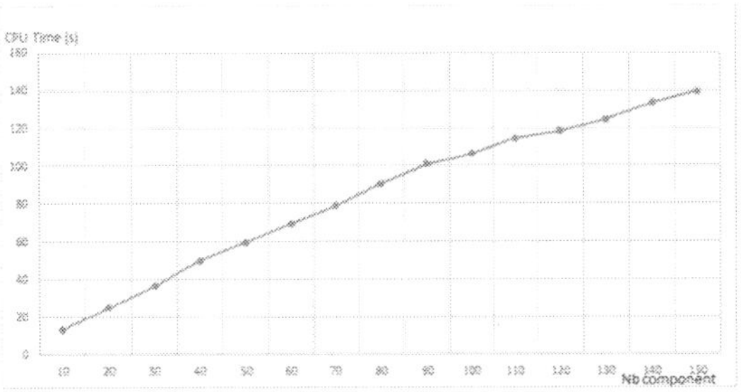

Figure 5: Time CPU evolution

the components of the system by using sensors and embedded solutions while taking advantage of the internet of things. Other MOEAs will be also investigated and compared with SMPSO, and, more importantly, several different instances. One crucial issue is how well this algorithm scales with the problem complexity (number of vehicle and number of components).

References

[1] B. Abou El Majd, J.A. Désidéri and A. Habbal, " Aerodynamic and structural optimization of a business-jet wingshape by a Nash game and an adapted split of variables". *Mécanique Industries*, 11(3-4), 209-214, 2010.

[2] Ab-Samat, Hasnida, and S. Kamaruddin. "Opportunistic maintenance (OM) as a new advancement in maintenance approaches: A review".*Journal of Quality in Maintenance Engineering 20.2*, 98-121, 2014.

[3] S. Alaswad and Y. Xiang," A review on condition-based maintenance optimization models for stochastically deteriorating system", *Reliab. Eng. Syst. Saf.*, vol. 157, p. 54-63, janv. 2017.

[4] P. Baraldi, F. Cadini, F. Mangili, et E. Zio," Model-based and data-driven prognostics under different available information", *Probabilistic Eng. Mech.*, vol. 32, p. 66–79, 2013.

[5] K. Bouvard, S. Artus, C. Bérenguer, et V. Cocquempot," Condition-based dynamic maintenance operations planning and grouping. Application to commercial heavy vehicles", *Reliab. Eng. Syst. Saf.*, vol. 96, no 6, p. 601-610, juin 2011.

[6] R. Dawid, D. McMillan, and M. Revie," Review of Markov models for maintenance optimization in the context of offshore wind", 2015.

[7] R. Duvigneau, B. Abou El Majd and J.A. Désidéri. " Towards a self-adaptive parameterization for aerodynamic shape optimization". *In ESAIM: Proceedings (Vol. 22, pp. 169-174)*. EDP Sciences, 2008.

[8] K .C. Dey, A. Mishra, and M. Chowdhury, " Potential of intelligent transportation systems in mitigating adverse weather impacts on road mobility: a review", *IEEE Transactions on Intelligent Transportation Systems*, vol. 16, no. 3, pp. 1107–1119, 2015.

[9] J.A. Désidéri, R. Duvigneau, B. Abou El Majd and Z. Tang, " Algorithms for efficient shape optimization in aerodynamics and coupled disciplines". *In 42nd AAAF Congress on Applied Aerodynamics*, Sophia-Antipolis, France, 2007.

[10] R. Fadil, B. Abou El Majd, H. Rahil, H. El Ghazi and N. Kaabouch, " Multi-objective Optimization Approach for Air Traffic Flow Management". *In: MATEC Web of Conferences*, 105, 0005, 2017.

[11] W.F. Fihri, Y. Arjoune, H. El Ghazi, N. Kaabouch and B. Abou El Majd, " A particle swarm optimization based algorithm for primary user emulation attack detection". *In: IEEE consumer communications and networking conference*, p. 1–6, 2018.

[12] K. He, L.M. Maillart, and O.A. Prokopyev, " Scheduling Preventive Maintenance as a Function of an Imperfect Inspection Interval", *IEEE Transactions on Reliability*, vol. 64, no. 3, pp. 983–997, Sep. 2015.

[13] K.T. Huynh, A. Barros, and C. Bérenguer," Maintenance decision-making for systems operating under indirect condition monitoring: value of online information and impact of measurement uncertainty", *IEEE Trans. Reliab.*, vol. 61, no 2, p. 410–425, 2012.

[14] S. K. Kinnunen et al., " Decision making situations define data requirements in fleet asset management", *in Proceedings of the 10th World Congress on Engineering Asset Management (WCEAM 2015)*, 2016, p. 357–364.

[15] A. Le Mortellec, J. Clarhaut, Y. Sallez, T. Berger and D. Trentesaux, " Embedded holonic fault diagnosis of complex transportation systems". *Engineering Applications of Artificial Intelligence*, 26(1), 227-240, 2013.

[16] R. Lesobre, " Modélisation et optimisation de la maintenance et de la surveillance des systèmes multi-composants-Applications à la maintenance et à la conception de véhicules industriels", *PhD Thesis*, Université Grenoble Alpes, 2015.

[17] A.J. Nebro, J.J. Durillo, J. García-Nieto, C.A. Coello Coello, F. Luna, and E. Alba. " Smpso: A new pso-based metaheuristic for multi-objective optimization ". In *2009 IEEE Symposium on Computational Intelligence in Multicriteria Decision-Making (MCDM 2009)*, pages 66-73. IEEE Press, 2009.

[18] J. Nash. " Two-person cooperative games". *Econometrica: Journal of the Econometric Society*, pages 128-140, 1953.

[19] R.P. Nicolai and R. Dekker, " A review of multi-component maintenance models", *In : Proc. of European Safety and Reliability Conference ESREL*. 2007.

[20] M. Reyes and C.A. Coello Coello. " Improving PSO-based multi-objective optimization using crowding, mutation and ϵ-dominance ". In *C.A. Coello, A. Hernández, and E. Zitler, editors, Third International Conference on Evolutionary MultiCriterion Optimization, EMO 2005*, volume 3410 of LNCS, pages 509-519. Springer, 2005.

[21] H. Rahil, B. Abou El Majd, and M. Bouchoum, " Optimized Air Routes Connections for Real Hub Schedule Using SMPSO Algorithm". *In : Recent Developments in Metaheuristics*, Springer, Cham, p. 369-384, 2018.

[22] Y. Sallez, T. Berger, D. Deneux and D. Trentesaux, " The lifecycle of active and intelligent products: The augmentation concept", *International Journal of Computer Integrated Manufacturing*, 23, 905–924, 2010.

[23] Y. Sallez, T. Berger and D. Trentesaux, " A stigmergic approach for dynamic routing of active products in fms", *Computers in Industry*, 60, 204–216, 2009.

[24] O. Senechal and D. Trenteseaux," Spécification d'une méthodologie pour l'aide à la décision dans le cadre de la maintenance basée sur la performance environnementale Application aux systèmes ferroviaires.", *Congrès International de Génie Industriel (CIGI)*, Compiègne, France, mai 2017.

[25] O. Svensson, S. Thelin, S. Byttner, et Y. Fan," Indirect Tire Monitoring System - Machine Learning Approach", *IOP Conf. Ser. Mater. Sci. Eng.*, vol. 252, p. 012018, oct. 2017.

[26] H. C. Vu, P. Do, A. Barros, and C. Bérenguer," Maintenance grouping strategy for multi-component systems with dynamic contexts", *Reliability Engineering and System Safety*, vol. 132, pp. 233–249, Dec. 2014.

[27] H. Wang, L. Jiao, and X. Yao, "s Two Arch2: An Improved Two-Archive Algorithm for Many-Objective Optimization ", *IEEE Transactions on Evolutionary Computation*, vol. 19, no. 4, pp. 524–541, Aug. 2015.

[28] L. Xiao, S. Song, X. Chen, and D. W. Coit," Joint optimization of production scheduling and machine group preventive maintenance", *Reliability Engineering and System Safety*, vol. 146, pp. 68–78, Feb. 2016.

[29] M. Yildirim, X. A. Sun, et N. Z. Gebraeel," Sensor-driven condition-based generator maintenance scheduling—Part I: Maintenance problem", *IEEE Trans. Power Syst.*, vol. 31, no 6, p. 4253–4262, 2016.

FUZZY PERFORMANCE MEASUREMENT SYSTEM FOR THE MAINTENANCE FUNCTION

NAJI Mehdi Amine[1, a], MOUSRIJ Ahmed[1]

[1] *Industrial Management and Innovation Laboratory; Faculty of Science and Technology, Hassan 1st Settat University, Morocco*

Abstract. Maintenance aims to sustain the manufacturing process by ensuring that the production tools are well managed. However, modern production means are more automated and therefore more complex, so they require a performing maintenance for which planning and implementation are increasingly difficult. This paper, therefore, emphasizes the elaboration of a maintenance performance measurement system. Therefore, for an appropriate and reliable performance evaluation, we adopt a multi-level multi-criteria approach where every criterion influencing the maintenance performance either indirectly or directly are specified. Thereafter, indicators for the specified criteria would be measured using fuzzy logic to overcome limitations due to lack of available resources and data. For a global maintenance measurement, the elementary measurements will be aggregated into a total measurement through a multiple-criteria decision-making method such as MACBETH.

1 Introduction

To remain in a competitive position, the manufacturing systems must be more productive, effective and economical [1]. However, the complexity of these manufacturing systems, the large diversity of manufactured products and the uncertainty of the production environment are behind critical dysfunctions. These malfunctions and failures negatively affect production capacity, production quality and cost, as well as potential health and safety risks and accidents to personnel, plants and the environment [2]. This is the reason behind maintenance, which is no more restricted to repair facilities, but rather involves preventing their dysfunction [3-4]. Consequently, maintenance is now a vital element in the global business success for any manufacturing company [5]. Therefore, a performing maintenance became a necessity and not a possibility.

Achieving the required performance begins with its measurement as without it there are practically any indications of improvement [6]. A performance measurement system is a management tool that allows a company to evaluate how its activities are effective and efficient [7]. This tool is structured in a multi-criteria approach, including a combination of performance indicators selected according to the company's objectives [8]. The main challenge when measuring maintenance performance is defining and choosing indicators that reflect the corporate strategy and provide quantitative feedback on the maintenance strategies performance [9]. Indeed, a suitable indicator would track and control performance, assist identifying performance gaps, and continuous improvement [5]. Therefore, to make this system more useful, there is an obligation in defining the indicators to be measured, the way in which they should be evaluated and who is responsible for their implementation [10].

However, in an environment characterized by cost, time, resources and skill limitations, measuring the elementary performance is a big issue. For this reason, we propose that the fuzzy logic is used to quantify this elementary performance. Subsequently, by using the multi-criteria decision-making method MACBETH, the elementary performance measurements are aggregated into an overall measure that provides the total maintenance performance. Consequently, the suggested approach allows measuring the global performance and the local performances of the maintenance, as well as defining the necessary actions for a continuous maintenance performance enhancement.

2 Methodology

2.1. The identification of the elementary maintenance performance measurement indicators

In order, to measure maintenance performance correctly, the strategic aspects are specified and aligned with the company's strategy. These strategic aspects are the financial aspect, the technical aspect, the internal process aspect, the support aspect, the human resources aspect and the health, safety and environment (HSE) aspect. Afterwards, they will be disaggregated until reaching elementary indivisible sub-criteria. This paper will focus

[a] NAJI Mehdi Amine: naji.mehdi.amine@gmail.com

on financial performance measurement, which is presented in Figure 1. A similar methodology is used when assessing the other dimensions of the maintenance performance.

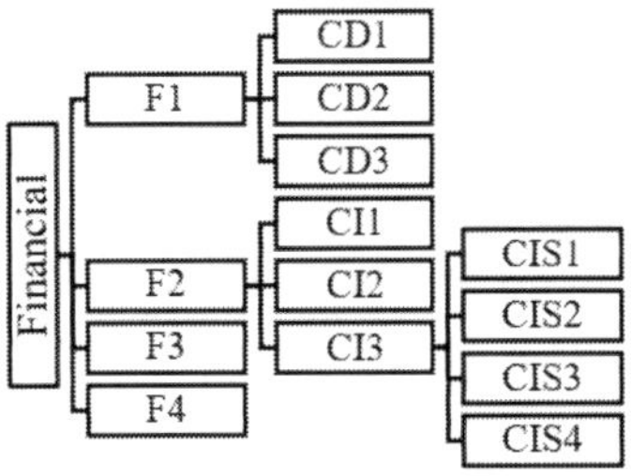

Figure 1 Financial maintenance performance indicators.

*With **F1** direct cost, **F2** indirect Cost, **F3** maintenance intensity and **F4** capital replacement value. **CD1** Internal labor cost, **CD2** Subcontract cost, **CD3** Maintenance material purchasing cost (spare part + consumable). **CI1** training and learning costs, **CI2** HSE related cost and **CI3** unavailability cost. **CIS1** palliative cost, **CIS2** contractual penalty, **CIS3** production losses, **CIS4**: Intermediate inventory cost.*

2.2. The elementary Performance Measurement using the Fuzzy Logic:

The fuzzy logic has been inserted by Zadeh (1965) [11], in order to overcome the complexity and fuzziness of the real systems, by representing them through linguistic values that will follow a mathematical treatment transformed into numerical ones [12].

2.2.1. Linguistic terms and triangular fuzzy number identification

First, to measure elementary performance, it is necessary to construct the appropriate rating scale. This performance is from 0 to 1. The 0 is a complete dissatisfaction of the target, and the 1 is a complete satisfaction of the target. Then the interval [0, 1] will be subdivided into six sub-intervals. The interval [0, 0.45] denotes a very poor satisfaction (VP), [0.35, 0.5] denotes a poor satisfaction (P), [0.45, 0.65] a medium satisfaction M, [0.6, 0.8] a good satisfaction (G), [0.75, 0.92] a very good satisfaction (VG) and [0.9, 1] indicates an excellent satisfaction (E).

Then for each criterion C_i represented in Figure 1, decision-makers will linguistically express their satisfaction degrees regarding to the objective O_i using a specific term. The result is reported in Table 1.

Table 1 Satisfaction degrees

	CD1	CD2	CD3	CI1	CI2	CIS1	CIS2	CIS3	CIS4	F3	F4
K1	**VG**	**G**	**G**	**P**	**P**	**P**	**VP**	**P**	**P**	**G**	**G**
K2	**G**	**M**	**G**	**M**	**M**	**P**	**VP**	**VP**	**P**	**M**	**G**
K3	**G**	**G**	**G**	**M**	**P**	**M**	**P**	**P**	**M**	**M**	**M**
K4	**VG**	**G**	**G**	**M**	**M**	**P**	**P**	**P**	**M**	**G**	**G**

After that, every linguistic term is assigned to the appropriate fuzzy number. In this study, the triangular fuzzy numbers are chosen for their simplicity [13-14-15].

$\widetilde{p_K}$ is the triangular fuzzy number of a specific criterion C_i. $\widetilde{p_K} = (L_K, M_K, R_K)$ with K = VP, P, M, G, VG and E where L and R are two points on the left and right of the triangular fuzzy number representing the interval $[L_K, R_K]$. Mk is the most possible value. The result is in the table 2.

Table 2 Triangular fuzzy number TFN

Performance Satisfaction Degree	Triangular Fuzzy number
Very poor VP	(0, 0.35, 0.45)
Poor P	(0.35, 0.45, 0.5)
Medium M	(0.45, 0.6, 0.65)
Good G	(0.6, 0.75, 0.8)
Very Good VG	(0.75, 0.85, 0.92)
Excellent E	(0.9, 1, 1)

Then, each linguistic parameter is assigned with the corresponding triangular fuzzy number. The result is shown in Tables 3 and 4.

2.2.2. Aggregation of the fuzzy numbers

In order to get a consensus among the various evaluators, the satisfaction degrees $\widetilde{P}_i^K = (L_i^K, M_i^K, R_i^K)$ K= 1,.....,n expressed by each evaluator on the evaluation criterion C_i should be aggregated into a unique TFN $\widetilde{P}_i = (L_i, M_i, R_i)$ using equation (1) [16]:

$$\begin{cases} L_i = \min\limits_K (L_i^K) \\ M_i = \frac{1}{k}\sum\limits_{k=1}^{n} M_i^K \\ R_i = \max\limits_K (R_i^K) \end{cases} \quad (1)$$

By using equation (1), the various fuzzy numbers expressed by the decision-makers will be aggregated into a single fuzzy number reflecting the satisfaction degree associated to each elementary sub criterion Cij. The result is shown in Tables 5 and 6.

Table 3 the associated triangular fuzzy numbers

	CD1	CD2	CD3	CI1	CI2
K1	(0.75, 0.85, 0.92)	(0.6, 0.75, 0.8)	(0.6, 0.75, 0.8)	(0.35, 0.45, 0.5)	(0.35, 0.45, 0.5)
K2	(0.6, 0.75, 0.8)	(045, 0.6, 0.65)	(0.6, 0.75, 0.8)	(045, 0.6, 0.65)	(045, 0.6, 0.65)
K3	(0.6, 0.75, 0.8)	(0.6, 0.75, 0.8)	(0.6, 0.75, 0.8)	(045, 0.6, 0.65)	(0.35, 0.45, 0.5)
K4	(0.75, 0.85, 0.92)	(0.6, 0.75, 0.8)	(0.6, 0.75, 0.8)	(045, 0.6, 0.65)	(045, 0.6, 0.65)

Table 4 the associated triangular fuzzy numbers

	CIS1	CIS2	CIS3	CIS4	F3	F4
K1	(0.35, 0.45, 0.5)	(0, 0.35, 0.4)	(0.35, 0.45, 0.5)	(0.35, 0.45, 0.5)	(0.6, 0.75, 0.8)	(0.6, 0.75, 0.8)
K2	(0.35, 0.45, 0.5)	(0, 0.35, 0.4)	(0, 0.35, 0.4)	(0.35, 0.45, 0.5)	(045, 0.6, 0.65)	(0.6, 0.75, 0.8)
K3	(045, 0.6, 0.65)	(0.35, 0.45, 0.5)	(0.35, 0.45, 0.5)	(045, 0.6, 0.65)	(045, 0.6, 0.65)	(045, 0.6, 0.65)
K4	(0.35, 0.45, 0.5)	(0.35, 0.45, 0.5)	(0.35, 0.45, 0.5)	(045, 0.6, 0.65)	(0.6, 0.75, 0.8)	(0.6, 0.75, 0.8)

Table 5 Aggregation of TFNs

CIS1	CIS2	CIS3	CIS4	F3	F4
(0.35, 0.48, 0.65)	(0,0.37, 0,5)	(0,0.41, 0,5)	(0.35, 0.52, 0.65)	(0.45, 0.68, 0.8)	(0.45, 0.71, 0.8)

Table 6 Aggregation of TFNs

CD1	CD2	CD3	CI1	CI2
(0.6, 0.8, 0.92)	(0.45, 0.71, 0.8)	(0.6, 0.75, 0.8)	(0.35, 0.56, 0.65)	(0.35, 0.52, 0.65)

2.2.3. Defuzzification

In order to convert the TFN $\tilde{\bar{P}}_i$ related to each evaluation criterion Ci into crisp value. The $\tilde{\bar{P}}_i$ will be deffuzzify by the CFCS method (Converting the Fuzzy data into Crisp Scores) proposed by Opricovic [17].

In the presented study, the fuzzy logic is utilized for quantifying the elementary maintenance performance related to the criterion C_i. So $\tilde{P}_i = (L_i, M_i, R_i)$ is the membership function of the elementary performance for the criterion C_i.

Then:

The normalization of the TFN:

$$r_i^{max} = R_i \quad , \quad l_i^{min} = L_i$$

$$\Delta_{min}^{max} = R_i - L_i$$

$$x_l = \frac{L_i - l_i^{min}}{\Delta_{min}^{max}} = 0$$

$$x_m = \frac{M_i - l_i^{min}}{\Delta_{min}^{max}} = \frac{M_i - L_i}{R_i - L_i}$$

$$x_r = \frac{R_i - l_i^{min}}{\Delta_{min}^{max}} = 1$$

The Left and Right normalized value x^{ls} and x^{rs}:

$$x^{ls} = \frac{x_m}{1 + x_m - x_l} = \frac{M_i - L_i}{r_i - 2L_i + M_i}$$

$$x^{rs} = \frac{x_r}{1 + x_r - x_m} = \frac{R_i - L_i}{2R_i - L_i - M_i}$$

The total normalized crisp value:

$$x^{crisp} = [x^{ls}(1 - x^{ls}) + x^{rs}x^{rs}]/[1 - x^{ls} + x^{rs}]$$

Finally, the crisp values P_i, which refer to the elementary performance of the criterion C_i:

$$P_i = L_i + x^{crisp}(L_i - R_i) \qquad (2)$$

By using this defuzzification method, the aggregated fuzzy numbers will be converted into a performance expression. Therefore, we will get the elementary performance Pi for each Ci criterion. The result is displayed in figure 2.

According to Figure 2, the best performances are P_{CD1} = 0.78 and P_{DC3} = 0.73. On the other hand, the poor performances are P_{CIS3} = 0.35 and P_{CIS2} = 0.32.

However, the elementary performance measurements are not suitable to provide a holistic maintenance performance assessment. Because of this, they must be aggregated to form a unified measure.

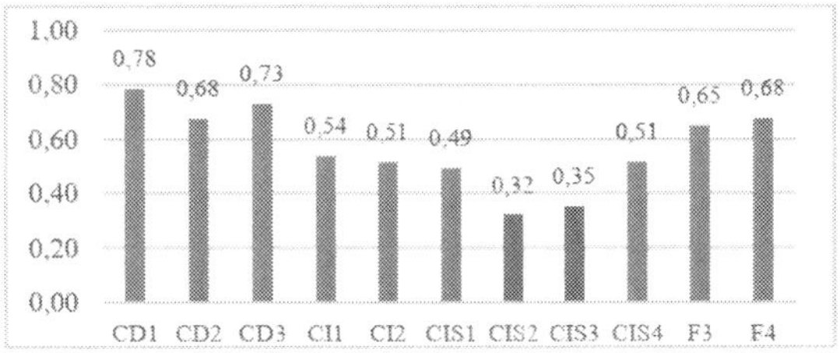

Figure 2 Financial elementary performances

2.3. The overall maintenance performance measurement

The evaluation criteria and aspects have a different effect on the performance. As a result, to identify the relative importance of each evaluation criterion and aspect, a multi-criteria decision-making MCDA method such as MACBETH is needed. Indeed, this method allows quantifying the relative weight of each criterion by using the comparative attractiveness of each criterion as expressed linguistically by the decision-makers [18].

To do this, the decision-makers will express the attractiveness relating to the evaluation criteria of the same hierarchical level. By using the MACBETH method and the M-MACBETH software, each criterion and sub-criterion will be weighted. The weights W_i of F1, F2, F3, and F4 are shown in Figure 3.

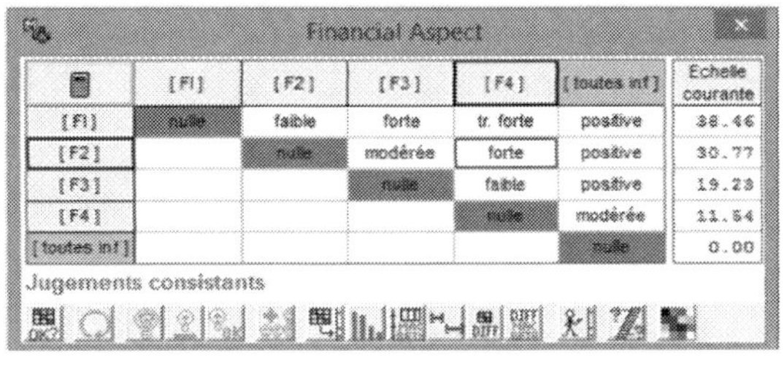

Figure 3 Financial criteria weighing using M-MACBETH.

After that, the overall maintenance performance is calculated through the aggregation of the elementary performance using the weighted arithmetic mean as in equation (3).

$$P_{ag} = Ag(P_1, \ldots, P_n) = \sum_{i=1}^{n} W_i P_i \qquad (3)$$

The result is displayed in figure 4.

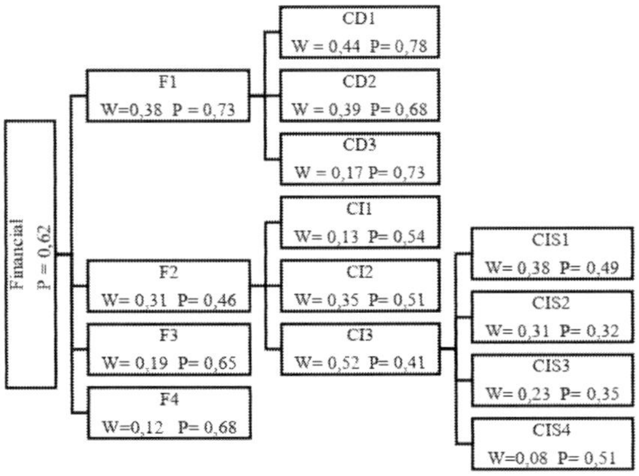

Figure 4 Aggregation of the elementary performances

As a result, the overall maintenance financial performance is 0.62. In other words, the company has achieved 62% of its maintenance financial target. In addition, the weighting of the criteria by MACBETH allows to precisely identify the most critical improvement areas, in this study the performance of the direct cost and that of the indirect cost are the most critical. Therefore, these two criteria must be prioritized for further improvement.

4 Conclusion

The performance measurement system of the maintenance function introduced in this paper is elaborated through three sections. Since the first part defines and prioritizes the performance measurement indicators that are the most appropriate. The second section, by using the fuzzy logic the elementary performances are quantified trough a qualitative evaluation. Subsequently, in the third section, these elementary measures are aggregated to an overall maintenance performance measure. All of this will give the company a synthesized and further detailed visibility on both local and global performance for an optimal maintenance improvement.

References

1. C. Wang, S. Hwang, A stochastic maintenance management model with recovery factor. *Journal of Quality in Maintenance Engineering*, **10**, 2: 154-65 (2004).
2. A. Parida, U. Kumar, D. Galar, C. Stenström. Performance measurement and management for maintenance: a literature review. *Journal of Quality in Maintenance Engineering*, **21**, 1: 2-33 (2015).
3. A. Tsang. A strategic approach to managing maintenance performance. *Journal of Quality in Maintenance Engineering*, **4**, 2: 87-94 (1998).
4. U. Kumar, D. Gal, A. Parida, C. Stenström, L. Berges, Maintenance performance metrics: a state-of-the-art review. *Journal of Quality in Maintenance Engineering*, **19**, 3: 233-277 (2013).
5. P. Muchiri, L. Pintelon, L. Gelders, H. Martin. Development of maintenance function performance measurement framework and indicators. *International Journal of Production Economics,* **13**, 1: 295–302 (2011).
6. A. Parida, T. Åhrén, and U. Kumar, U. Integrating maintenance performance with corporate balanced scorecard. *Proceedings of COMADEM 2003:* 53-59 (2003).
7. A. Neely, M. Gregory, K. Platts. Performance measurement system design: a literature review and research agenda. *International journal of operations & production management*, **15**, 4: 80-116, (2005).
8. L. Berrah, G. Mauris, A. Haurat, L. Foulloy. Global vision and performance indicators for an industrial improvement approach. *Computers in Industry*, **43**, 3: 211-225, (2000).
9. C. Stenström, A. Parida, D. Galar. Performance indicators of railway infrastructure. *The international Journal of railway technology*, **1**, 3: 1-18 (2012).
10. K.Y, Kutucuoglo. J, Hamali. Z, Irani, J. M, Sharp. A framework for managing maintenance using performance measurement systems, *International Journal of Operation and Production Management*, **21**, 1: 173-194 (2001).
11. L. A, Zadeh. Information and control. *Fuzzy sets*, **8**, 3: 338-353 (1965).
12. B. Al-Najjar, I. Alsyouf. Selecting the most efficient maintenance approach using fuzzy multiple criteria decision-making. *International journal of production economics*, **84**, 1: 85-100 (2003).
13. M. Grabisch. The application of fuzzy integrals in multicriteria decision-making. *European journal of operational research*. **89**, 3: 445-456 (1996).
14. M. L. Tseng, J. H. Chiang, L. W. Lan. Selection of optimal supplier in supply chain management strategy with analytic network process and choquet integral. *Computers & Industrial Engineering*, **57**, 1: 330-340 (2009).
15. N. Yalcin, A. Bayrakdaroglu, C. Kahraman. Application of fuzzy multi-criteria decision-making methods for financial performance evaluation of Turkish manufacturing industries. *Expert Systems with Applications*. **39,** 1: 350-364, (2012).
16. A. Shemshadi, H. Shirazi, M. Toreihi, and M.J Tarokh. A fuzzy VIKOR method for supplier selection based on entropy measure for objective weighting. *Expert Systems with Applications*. **38**, 10: 12160-12167 (2011).
17. S. Opricovic, G.H. Defuzzification within a multicriteria decision model. *International Journal of Uncertainty, Fuzziness and Knowledge-Based Systems*, **11**, 5: 635-652 (2003).
18. C.A. Bana e Costa, J.C. Vansnick. Applications of the MACBETH approach in the framework of an additive aggregation model. *Journal of Multi-Criteria Decision Analysis*, **6**, 2: pp. 107-114 (1997).

A literature review on multi-echelon inventory management: the case of pharmaceutical supply chain

Nouçaiba Sbai[1,a], *Abdelaziz* Berrado[1]

[1]Equipe AMIPS-Ecole Mohammadia d'Ingénieurs, Université Mohammed 5, Rabat, Maroc

Abstract. Inventory management remains a key challenge in supply chain management. Many companies recognize the benefits of a good inventory management system. An effective inventory management helps reaching a high customer service level while dealing with demand variability. In a complex supply chain network where inventories are found across the entire system as raw materials or finished products, the need for an integrated approach for managing inventory had become crucial. Modelling the system as a multi-echelon inventory system allows to consider all the factors related to inventory optimization. On the other hand, the high criticality of the pharmaceutical products makes the need for a sophisticated supply chain inventory management essential. The implementation of the multi-echelon inventory management in such supply chains helps keeping the stock of pharmaceutical products available at the different installations. This paper provides an insight into the multi-echelon inventory management problem, especially in the pharmaceutical supply chain. A classification of several multi-echelon inventory systems according to a set of criteria is provided. A synthesis of multiple multi-echelon pharmaceutical supply chain problems is elaborated.

1 Introduction

Controlling the material flow from suppliers of raw material to final customers has always been a critical problem. Top management of all organizations in any sector of the economy have recognized the strategic importance of supply chain management [1]. Viewing the supply chain as a whole, adopting a strategic orientation with cooperative efforts and having a customer focus have always been the main characteristics of the supply chain management [2].

Innovation leaders focused on effective inventory management as an essential part of supply chain management. In fact, it leads to enable shorter lead times to customers, makes working capital available for sustained profitability and helps dealing with demand variability.

The complexity of supply chains nowadays and the high interaction between all its nodes, make the determination of optimal inventory policy so difficult [3]. Adopting local optimization in each site of the supply chain cannot always have good results. In fact, single-echelon inventory management assumes that inventory is located in only one storage point where demand is met and replenishments are added. Actually, inventory is located in many storage points that belong to the same system which made researchers think about integrated approaches and modeling the system as a multi-echelon inventory system [4].

The high responsibility of the multi-echelon pharmaceutical supply chain to ensure the delivery of medicine in the right conditions makes the inventory management of this system more complex [5].

In this paper, we focus our attention on multi-echelon inventory systems. We consider specifically the multi-echelon inventory management in the pharmaceutical supply chain.

This paper is organized in three sections. The next section presents the multi-echelon inventory system. Then, we present in Section 3 a literature review on multi-echelon inventory management and especially in the pharmaceutical supply chain.

2 Multi-echelon inventory system

The multi-echelon supply chain has multiple stages and many players at each stage [6]. The products are first sent to a central warehouse then shipped to the distribution centers that make the delivery to the final customer.

A multi-echelon system is a network composed of stages that are grouped into echelons. Stages can represent physical locations, items in stock or in processing activities [7]. Figure 1 presents a clear illustration of a multi-echelon inventory system.

The echelon inventory is defined as the inventory between a stage in the supply chain and the final customer. That means, the inventory position added to all downstream installations inventory [8].

[a] Corresponding author: noucaibasbai@research.emi.ac.ma

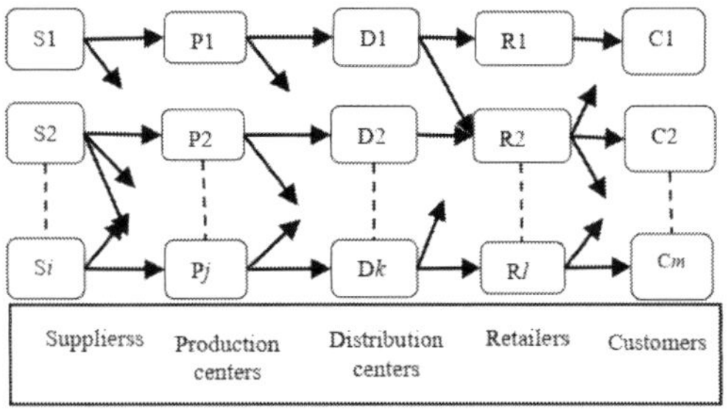

Fig. 1. An illustration of a multi-echelon supply chain with *i* suppliers and *m* customers.

2.1 Structures of multi-echelon inventory system

A multi-echelon inventory system may involve multiple suppliers and multiple customers for each inventory location in the system. On the basis of their network structure, multi-echelon inventory systems can be classified into serial systems, assembly systems, distribution systems and general systems.

In serial systems illustrated in Figure 2, each inventory location has at most one predecessor and one successor. In general, dealing with serial systems could be easier than with other types of multi-echelon systems.

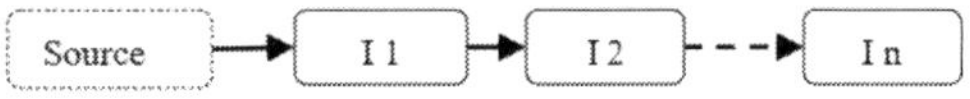

Fig. 2. Serial System of *n* inventory locations

The second type is assembly systems. As illustrated in Figure 3, each installation has one immediate successor but multiple components can be assembled into a single item or final product which can result in many suppliers for a stock.

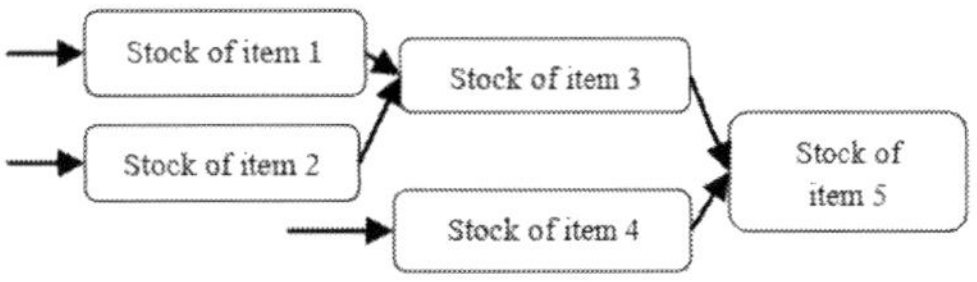

Fig. 3. Example of assembly system

A third structure is the distribution system in Figure4. This type is characterized by the increase of the number of parallel installations along the material flow, and this because a supplier may have multiple customers.

Finally, the general structure presented in Figure 5, can include any of the previous three structures as parts of the whole system.

2.2 Characteristics of multi-echelon inventory system

A multi-echelon inventory system has many characteristics that differentiate it from other systems and make it present many challenges in supply chain management. On the one hand, a multi-echelon inventory system is composed of multiple echelons that contain multiple stocks with physical locations.

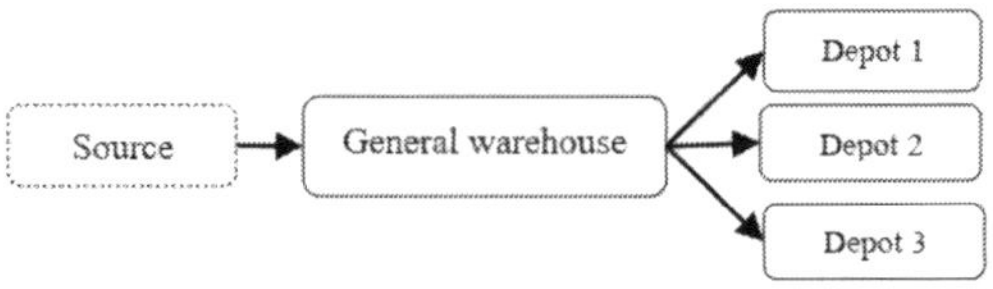

Fig. 4. Distribution system [4]

Each stock at a downstream position is a customer of one or multiple stocks at an upstream position [9]. On the other hand, this system can be characterized by the higher levels of average inventory which lead to have some trade-offs like having lower transportation costs, lower pipeline inventory, and faster response in meeting the demand at the end-customer level [4].

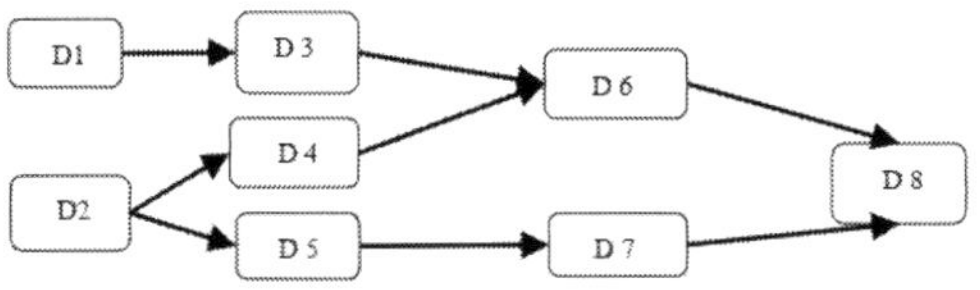

Fig. 5. Illustration of a general multi-echelon inventory system of 8 depots [1]

2.3 Multi-echelon inventory management

Inventory management approach has a direct impact on the overall inventory levels in the supply chain. Many companies used to manage their inventories in isolation or independently. Adopting a single-stage optimization can lead to have the sizing of inventories by applying the appropriate inventory models but it doesn't take into account the interdependencies between the different stages. Thus, multiple actors in the supply chain started to pay more attention to multi-echelon inventory management as a global optimization approach [4].

Multi-echelon inventory management or "the basic multi-activity inventory control problem", is establishing rules or policies which cause flows of the product through the network and which satisfy a prescribed performance objective [7]. As information is becoming centrally available for many companies, a more effective and efficient handling of inventories is possible. Multi-echelon inventory management as an integrated approach, considers completely the factors related to inventory optimization: cost, variability, service and complexity. It sizes correctly the stock across the entire supply chain while taking into account the complex interdependencies between stages as well as variables that cause excess of inventory.

3 Literature review

In this section, we present a review on the models proposed for the optimization of multi-echelon inventory systems in general with a focus on pharmaceutical supply chain. We also define the problem of selecting the appropriate multi-echelon inventory system.

3.1. Literature review on multi-echelon inventory management

The multi-echelon inventory system was discussed first by Clark and Scarf [10]. Before this research work, all papers treated the inventory problem by determining the optimal purchasing quantities at a single installation only. This paper presented the problem as a multi-installation model. Since that time, many authors tried to develop new approaches and ideas based on the model given by Clark and Scarf [3, 11, 12, 13, 14, 15, 19].

In Table 1, we classified several models that were formulated by a lot of researchers in order to solve the multi-echelon inventory problem. Based on many assumptions and aspects considered in the single-stage optimization as well as those mentioned in the papers as characteristics of the models, we chose the following criteria as the ones that were assumed by the authors as necessary features in their formulations:

• Demand type: Deterministic or stochastic, static or non-static demand and the capability of the system to offer the option of emergency deliveries. In a deterministic model, external demands at each installation of the system are known with certainty. In a stochastic model, the demands obey a given probability distribution. For the case of static demand, parameters used to define external demand don't vary over time.

•Inventory/ Review policy: There are two replenishment policies often used in practice. The continuous review policy and the periodic review policy. In a continuous review model, the inventory position of the system is reviewed continuously over time and a quantity Q is ordered when the level of inventory reach the reorder point. In a periodic review model, the inventory status is reviewed at regular periodic intervals and the inventory level is raised to a predefined point. Besides that, this criteria includes the replenishment decisions (centralized or decentralized). In the centralized case, the objective is to find the optimal cost for the whole system. In the decentralized case, every installation in the system intends to optimize its own costs independently.

•Product: This criteria helps to classify the models into systems that are manufacturing a single product and those producing multiple products.

•Number of echelons: The number of echelons contained in the system.

3.2 The case of pharmaceutical supply chain

In the past, pharmaceutical supply chains were considered as a tool to deliver products to markets with taking into account security of supply. Recently, pharmaceutical industries are looking for new approaches in order to create additional benefits. There are a lot of actors in the pharmaceutical supply chain: primary manufacturing installations, distribution centers, wholesalers, hospitals and many others. The interdependencies between those components and the sensitivity of the pharmaceutical products make the call for sophisticated supply chain optimization techniques necessary.

The multi-echelon inventory problem in pharmaceutical supply chains was investigated by many authors. Tamil [5] treated a two-echelon supply chain inventory model for multiple pharmaceutical products under realistic problems in healthcare industries. He used a continuous review integrated production-distribution inventory model for a pharmaceutical company and a hospital supply chain , considering multiple pharmaceutical products, variable lead time and many constraints related to space availability and customer service level. Guerrero [15] presented a methodology to find near-optimal joint inventory control policies for a real case of a one-warehouse, n-retailer distribution system of infusion solutions at a University Medical Center in France. They considered emergency replenishments and target service level constraints under stochastic demand. The objective of the model was to optimize the inventory policy of the supply chain including the special constraints and operational differences that the healthcare industry demands. Kim [17] designed and developed an integrated supply chain management system for optimizing inventory control and reducing material handling costs of pharmaceutical products in healthcare sector. The supply chain studied was composed of pharmaceutical companies, a wholesaler and hospitals. The real time information sharing providing by the SCM system proposed helped forecasting demand more accurately by the wholesaler and reduce the total supply chain cost of pharmaceutical products by more than 30 %. Noorfa and Andrew [18] evaluated inventory management in the private healthcare sector in Malaysia. The paper focused on the inventory and delivery management process. After investigating a leading healthcare company in Malaysia who owns a wholesaler and a chain of medical clinics, the authors found two issues related to urgent orders due to poor inventory control methods and stock availability at the wholesaler problems. In order to reduce the operating cost and increase the customer service level, they suggested implementing a VMI approach based on the organization's current inventory management [18].

3.3 Definition of the problem of selecting the appropriate multi-echelon inventory system

The main problems faced by pharmaceutical supply chain sector are the need to balance future capacity with anticipated demand at the designed stage, and the difficulty to ensure responsiveness at the operation stage [16]. Nowadays, adopting supply chain management practices by applying techniques and methods developed in the industrial settings takes an important consideration by many health care organizations. In this context, our objective is to develop an approach to help decision makers choose the appropriate multi-echelon inventory system that satisfies the needs of the system and corresponds to their priorities.

4 Conclusion

In this paper, we presented a literature review on multi-echelon inventory problem. Depending on the network structure, the multi-echelon inventory system may involve multiple suppliers and multiple costumers. Multiple multi-echelon inventory models were classified according to a set of criteria. Demand type, review policies, manufacturing single product or multiple products and the number of echelons involved in the system are necessary features and relevant parameters required for formulating and controlling a multi-echelon inventory system. A literature review on the multi-echelon pharmaceutical supply chains was elaborated.

Table 1. Classification of multi-echelon inventory models according to criteria used in the literature

Network structure/ Criteria		**Serial system**	**Assembly system**	**Distribution system**
Demand	Stochastic	[10]		[11],[12],[15], [3],[10],[14]
	Deterministic	[19]		[19],[12]
	Static			[11],[12],[14]
	Emergency deliveries			[11],[15]
Inventory	Continuous Review policy		[3]	[14], [3]
	Periodic review policy			[11],[15],[13]
	Centralized replenishment decisions	[19]		[14],[19]
	Decentralized replenishment decisions	[19]		[14],[19]
Product	Single product	[10]		[3],[10]
	Multiple products	[19]		[11],[12],[15], [19]
Number of echelons	2	[19]		[15],[3],[14], [19]
	3			[11], [12]
	> 3	[10]		[13],[10]

Lastly, the literature review on multi-echelon inventory management presented in this paper will be helpful to build alternatives for decision makers in order to choose the appropriate multi-echelon inventory system that corresponds to their priorities and responds to the needs of the whole system. This could be treated in our future work.

This work was conducted within the research project RSCM2015-2018. The authors would like to thank the Moroccan MS, MESRSFC and CNRST for their support.

References

1. S. Axsäter, *Inventory Control* (Springer International Publishing, 2015)
2. J.T.Mentzer, W.Dewitt, J.S.Keebler, S.Min, N.W.Nix, C.D.Smith, Z.G.Zacharia. *"Defining supply chain management."* J. Bus. Logist.,22.2 (2001)
3. R.Ganeshan, *"Managing supply chain inventories: a multiple retailer, one warehouse, multiple supplier model."* IJPE, 59 (1999): 341-354
4. P.Vrat, *Materials Management* (Springer India, 2014)
5. R.Uthayakumar, S.Priyan, *"Pharmaceutical supply chain and inventory management strategies for optimization: a study on pharmaceutical company and hospital."* ORHC, 2.3 (2013)
6. P.Meindl, S.Chopra, *Supply Chain Management, Strategy, Planning, and Operation,* (Pearson Education, Inc., 2013).
7. A.J.Clark, *"An informal survey of multi-echelon inventory theory."* NRL (1972)
8. S. Axsäter, K.Rosling, *"Notes: Installation vs. echelon stock policies for multilevel inventory control."* MS, 39.10 (1993)
9. P.LL, *Optimization of (R,Q) policies for multi-echelon inventory systems with guaranteed service* (LOSI, UTT,2013)
10. H.Scarf, A.J.Clark, *"Optimal policies for a multi-echelon inventory problem."* MS, 6.4 (1960)
11. W.Zhou, L.Chen, H. Ge, *"A multi-product multi-echelon inventory control model with joint replenishment strategy."* Appl. Math. Model., 37 (2013): 2039-2050
12. P.Tsiakis, N.Shah , C.C.Pantelides, *"Design of multi-echelon supply chain networks under demand uncertainty."* Ind. Eng. Chem. Res., 40.16 (2001): 3585-3604
13. E.B.Diks, A.G. de Kok, *"Optimal control of a divergent multi-echelon inventory system."* Eur. J. Oper. Res., 111 (1998): 75-97
14. K.Moinzadeh, *"A multi-echelon inventory system with information exchange."* MS, 48.3 (2002)
15. W.J.Guerrero, T.G.Yeung, C.Guéret, *"Joint-optimization of inventory policies on a multi-product multi-echelon pharmaceutical system with batching and ordering constraints."* Eur. J. Oper. Res., 231 (2013)
16. N.Shah, Comput.*" Pharmaceutical supply chains: key issues and strategies for optimization."* Chem. Eng., 28 (2004): 929-941
17. D.Kim, *"An integrated supply chain management system: a case study in healthcare secotr."* EC-Web, LNCS, 3590 (2005): 218-227

18. N.H.Mustaffa, A.Potter, *"Healthcare supply chain management in Malaysia: a case study."* Supply chain manag. 14. 3 (2009)
19. A.M.M.N.Ahsan, K.A.Zamam, M.N.Sultana, *"An optimal joint multi-item replenishment policy for a two level supply chain: centralized versus decentralized decision."* AJOR, 3.2 (2012)

Optimizing the Multi-Objective Deployment Problem of Mlat System

Rabie Fadil[1], *Badr* Abou El Majd[1], *Hassan* El Ghazi[2], and *Hicham* Rahil[3]

[1]CeReMAR, LMSA Lab, FSR, Mohamed V University, Rabat, Morocco
[2]STRS Lab, National Institute of Posts and Telecommunication, Rabat, Morocco
[3]LIMSAD Lab, FSAC, Hassan II University of Casablanca, Casablanca, Morocco

Abstract. Multilateration (MLAT) systems are powerful means for air traffic surveillance. These systems aim to extract, and display to air traffic controllers identification of aircrafts or vehicles equipped with a transponder. They provide an accurate and real-time data without human intervention using a number of ground receiving stations, placed in some strategic locations around the coverage area, and they are connected with a Central Processing Subsystem (CPS) to compute the target (i.e., aircraft or vehicle) position. The MLAT performance strongly depends on system layout design which consists on deploying the minimum number of stations, in order to obtain the requested system coverage and performance, meeting all the regulatory standards with a minimum cost. In general, choosing the number of stations and their locations to cope with all the requirements is not an obvious task and the system designer has to make several attempts, by trial and error, before obtaining a satisfactory spatial distribution of the stations.

In this work we propose a new approach to solve the deployment of Mlat stations problem by focusing on the number of deployed stations and the coverage as the main objectives to optimize. The Non-dominated Sorting Genetic Algorithm II(NSGA-II) was used in order to minimize the total number of stations required to identify all targets in a given area, with the aim to minimize the deployment cost, accelerating processes, and achieve high availability and reliability. The proposed approach is more efficient and converge rapidly which makes it ideal for our research involving optimal deployment of Mlat station.

1 INTRODUCTION

Nowadays, multilateration (MLAT) systems are a feasible option to be used in the air traffic control (ATC) technological infrastructures. These systems were widely deployed over a high number of airports around the world, for the aircraft surveillance and control in all flight phases, and in many cases, they are replacing the classical secondary surveillance radar. A multilateration system aim to detect, locate, and identify cooperating targets (i.e., aircraft or vehicle) position by receiving and processing suitable signals emitted by onboard transponder devices, according to the Secondary Surveillance Radar (SSR) international standards (e.g., the Mode A/C and Mode S signals).

In these systems, a number of ground receiving stations, with capabilities to measure some physical characteristics of signals emitted by transponders, such as Time of Arrival (TOA), are placed in some strategic locations around the coverage area, and they are connected with a Central Processing Subsystem (CPS to compute the target (i.e., aircraft or vehicle) position as shown in figure 1.

In the standard, and most widely extended, configuration of MLAT systems, the Mode S transmissions and the unsolicited transponder emissions (i.e., the squitter) as well as the responses to interrogations elicited by the MLAT system, can be used. These signals are received by the ground stations and their TOAs are measured and sent to the CPS, where the transponder position is calculated [1] [2]. This calculation is based on the Time Difference of Arrival (TDOA) measurements. They are the time of arrival (TOA) differences between the ith receiving station and a reference one (normally designated as the number 1). This is sketched in figure 2, which, for the sake of simplicity, refers to the two-dimensional case. It

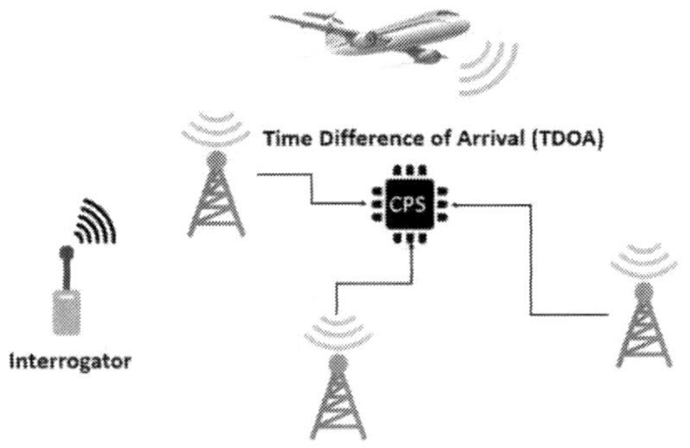

Figure 1: Principle of Mlat system

can be observed that the target is located at the intersection of the two hyperbolas.

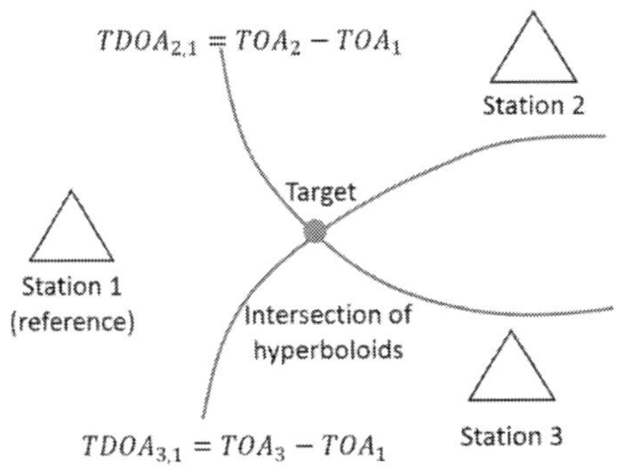

Figure 2: An illustrated (2D) view of TDOA Localization

The TDOA function geometrically represents an hyperboloid (or hyperbola), and it can be expressed as follows:

$$TDOA_{i,1} = \frac{1}{c}\Big(\sqrt{(x-x_i)^2+(y-y_i)^2+(z-z_i)^2} - \sqrt{(x-x_1)^2+(y-y_1)^2+(z-z_1)^2}\Big) + n_{i,1} \quad (1)$$

Where c is the speed of light, (x, y, z) the unknown target position (aircraft position), and (x_i, y_i, z_i) is the known position of the ith station ($i = 1$ denotes the reference station). $n_{i,1}$ is a TDOA random noise term, which generally is assumed to be zero mean Gaussian distributed [3].

In some scenarios, it is common to find a numerical problem when solving the system of hyperbolic equations. This problem is defined in the literature as an ill-conditioned problem and the consequence of this is that, when the system of equations is solved, the solution is not correct or it has a big error. The mathematical interpretation of this problem goes back to the three conditions of Jacques Hadamard [4], namely, the solution exists, the solution is unique, and the solution depends continuously on the data. If at least one of these conditions is not satisfied the problem becomes ill-conditioned. On the other hand, the effects of this problem in the multilateration systems accuracy have been studied in [5] [6].

Several methods were adopted highlighting the challenges existing in target lacalization. Most of them aim to construct and solve the system of hyperbolic equations. Linearizing the system of hyperbolic equations by Taylor series expansion [7] [8] is the most accepted strategy to solve these hyperbolic equations, for estimating the target position [9]. Numerical approaches were also used to solve the localization problem [9] [10] [11]. These methods assume certain numerical approximations between the target position and its derived parameter in order to simplify the solution. The most common assumed approximation is that of mutual numerical independence between them. Contradictory to the numerical approaches, the algebraic approaches [12] [13] [14] do not use any statistical assumptions nor numerical approximations. They algebraically manipulate the hyperbolic equations until directly set an inverse problem that linearly relates the unknown target position with the known parameters. Some techniques add new measurement capabilities to the system, such as the angle of arrival [15] or the round-trip-delay [14] [16] in order to change the ill-conditioned Jacobian matrix into a well-conditioned Jacobian matrix. However all these solutions, although are efficient options, require in much cases significant economic investments.

The MLAT performance strongly depends on system layout design, which consists on deploying the minimum number of stations, in order to obtain the requested system coverage and performance, meeting all the regulatory standards with a minimum cost. In general, choosing the number of stations and their locations to cope with all the requirements is not an obvious task and the system designer has to make several attempts, by trial and error, before obtaining a satisfactory spatial distribution of the stations.

This paper introduces an original methodology to determine the number of stations that should be deployed and their position in order to obtain the requested system coverage and performance, meeting all the regulatory standards with a minimum cost. To the best of our knowledge, this article is the first to tackle the problem of Mlat system deployment with a multi-objective optimization approach.

The paper is organized as follows: Section 1 presents the general aspects and the classical solution for the MLAT localization problem. In Section 2 the mathematical formulation of the model and the optimization algorithm are defined. Section 3 presents some experimental results that validate the model and the optimisation process. Finally, Section 4 concludes the paper and states some open questions.

2 MATHEMATICAL MODEL

2.1 Multi-objective optimization

Multi-objective optimization involves the simultaneous optimization of multiple objective functions. In many cases, the objectives are defined in incomparable units, and they present some degree of conflict among them (i.e., one objective cannot be improved without deterioration of at least another objective).

In principle, multi-objective optimization is very different than the single-objective optimization. In single objective optimization, one tries to obtain the best design or decision, which is usually the global minimum or the global maximum depending on the optimization problem. In the case of multiple objectives, there may not exist one solution which is best (global minimum or maximum) regarding all objectives. In a typical multi objective optimization problem, there exists a set of solutions which are superior to the rest of solutions in the search space when all objectives are considered but are inferior to other solutions in the space in one or more objectives. These solutions are known as Pareto-optimal solutions or nondominated solutions. The rest of the solutions are known as dominated solutions.

One way to solve multi-objective optimization problems is to reformulate the problem as a single objective problem. The main advantage of this approach is that any generic optimization algorithm can be used to optimize this single objective function. However, this approach also suffers from several drawbacks: it requires a prior knowledge of the trade-off the decision maker is willing to make between the different objectives (the relative importance or weights of the objectives). The second approach requires that the decision maker interacts with the optimization procedure typically by specifying preferences between pairs of presented solutions. Interactive approaches consider only a small set of non-dominated solutions due to the effort required.

An interesting approach based on game theory was proposed by the authors of [17] [18] [19] to solve multi-objective optimization problems. This approach considers the multi-objective optimization problem as a multi-player co-operative game where each objective function to be optimized is a player in the game. A game is said to be co-operative if the players are able to reach an agreement on strategies. The players are the objective functions, which are ultimately controlled by the decision maker and so can be expected to reach an 'agreement', meaning the game is co-operative. Using the fundamental text on co-operative games [20].

Pareto optimization, finds a representative set of non-dominated solutions approximating the Pareto front. Evolutionary Algorithms (EAs) [21] [22], bio-inspired optimization algorithms crudely mimicking natural evolution by implementing stochastic optimization through "natural selection" and "blind variations", can easily be turned into multi-objective optimizers by replacing the "natural selection", that favors the best value of the objective function, by some 'Pareto selection' based on the Pareto dominance relation. Some diversity criterion is generally used, ensuring a wide spread of the population over the Pareto front. The resulting algorithms, Multi-Objective Evolutionary Algorithms (MOEAs), have demonstrated their ability to do in a flexible and reliable way [23] [24].

Several MOEAs have been proposed in the literature, based on different implementation of the Pareto dominance selection and the diversity criterion [25] [26] [28]. In particular, many MOEAs use an archive of solutions, where they maintain the non-dominated solutions ever encountered during the search. Pareto optimization methods, such as evolutionary multi-objective optimization algorithms, allow decision maker to check the potential solutions without a priori judgements regarding the relative importance of objective functions.

In single-objective optimization, it is possible to determine between any given pair of solutions if one is better than another one. As a result, we usually obtain a single optimal solution. However, in multi-objective optimization there is not a straightforward method to determine if a solution is better than another one. The method most commonly adopted in multi-objective optimization to compare solutions is the one called Pareto dominance relation, which, instead of a single optimal solution, leads to a set of alternatives with different trade-offs among the objectives. Given two solutions **X** and **Y** of a multi-objective problems, **X** is obviously to be preferred to **Y** in the case where all objective values for **X** are better than those of **Y**, one at least being strictly better: in such case, **X** is said to Pareto-dominate **Y**. However, Pareto-dominance is not a total order, and most solutions are not comparable for this relationship. The set of interest when facing a multi-objective problem is the so-called Pareto set of all solutions of the search space that are not dominated by any other solution: such non-dominated solutions are the best possible trade-offs between the contradictory objectives, i.e., there is no way to improve any of them on one objective without degrading it on at least another objective. In multi-objective optimization, the search space is generally called the design space, by contrast with the objective space, where the Pareto front is the image of the set of non-dominated solutions. the Pareto set can contains a large number of solutions. At this stage, the decision-maker selects feasible solutions according to criteria which depend of explicit objective function preferences. In order to make the section of a feasible solution a manageable task, the set of Pareto solutions has to be filtered or reduced to a small number of representative solutions [28].

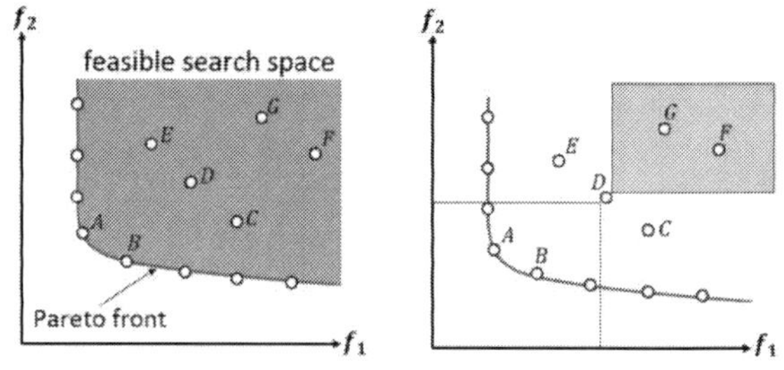

Figure 3: Pareto dominance

Figure 3 illustrates these ideas for a two-objective minimization problem. Solution D dominates F and G because their objective values for both f_1 and f_2 are greater than those for D. Meanwhile, the objective values for A and B are smaller than those for solution D, and so both A and B dominate D. However neither C nor E dominate, or are dominated by D.

2.2 Mathematical formulation

We consider a set of N_s stations that will be used to locate M targets, the stations are located at positions x_i with $i = 1, ..., N_s$ and the targets are located at θ_j with $j = 1, ..., M$. Each station is characterised by an interrogation range r_i which define the coverage area of the station i. The distance between the station i and the target j is given by :

$$d_{ij} = \|x_i - \theta_j\| \quad (2)$$

For each station we associate a binary variable.

$$a_i = \begin{cases} 1 & \text{if the station } i \text{ is deployed} \\ 0 & \text{otherwise.} \end{cases}$$

Then, the number of deployed stations can be stated as follows:

$$f_1 = \sum_{i=1}^{N_s} a_i \quad (3)$$

Now, we need to calculate the number of covered targets for a station i.

$$b_j^i = \begin{cases} 1 & \text{if } \exists! i \in \{1, ..., N_s\} : d_{ij} \le r_i, j \in \{1, ..., M\} \\ 0 & \text{otherwise.} \end{cases}$$

Then, the covered area for station i is expressed as follows:

$$f_2 = M - \sum_{j=1}^{M} b_j^i, \forall i \in \{1, ..., N_s\} \quad (4)$$

The multi-objective optimization problem can be formally stated as follows:

$$\begin{cases} \min \ \sum_{i=1}^{N_s} a_i \\ \max \ M - \sum_{j=1}^{M} b_j^i \\ \left(x, b, a\right) \in \left\{\mathbf{R}_+^{N_s} \times \mathbf{R}_+^{N_s}, \{0,1\}^{N_s} \times \{0,1\}^{M}, \{0,1\}^{N_s}\right\}. \end{cases}$$

2.3 Algorithm

This section introduces the evolutionary optimization algorithm that has been used to validate numerically the mathematical model described above, namely the Non-dominated Sorting Genetic Algorithm (NSGA-II) [27].

The algorithm generates a first initial population and evaluates it. On this initial population, it performs crossover and mutation to obtain the offspring population. The offspring population is evaluated. The two populations are composed into a single one and sorted according to the domination relationship. The individuals, which are non-dominated with respect to one another, are further sorted by their crowding measurement (density computation). Then the best individuals are selected according to their fitness assigned after the sorting procedure. These selected individuals form the next population.

3 EXPERIMENTS

In this section, we present results from MATLAB simulation on the multi-objective optimization of the Mlat system design layout is described, the main goal is to choose the suitable number of deployed stations and where to set them up in order to obtain the requested system coverage and performance. The NSGA-II algorithm was used to solve an instance including $N_s = 20$ stations and $M = 50$ targets, all the stations have an interrogation range $r = 40m$. The targets are randomly distributed from (x, y) positions $(1,5)$ to $(100,120)$. NSGA-II has been chosen here for its long record of successes, demonstrating its robustness to find a good approximation of the Pareto front. The chosen variation operators for NSGA-II algorithms are shown in the table below:

NSGA-II operators

Encoding: Binary
Population size : 100
Maximal number of evaluations : 100
Selection : Binary tournament
Crossover : Intermediate Crossover , $P_c = 0.9$
Mutation : Gaussian, $P_m = 0.01$

All the results were computed on an Intel® Core™ i5-4200M CPU with 4× 2.5 GHz and 4Go of memory.

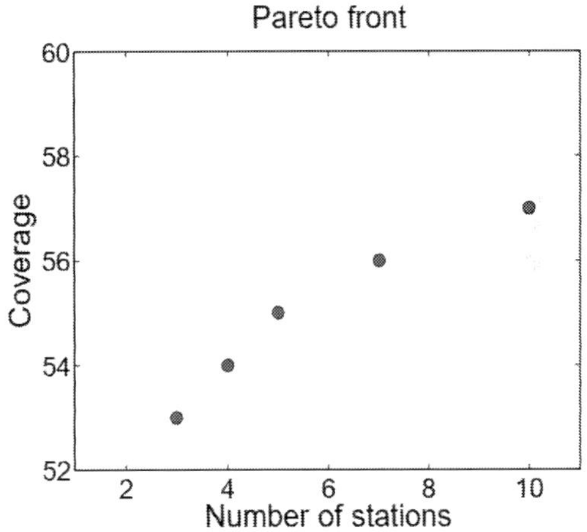

Figure 4: Pareto front obtained with NSGA-II

The Pareto front depicted in figure 4 shows the trade-off between minimizing the number of deployed stations and maximizing the number of lactated targets. If the decision maker wants to minimize the number of stations no matter how many targets are located she/he shall select a solution toward the lower left corner of the graph. Otherwise, if she/he thinks that the number of estimated targets matters, she/he can choose a compromise in the top right corner of the graph, resulting in higher number of located targets, at the price of a higher number of stations. Figure 5 shows a representation of a solution in the decision variable space. The solution corresponds to a number of 4 stations that were used to lactate 54 targets.

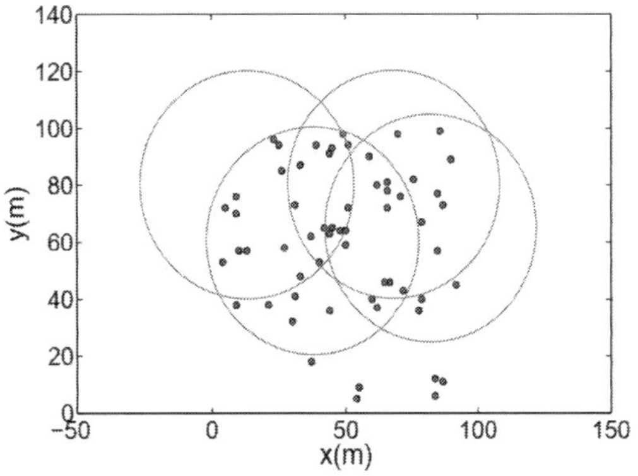

Figure 5: Representation of a solution in the decision variable space. A number of 4 stations were used to lactate 54 targets

4 Conclusion

Multilateration systems are a powerful means for the surveillance function of Air Traffic Control (ATC) operations. The MLAT performance strongly depends on system layout design which consists on deploying the minimum number of stations, in order to obtain the requested system coverage and performance, meeting all the regulatory standards with a minimum cost. In this paper we have presented a new approach for optimizing the system design layout by minimizing the number of stations while maximizing the number of lactated targets. The approach is based on multi-objective optimization and it is intended to find the optimal spacial distribution of the ground stations and also their number in order to have a maximum system coverage. Numerical results demonstrated the ability of the propose approach to obtain the most efficient localization strategy. The new technique can be applied to design, deploy and operate the MLAT systems for airport surface surveillance as well as for take-off-landing, approach and en-route control.

References

[1] EUROCAE-WG-41: ED-117, *MOPS for Mode S multilateration systems for use in advanced surface movement guidance and control systems (A-SMGCS)*. In: The European Organisation for Civil Aviation Equipment (EUROCAE) (November 2003).

[2] EUROCAE-WG-70: ED-142, *technical specification for wide area multilateration (WAM) systems*. In: The European Organisation for Civil Aviation Equipment (EUROCAE) (September 2010).

[3] W.R. Hahn, S.A. Tretter, *Optimum processing for delay-vector estimation in passive signal arrays*. IEEE Trans. Inf. Theory IT–19(5), 608–614 (1973)

[4] J. Hadamard, *Lectures on Cauchy's Problem in Linear Partial Differential Equations*, Yale University Press, New Haven, 1923.

[5] M. Leonardi, A. Mathias, G. Galati, *Two efcient localization algorithms for multilateration*. Int. J. Microw. Wirel. Technol, 1 (2009), 223–229.

[6] G. Galati, M. Leonardi, M. Tosti, *Multilateration (local and widearea) as a distributed sensor system: lower bounds of accuracy*, in European Radar Conf, EuRAD, Amsterdam, 30–31 October 2008.

[7] W.H. Foy, *Position-location solution by Taylor-series estimation*. IEEE Trans. Aerosp. Electron. Syst., AES-12 (1976), 187–194.

[8] D.J. Torrieri, *Statistical theory of passive location systems*. IEEE Trans. Aerosp. Electron. Syst., AES-20 (1984), 183–198.

[9] I.A. Mantilla-Gaviria, M. Leonardi, G. Galati, J.V. Balbastre-Tejedor and E.D.L.R. Davó, *Efficient location strategy for airport surveillance using Mode-S multilateration systems*. International Journal of Microwave and Wireless Technologies, 4(2), 209-216, 2012.

[10] K. Yang, J. An, X. Bu, G. Sun, *Constrained total least-squares location algorithm using time-difference-of-arrival measurements*. IEEE Trans. Veh. Technol. 59(3), 1558–1562 (2010)

[11] I.A. Mantilla-Gaviria, M. Leonardi, J.V. Balbastre-Tejedor, Edl. Reyes, *On the application of singular value decomposition and Tikhonov regularization to ill-posed problems in hyperbolic passive location*. Math. Comput. Model. 57(7–8), 1999–2008 (2013).

[12] M. Geyer and A. Daskalakis, *Solving passive multilateration equations using Bancroft's algorithm*, in Digital Avionics Systems Conference, Bellevue, WA, USA, 1998, pp. F41/1 - F41/8.

[13] S. Bancroft, *An algebraic solution of the GPS equations*, IEEE Transactions on Aerospace and Electronic System, vol. AES-21, pp. 56-59, January 1985.

[14] R.O. Schmidt, *A new approach to geometry of range difference location*, IEEE Transactions on Aerospace and Electronic System, vol. AES-8, pp. 821-835, Nov. 1972.

[15] G. Galati, M. Leonardi, M. Tosti, *Multilateration (local and wide area) as a distributed sensor system: lower bounds of accuracy, in European Radar Conf.* EuRAD, Amsterdam, 30–31 October 2008.

[16] E. Perl, M.J.Gerry, *Target localization using TDOA distributed antenna*, US 2005/0035897 A1, USA, 17 February 2005.

[17] B. Abou El Majd, J.A. Désidéri and A. Habbal, *Aerodynamic and structural optimization of a business-jet wingshape by a Nash game and an adapted split of variables*. Mécanique Industries, 11(3-4), 209-214, 2010.

[18] R. Duvigneau, B. Abou El Majd and J.A. Désidéri. *Towards a self-adaptive parameterization for aerodynamic shape optimization.* In ESAIM: Proceedings (Vol. 22, pp. 169-174). EDP Sciences, 2008.

[19] J.A. Désidéri, R. Duvigneau, B. Abou El Majd and Z. Tang, *Algorithms for efficient shape optimization in aerodynamics and coupled disciplines.* In 42nd AAAF Congress on Applied Aerodynamics, Sophia-Antipolis, France, 2007.

[20] J. Nash. *Two-person cooperative games.* Econometrica: Journal of the Econometric Society, pages 128-140, 1953.

[21] A. Eiben and J. Smith, *Introduction to Evolutionary Computing.* Springer Verlag, 2003.

[22] W.F. Fihri, Y. Arjoune, H. El Ghazi, N. Kaabouch and B. Abou El Majd, *A particle swarm optimization based algorithm for primary user emulation attack detection.* In: IEEE consumer communications and networking conference, p. 1–6, 2018.

[23] C.A.C. Coello, D.A.V. Veldhuizen, and G.B. Lamont, *Evolutionary Algorithms for Solving Multi-Objective Problems.* Kluwer Academic Publishers, 2002.

[24] K. Deb, *Multi-Objective Optimization Using Evolutionary Algorithms.* John Wiley, 2001.

[25] A. Raghib, B. Abou El Majd, and B. Aghezzaf, *An Optimal Deployment of Readers for RFID Network Planning using NSGA-II.* In Recent Developments in Metaheuristics, Operations Research/Computer Science Interfaces Series, Springer, vol. 62, p. 463–476, 2018.

[26] H. Rahil, B. Abou El Majd, and M. Bouchoum, *Optimized Air Routes Connections for Real Hub Schedule Using SMPSO Algorithm.* In Recent Developments in Metaheuristics, Operations Research/Computer Science Interfaces Series, Springer, vol. 62, p. 369-384, 2018.

[27] K. Deb, A. Pratap, S. Agarwal, and T. Meyarivan, *A Fast Elitist Multi-Objective Genetic Algorithm: NSGA-II,* IEEE Transactions on Evolutionary Computation, vol. 6, pp. 182-197, 2000.

[28] R. Fadil, B. Abou El Majd, H. Rahil, H. El Ghazi and N. Kaabouch, *Multi-objective Optimization Approach for Air Traffic Flow Management.* In: MATEC Web of Conferences 105, 0005, 2017.

Machine Learning for Supply Chain's Big Data: State of the art and application to Social Networks' data

Radouane El-Khchine[1,a], *Amine* Amar[2], *Zine Elabidine* Guennoun[2], *Charaf* Bensouda[1] and *Youness* Laaroussi[2]

[1] *Ibn Tofail University, Department of Mathematics, Kenitra, Morocco*
[2] *Mohamed V University, Department of Mathematics, Rabat, Morocco*

Abstract. In the context of today's pattern of globalization and a huge amount of information, a smart supply management chain is required. Naturally, statistics and operations research are used for optimizing supply and demand objectives. However, the new context brings out new opportunities at descriptive, predictive and prescriptive levels for supply chain network design, logistics and distribution and strategic sourcing. The key question is still how to capture and to use information. One striking example can be taken from social media, where their use allow to gain insight into the perception of consumers and to capture a real time overview of consumer reactions, regarding one or more specific events. In this regard, different modern approaches, such as IoT or Quantum neural network, are developed. In the same line of thought, we propose an analytic approach, based on KNN, Logistic Regression and SVM with the use of Twitter data in chicken supply chain management. Results identify the main concerns related to chicken products and allow to the development of a consumer-centric supply chain. The proposed approach can be extended to other topics such as anomaly detection and codification of customer intelligence.

1 Introduction

Supply chain management is presented as the field which consists, in parts, to provide the right quantities of goods most efficiently at the right place in the right order within the right time. Meeting these demands requires planning the inbound logistics. The process of planning depends on frequently changing information of product development, assembly line planning and purchasing. Currently, a high amount of time is spent for gathering information during planning and existing knowledge from previous planning processes. Planning process can be separated into strategic (long-term) planning which generates an initial evaluation for feasibility of different plant and supplier locations to integrate new products into production network, tactical (mid-term) inbound logistics planning which focuses on the engineering of logistics process alternatives and their evaluation and operational (short-term) planning of logistics before start of production and where all preselected logistics process and resources will be continuously detailed and integrated into the production plant by pre-series processes during the ramp up.

Analytics in supply chain management is not a new task[1]. For a long time, supply chain management has use statistics and operation research for optimizing the objectives of matching supply and demand. Business analytics using information system support has a strong relationship to supply chain performance [2]. However, the development of big data indeed brings out new opportunities. The term supply chain analytics can be used to define the advanced big data analytics in supply chain management [3]. This analytics can be categorized into descriptive, predictive and prescriptive analytics [4].

Big data analytics permits users to capture, store, and analyse huge amount of data, from internal as well as external of the organization, from multiple sources such as corporate database, sensor-captured data such as RFID, mobile-phone records and locations, and internet in order to understand the meaningful insights [5]. Akter et al. [6] suggested that big data analytics has a big impact to enhance firm performance. By improving big data analytics capability, a firm could create new products and services, provides better customer service, increases sales and revenue, and expands into the new market. Zhong et al. [7] provided the discussion on the big data applications in various sectors such as financial services sector, healthcare, logistics, and manufacturing. However, present research reveals that there is a limited agreement regarding the performance of big data that support supply chain management, especially for strategic sourcing, supply chain network design, product design and development; demand planning, procurement, production, inventory, logistics and distribution, supply chain agility and sustainability.

Consequently, this article aims to explore the application of social media big data and its analysis in supply chain management, especially for chicken supply chain.

In this regard, we propose an analytic approach, based on a comparison between three machine learning models: K-Nearest Neighbours (KNN), Logistic Regression, and Support Vector Machine (SVM) with the use of Twitter data related to chicken supply chain. Results identify the main concerns for chicken products and allow to the development of a consumer-centric supply chain. The proposed approach can be extended to other topics such as anomaly detection and codification of customer

[a] Corresponding author: radouane.elkhchine@gmail.com

https://doi.org/10.1051/matecconf/201820000015

intelligence. The rest of this paper is organized as follow: the next section expose several works related to the analyse of supply chain using big data. The third part explains preparation of data extracted from Twitter. In the fourth section, we present theoretical overview of methods and algorithms used, and the processing steps, when the fifth part presents the obtained results. This document ends with conclusion and an overview of further works.

2 Related works

Food products supply chain, such as chicken products pass through several steps, from production (farmer, abattoir and processor), to retailer and consumer. To have a constantly improved supply chain, continuous information should be analysed in every step, especially the one provided by consumer [8, 9]. Indeed, more and more industries are nowadays pursuing a consumer-driven supply chain [10, 11].

To meet aforementioned goals, researchers have employed social media information to retrieve valuable insights [12]. In fact, consumer opinion, given by comments on online social media provide management with unprecedented opportunities to leverage collective consumer intelligence for enhancing supply chain management and sales forecasting [13]. Several methods could be used for extraction of intelligence from tweets, such as Naïve Bayesian classifier, support vector machine, or artificial neural networks [10]. The sentiment analysis and social media are used to study several phenomena in different filed such as finance, medicine and others [14, 15].

N. Shukla et al. [16] propose a big data analytics based approach, which considers Twitter data for identifying supply chain management issues in beef supply chain. The results indicated that the proposed text analytic approach can be helpful to identify crucial customer feedback for supply chain management.

Other research works are conducted to investigate the effectiveness of e-retailers' logistics-related customer service interactions on Twitter with a view towards identifying effective and ineffective social media customer service strategies [17]

Social media have been used by many businesses, W. He et al. [18] apply text mining to analyze unstructured text content on Facebook and Twitter of three largest pizza chains. The results show the competitiveness of social media analysis and the power of text mining to extract business value from the vast amount of available social media data.

3 Data

To harvest Twitter posts, we developed a python script that uses Twitter API, we used a list of keywords to get data related to a specific topic. This list contains 22 keywords of the logistics and supply chain's jargon combined with the word "chicken". We then got all the 10 days of the available historical data for each keyword.

The script send as much queries as needed to get all available tweets, each query returns a JSON file with 100 posts and a huge amount of related information we call metadata. Each file should be parsed to keep only needed information. In our case, we've been interested in variables like: Twitter post's ID so we can follow the historical line while harvesting data, the post itself is the more important variable, the place so we can do some geographical analysis, the retweeted count, the keyword used to get the tweet, date, time, and time zone when the tweet was posted.

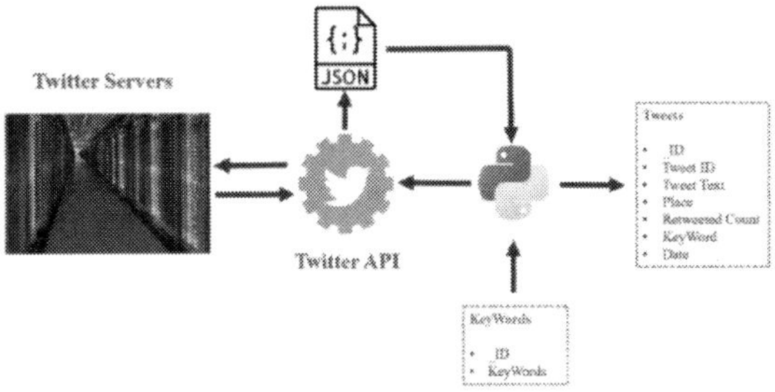

Figure 1. Data extraction process from Twitter servers to our databases.

The database where we store our data contains six tables: keywords, raw tweets table with different variables which will be cleaned and transformed to another table in order to train and test models, a backup table for raw tweets, a table for stop words, those are commonly used words we ignore, and a table of new tweets' extractions on which final classification model will be applied.

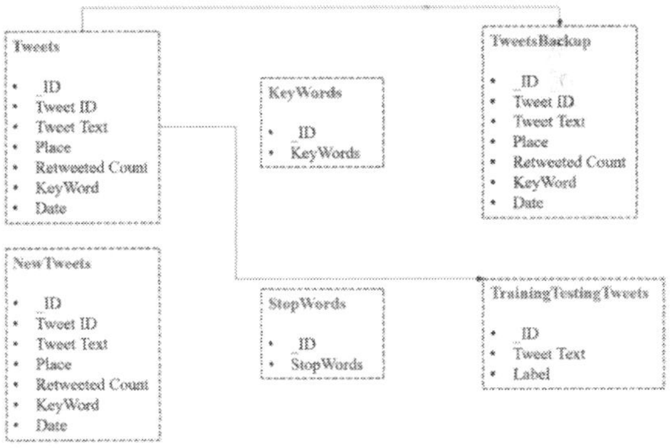

Figure2. Database schema

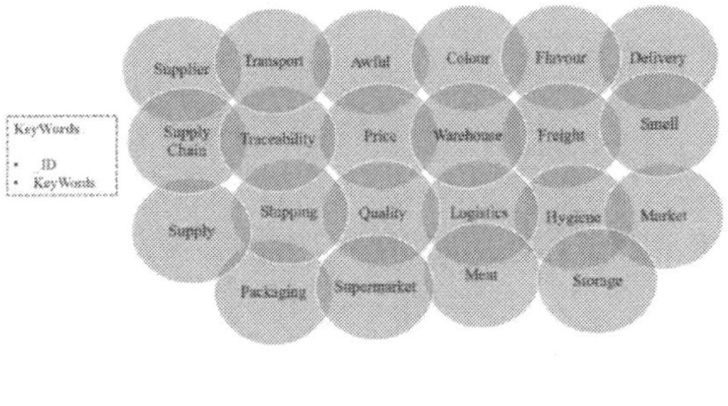

Figure3. Keywords list from the "KeyWords" table

As demonstrated by Morstatter et al.[19], the available free data collected using Twitter API is a good and sufficient representations of the full data on Twitter. This allows us to generate results obtained to the general consumer present on Twitter.

4 Proposed Methodology

4.1 Overview of used Models

The extracted data will be analyzed based on three main machine learning classification algorithms, K-Nearest neighbors (KNN), support vector machine (SVM) and logistic regression.

4.1.1 K -Nearest neighbors (KNN)

KNN algorithm is one of the most basic classification algorithms in Machine Learning. It belongs to the supervised learning domain and finds intense application in pattern recognition, data mining and intrusion detection. KNN is a non-parametric algorithm, i.e. it does not make any assumptions on the underlying data distribution and the model structure is completely determined from the data. Its purpose is to use a database in which the data points are separated into several classes to predict the classification of a new sample point, which will be classified according to the K (integer number) nearest points on the database [20].

4.1.2 Support Vector Machine (SVM)

A Support Vector Machine (SVM) is a discriminative classifier formally defined by a separating hyper-plane, it belongs to the supervised learning field. The SVM was developed in the 1990s and was extremely popular around the time and continue to be the high-performing algorithm with little tuning [21].

Given labelled training data, the algorithm outputs an optimal hyper-plane, dividing a plane in two parts where each class lay in a side, the hyper-plane maximizes the margin between classes.

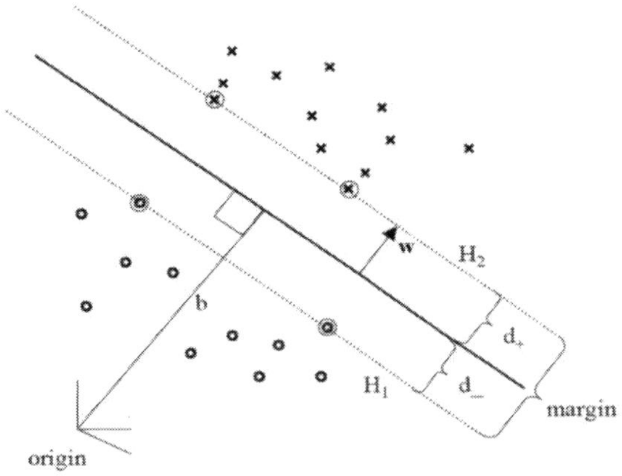

Figure4. Maximum margin hyper-plan for the SVM model

The margin is expressed by the following equations:

$$\vec{w}^T\vec{x}_i + b \geq +1 \quad for\; y_i = +1$$
$$\vec{w}^T\vec{x}_i + b \leq -1 \quad for\; y_i = -1$$

Both constraints can be combined into one set of inequalities:

$$\vec{y}_i(\vec{w}^T\vec{x}_i + b) - 1 \geq 0 \quad \forall i$$

Maximize the margine is equivalente to:

$$Max\frac{2}{\|\vec{w}\|} \Longrightarrow Min\|\vec{w}\| \Rightarrow Min\Phi(\vec{w}) = \frac{1}{2}\vec{w}^T\vec{w}$$

This is a classic optimization problem that could be resolved using Lagrangian method.

In some cases, data are not linearly separable, we use then a transformation to another space where it is easier to find the optimal hyper-plan.

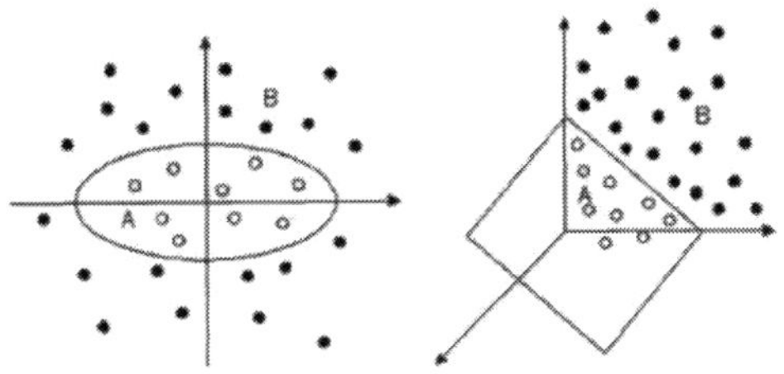

Figure5. Space transformation for non linear separable data

4.1.2 Logistic regression

Logistic regression is a statistical method for analysing a dataset in which there are one or more independent variables that determine an outcome, called dependent variable, which is a binary variable. The goal of logistic regression is to find the best fitting model to describe the relationship between the dependent variable and a set of independent variables.

Logistic regression predicts the probability of occurrence of an event by fitting a logistic function to data, called also, the sigmoid function and defined by:

$$S(x) = \frac{1}{(1 + e^{-x})}$$

We can remark that the curve given by S has a finite limit of 0 as x approaches $-\infty$ and 1 as x approaches $+\infty$ and when x=0 is S(x)= 0.5.

Thus, if the output is more than 0.5, we can classify the outcome as 1 (or YES) and if it is less than 0.5, we can classify it as 0(or NO) [22].

4.2 Processing Steps

The data on social media is highly unstructured. In fact, tweets contain text, URLs, hashtags, numbers, emoticons, and emoji. Therefore, appropriate text cleaning approach is required for effective knowledge gathering. Since there is no optimal data cleaning way, we develop our specific approach, depending on each case.

Indeed, after extracting the appropriate tweets that populate the main table, this table contains 13 693 entry. Since the models shouldn't be trained and tested on data with high similarity scores, the first step consists of deleting duplicates tweets and retweeted ones, next step is to perform a similarity test on the remaining tweets, so we reject tweets with similarity score higher than 80%, this operation took more than 5 hours, at this stage, we've got 4 906 tweets left. We used then a dictionary approach to label tweets; this generates labels between -5 and 5 depending on how many positive or negative dictionary words the tweet contains. Tweets with scores between -1 and 1 are dropped. Finally, we keep only 1 509 tweets with positive labels, and 1 509 tweets with negative labels. Those are the tweets populating the training and testing table of our database.

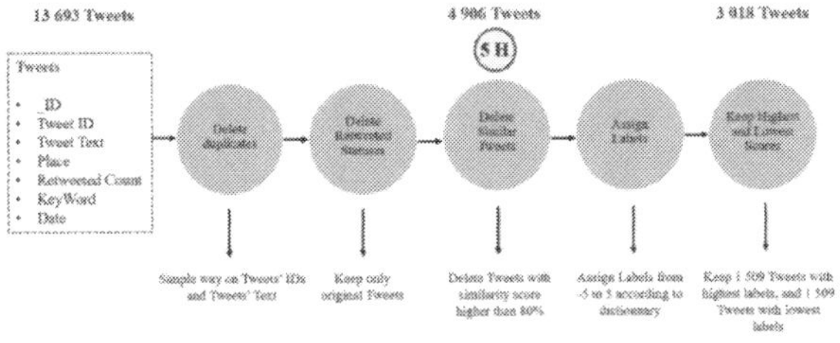

Figure6. Different steps for data cleaning process

The input for each used models should be under a matrix form, therefore, each tweet should pass through a "words to vectors" process. This consists in building a dictionary of all different words in all cleaned tweets, so each tweet could be represented by a vector.

First step is to remove "stop words", then delete words with length less than 3 characters, next step is to transform the words into their basic forms using python library NLTK, then transforming capital letters to small ones, then it is necessary to add new words to the dictionary. At this stage, a vector with dimension of dictionary size could represent a tweet by putting in each word on the dictionary the number of times it appears on the tweet, the values should be standardized, finally, we assign labels, 0 to negative and 1 to positive tweets.

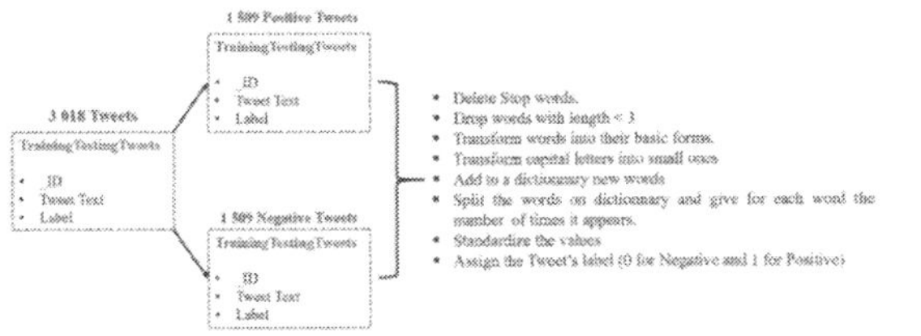

Figure7. Different steps of "Words to Vectors" process

	Word 1	Word 2	———	Word N	Label
Tweet 1	0	0.2	———	0.6	1
Tweet 2	0.23	0	———	0	1
Tweet 3	0.12	0	———	0.06	1
:	:	:	———	:	:
:	:	:	———	:	:
:	:	:	———	:	:
Tweet 1509	0.1	0.32	———	0.05	1
Tweet 1	0.3	0.15	———	0.1	0
Tweet 2	0.31	0	———	0	0
Tweet 3	0	0.01	———	0.02	0
:	:	:	———	:	:
:	:	:	———	:	:
:	:	:	———	:	:
Tweet 1509	0.32	0	———	0.08	0

Figure8. Final shape of input data after "Words to Vectors" process

In this study, three models are compared on same data for several times to measure which one performs the best to classify tweets. Before proceeding to this comparison, a previous study should be done to find best models' hyper-parameters to be used.

For SVM model, we used a grid search on several values of "Gamma" and "C" hyper parameters, for each combination, a corresponding SVM model is trained on data and tested three times after a random shuffle. Best scores are obtained for Gamma = 1 and C = 10, we choose a radian basis function as kernel.

We obtain the best hyper-parameter for KNN model by testing several values of K = 3, 5, 7, 10 on data. The best scores are shown for K = 5. The logistic regression model we used doesn't have any hyper-parameters.

Once all hyper-parameters known, we run the comparison of the three models on data. We first shuffle the matrix 20 times randomly, we split the data into training set of size 2 518 tweets and a testing set of 500 tweets, the three models are trained and tested on same data, and the scores are stored. This process is repeated 500 times, and took more than 15 hours.

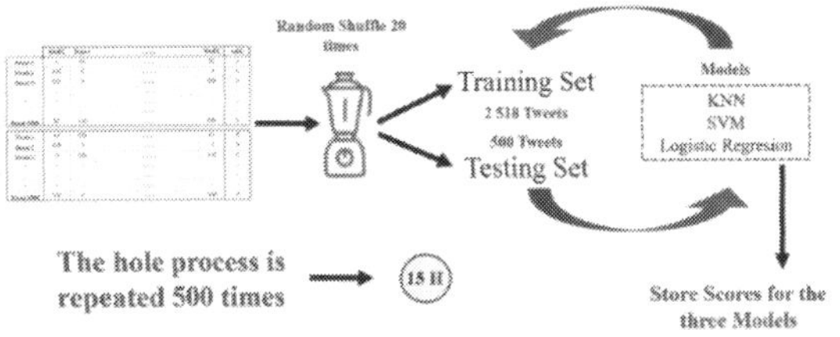

Figure9. Process for training and testing models to perform comparison between SVM, KNN, and Logistic regression

At this level, we could compare the three models and choose the best one.

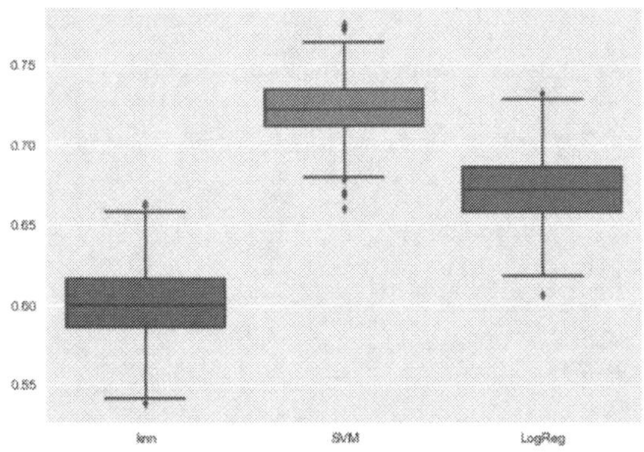

Figure10. Results of comparison between SVM, KNN, and Logistic regression

The KNN model made in average a score of 60% and may reach 66%, the logistic regression model records in average a score of 67% on testing data and can reach 73%, the SVM model scores on testing set in average around 72% and reaches 78%. Logistic regression model got the best score in 0.8% of cases and SVM is the best model in 99.2% of cases.

The SVM model with Gamma = 1, C = 10, and a radian basis function kernel is the one we'll be using next. We fit this model to the training data as much as needed until we got the SVM parameters that reach at least a score of 78% on testing data. The final SVM model made a score of 78.8% on testing set.

5Results

After performing the benchmark study which compares performance of KNN, Logistic regression, and SVM models, we do know the model that works the best on our kind of data. At this step, we perform a second massive extraction of Twitter data, respecting same keywords as before, during another 10 days. We obtained 11 032 of raw tweets, after a cleaning process (removing exact duplicates, removing URLs, removing emoticons and emoji, ...), we have a number of 6 417 tweets left on which our SVM model will be run.

Results of classification show sentiments related to the chicken supply chain based on consumer feedbacks on Twitter.

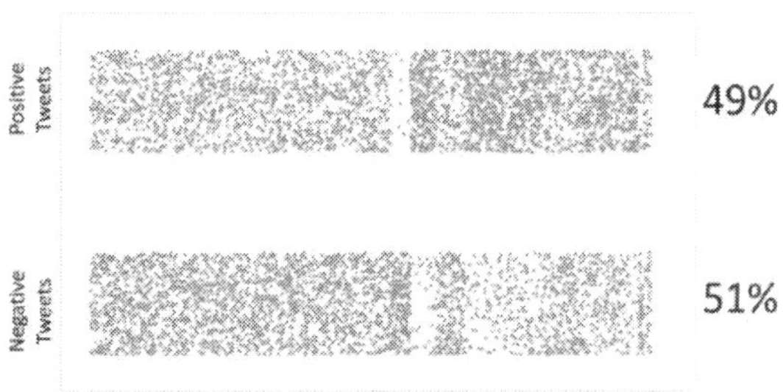

Figure11. Results of the SVM classifier on the 6 417 new tweets

Each point on the data visualization above is a one of the 6 417 tweets, 49% of customers on Twitter are expressing positive opinion about chicken supply chain. On the other side, 51% of feedbacks are negative.

It will be interesting to explore a geographical distribution of positive and negative feedbacks and how it is different from a country to another.

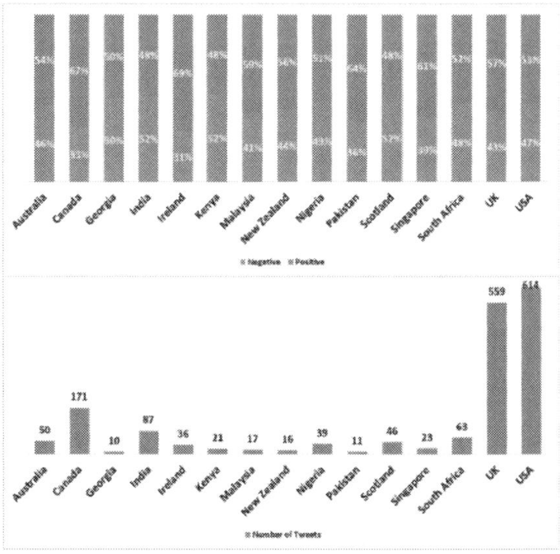

Figure12. Results of the SVM classifier on some chosen countries

Higher proportions of positive opinions related to chicken supply chain are shown in Ireland with 69%, followed by Canada where 67% of customers are happy with the quality of chicken supply chain, then United Kingdom with 57% of satisfied people on Twitter.

On the other hand, higher percentages of negative feedbacks arises from India and Scotland with 52%, then 48% for South Africa.

Another way to analyse results could be spreading the 6 417 tweets by keywords to figure out which one cumulate more positives or negatives consumers' feedbacks.

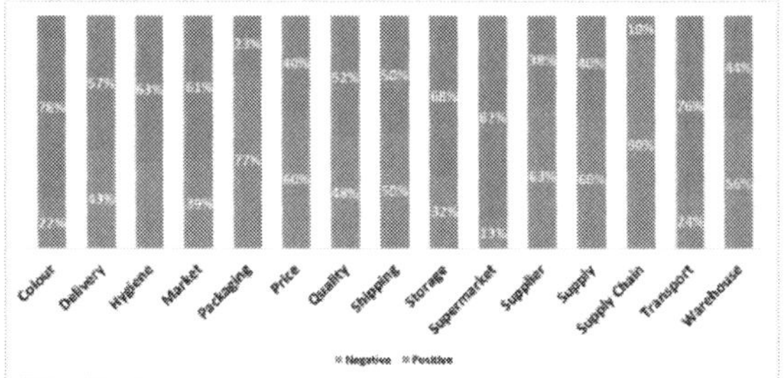

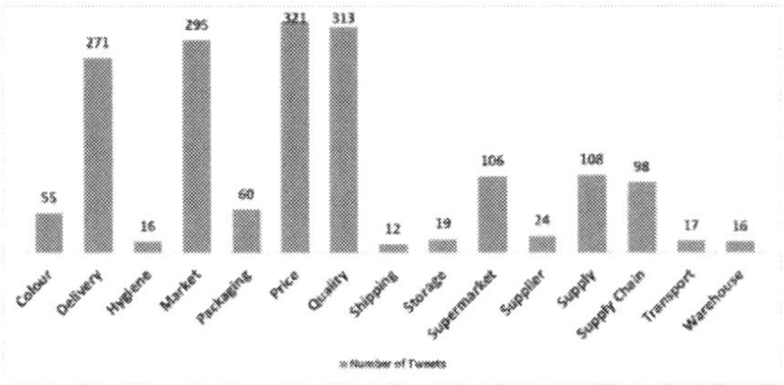

Figure13. Results of the SVM classifier on some chosen keywords

Higher parts of positive feedbacks are shown for keyword "Flavour" (90%), followed by the keyword "Supermarket" with 87%, and keyword "Colour" (78%), then 57% of opinions related to keyword "Delivery" are positives.

For negative opinions, 77% of consumers reacting on Twitter about "Packaging" are not happy, keyword "Price" has a proportion of 60%, then 57% of people speaking about "Meat" have a negative opinions.

6 Conclusion and further works

Analysis of the given information from social media, is used to understand the chicken supply chain, based on consumer feedback on Twitter. The study of such type of data conduct to understand reasons for positive and negative sentiments, identify communication nature and remarks of Twitter users discussing about several aspects related to chicken supply chain.

Consumers are using social media like Twitter to express their point of view on chicken meet quality (color, taste, price, packaging, ...), which generates plenty of useful available information to develop and improve supply chain strategy. This information is big in size considering its volume, variety and velocity and it is vague and unstructured in nature. In the proposed methodology, customers' tweets associated with chicken supply chain are being extracted, sorted and organized into tables with different variables which will be cleaned and transformed in order to train and test models. A new table of tweets is extracted on which final classification model will be applied.

Based on this analysis, results of classification showed interesting insights on how consumers are reacting on Twitter about topics related to chicken supply chain. Twitter provides several metadata that could be used for more advanced analysis like geographical data, date and time for time series analysis on how opinions polarity progresses, users, etc.

For businesses that adopt a consumer-centric strategy especially in supply chain management, this kind of studies could be very helpful to understand more and more consumers' needs.

In future, an enhanced list of keywords could be used for further analysis of the issue. Twitter analytics could be employed for longer time duration and, like time series analysis on tweets classifications could be performed. We also intend to use machine learning models to generate scores rather than binary classifications and using more advanced machine learning models such as deep artificial neural networks on larger datasets.

References

1. G.C. Souza, Supply chain analytics. Business Horizon, **57**, 595–605 (2014).
2. P. Trkman, K. McCormack, M.P.V. de Oliveira, and M.B. Ladeira, Decision Support System, **49**, 318–327 (2010).
3. G. Wang, A. Gunasekaran, E. W. Ngai, and T. Papadopoulos, International Journal of Production Economics, **176**, 98-110 (2016).
4. S. Tiwaria, H.M. Weeb, and Y. Daryanto, Computers & Industrial Engineering, **115**, 319-330 (2018).
5. J. Zakir, T. Seymour, K. Berg. Big data analytics. Issues in Information Systems, **16 (2)**, 81–90 (2015).
6. S. Akter, S.F. Wamba, A. Gunasekaran, R. Dubey and S.J. Childe, International Journal of Production Economics, **182**, 113-131 (2016).
7. R.Y. Zhong, S.T. Newman, G.Q. Huang, and S. Lan, Computers & Industrial Engineering, **101**, 572-591 (2016).
8. R. Handfield, T. Linton. The LIVING Supply Chain: The Evolving Imperative of Operating in Real Time; Wiley, ISBN: 978-1-119-30625-2, (2017).
9. M. Hugos. Essentials of Supply Chain Management. Wiley, ISBN: 978-0-470-94218-5 (2011).
10. J. Yoon, R. Narasimhan and M. K. Kim, International Journal of Production (2017).
11. B. Schulze-Ehlers and S. Anders, Renewable Agriculture and Food Systems, **33**, 73-85 (2018).
12. C. Dhaou. Social media sentiment analysis: lexicon versus machine learning, Journal of Consumer Marketing, **34 (6)**, (2107).
13. R. Y. K. Lau , W. Zhang, W. Xu. Parallel Aspect- Oriented Sentiment Analysis for Sales Forecasting with Big Data, Wiley Online Library (2017).
14. S. Krishnamoorthy. Sentiment analysis of financial news articles using performance indicators, Knowledge and Information Systems, **56 (2)**, 373-394 (2018).
15. J. Wei, X. Liao, H. Zheng, G. Chen, X. Cheng. Learning from context: A mutual reinforcement model for Chinese microblog opinion retrieval, Frontiers of Computer Science, **12 (4)**, 714-724 (2018).
16. A. Singha, N. Shukla, and N. Mishra, Transportation Research Part E: Logistics and Transportation Review, **114**, 398-415 (2018).
17. N. Shukla, N. Mishra and A. Singh, IMMM (2017).

18. J. Bhattacharjya, A. Ellison and S. Tripathi, International Journal of Physical Distribution & Logistics Management, **46**, 659-680 (2016).
19. W. He, S. Zha, L. Li, International Journal of Information Management, **33**, 464-472 (2013).
20. F. Morstatter, J. Pfeffer, H. Liu and K. M. Carley, In Proceedings the Seventh International AAAI Conference on Weblogs and Social Media, Boston, MA (2013).
21. O. Sutton, Introduction to k Nearest Neighbor Classification and Condensed Nearest neighbor Data Reduction, (2012).
22. S. R. Gunn, Technical report, University of Southampton, Faculty of Engineering, Science and Mathematics; School of Electronics and Computer Science (1998).

Supply Chain Performance Measurement: A Case Study about Applicability of SCOR® Model in Automotive Industry Firm

Radouane LEMGHARI[1,*], *Chafik* OKAR[2], and *Driss* SARSRI[1]

[1]The Innovative Technologies Laboratory, National School of Applied Sciences, 90000, Tangier, Morocco
[2]The Higher School of Technology of Berrechid, Hassan the 1st university, 26000, Settat, Morocco

Abstract. The evaluation of a supply chain is a major priority of companies; it is a task that remains difficult due to the complexity of these systems [1]. This evaluation involves a selection of performance measurement indicators, which are appropriate to the management of this chain. It is then necessary to have a structured approach and adequate methodological tools [2]. Indeed, we propose in this paper a practical method that will model in the first place a Moroccan automotive supply chain, according to the SCOR® model (Supply Chain Operations Reference), proposed by the Supply Chain Council. This method will also identify at each level the appropriate indicators for the performance evaluation depending on the strategic vision. In this context our research problem is made, it is interested in the contribution of the business modelling to improve logistics performance. To the best knowledge of the authors, this is the first work that proposes a case study believed to be easy to understand, practical and suitable for the automotive sector. In short, this study is a real application leap to resolve the problematic unanswered of practical SCOR® model using an industrial application in the Moroccan automotive sector.

1 Introduction

Over the past ten years, logistics sector in Morocco went through a strong growth. A development has been reflected in the improvement of the performance of the Logistics Performance Index (LPI) established annually by the World Bank. Thus, Morocco is currently at the 86th place worldwide in logistics performance after being at 94th in 2007 [3]. A national strategy has been implemented in Morocco to improve logistics performance and develop the competitiveness of Moroccan businesses. Indeed, the supply chain is a well-understood function and successfully deployed in large enterprises, this function remains misunderstood and underexploited in Moroccan automotive companies. The main assumption is that automotive firm requires tools specifically designed and tailored on their characteristics and needs [4]. The automotive industry is a highly globalized sector, where for many actors there is competition from around the world. Faced with the increase in supply and the strong pressure on prices, reinforced by the power of Asian groups (Toyota, Hyundai, etc.) and the arrival of new players from emerging countries (Geely, Tata Group, etc.), many companies seek to optimize their value chain in order to remain competitive [5]. In the supply chain of the automotive industry, many factories are working together to manufacture a product (car, motor, etc.). To meet customer needs, they must be designed and organized appropriately [6].

Logistics can be considered a key competitive factor in the automotive industry due to the increasing number of variants and options of the model. With the growing importance of logistics [7], the evaluation of logistics effectiveness and efficiency is gaining increased attention [4]. For this reason, the automotive sector characteristics and current performance measurement practice have been reviewed and analyzed to derive the main characteristics of an effective model for an automotive firm [4,7]. Moreover, there are many continuous improvement tools such as total quality management, six sigma, continuous process improvement and others that are available for companies to improve their operations. However, none of these improvements programs is dedicated to the logistics chain [8]. The SCOR® model was utilized because of its process orientation and growing use among professionals and academics who are directly involved with supply chains [9]. SCOR® is also becoming the common language for benchmarking and comparing supply chains and supply chain management practices [9].

Thus, in order to improve business systems, the return on investment, to deal with competitiveness and optimize the efficiency of their supply chain, some global companies apply the SCOR® model. In this context and for the same reasons, our company planned to model its supply chain based on the same model since it already has a quality management system based on the process approach. To meet these expectations, we followed the SCOR® methodology, which is based on

* Corresponding author: lemghari.radouane@gmail.com

performance measurement, benchmarking and implementation of best practices. Currently, our company should first seek to make a diagnosis of the performance of its supply chain to identify critical performance and improve their processes.

2 literature review

2.1 The Supply Chain Operations Reference (SCOR) model

In SCOR®, the integrated processes of PLAN, SOURCE, MAKE, DELIVER, RETURN and ENABLE from the supplier's supplier to the customer's customer represent supply chain management. Elements of business process engineering, metrics, benchmarking, leading practices, and people skills into a single framework are combined with SCOR® [10]. The version 4 was the first to include return process in the supply chain. The Revisions of the model are made by the members of the Council when it is determined that modifications need to be made to facilitate the use of the model in practice [11]. The release of version 11 for access, as well as the incorporation of the new material into all SCOR® training and certification programs, are announced by the Supply-Chain Council (SCC), the global not-for-profit organization supporting supply chain professionals and educators [12]. This major update includes Enable Process, Best Practices, and Redesign Costs. These updates bring the current model with the way these processes, practices, and measures are needed to practitioners of the supply chain that implement the model [12].

SCOR® describes the business activities related to all phases of the satisfaction of a customer request. The model itself is structured around primary management processes, and on SCOR® version 11, ENABLE is advanced to one of these processes, so there are six: PLAN, MAKE, SOURCE, DELIVER, RETURN, and ENABLE. Using these process definition blocks, SCOR® can be used to model supply chains that are very simple or very complex using a common set of definitions in disparate industries. In fact, public and private organizations and companies around the world use the model as a basis for projects to improve the global supply chain [12].

These processes focus on performance management, information, politics, the inventory strategy, capital, transportation, physical logistics network, regulatory and other management processes to enable Planning and execution of supply chain activities. SCOR® covers all customer relationships, product and market surrounding customer orders, purchase orders, work orders, return authorizations, forecasting and replenishment orders. It also includes physical movement of raw materials, work in progress, finished goods and merchandise return [10].

The SCOR® model has three levels of process detail. In practice, Level 1 describes the number of supply chains, how their performance is measured, and necessary competitive requirements. Level 2 presents the configuration of the planning and execution strategies in the material flow, involving standard categories such as "make-to-stock", "make-to-order" and "engineer-to-order". Level 3 considers the business processes and system functionality used to process sales orders, purchase orders, work orders, return authorizations, replenishment orders, and forecasts. Level 4 process details are not contained in SCOR® but must be defined to implement improvements and manage processes. Advanced users of the framework have defined process detail as far as Level 5, software configuration detail [10].

Organizations using the model SCOR® performance metrics can compare their performance levels compared to other organizations in the supply chain using a benchmarking tool called SCORmark. The SCORmark database contains historical data from over 1000 companies and 2000 supply chains. The benchmarking process using the SCORmark can be performed by the following steps: (1) defining the supply channels to be compared; (2) measuring the internal and external performances; (3) compares the performance to relevant industrial companies; (4) establish competitive demands; (5) calculate the opportunity value of improvement [18,19]. To facilitate the comparison, the SCORmark stratifies the process performance according to three points [14]:

- "Superior": is the performance (median value) on a specific indicator attained by 10% of the best classified SC's comparing to the total of the supply chains surveyed;
- "Advantage": is the performance (median value) among the top 10 companies and the median of all the supply chains considered;
- "Parity": is the performance (median value) of all the supply chains considered.

The APICS[a] the association announced the launch of the SCOR model version 12.0. Developed by a panel of international supply chain experts, this latest version of SCOR incorporates Omni-channel, metadata, blockchain and other emerging engines that supply chain professionals are using today. SCOR has been the global inter-sectorial standard for supply chain excellence over the last 20 years and, with this update, will continue to support ways to measure, improve and communicate the supply chain's business performance [15].

2.2 The performance measurement in the automotive industry

Lebas (1995) considers that performance is built by the management system and by managers. Performance management precedes performance measurement and makes sense of it. He says that performance measurement and performance management cannot be separated. In his opinion, all those who have focused exclusively on measurement, without understanding that the measures only speak of the consequences of the decisions that created the context of performance, have missed the opportunity to take control and master the process of creation of performance and success for the company or organizational unit examined [16].

[a] APICS is the association for supply chain management and the leading provider of research, education and certification programs that elevate supply chain excellence, innovation, and resilience.

The measurement mechanisms and performance evaluation have been incorporated into the commercial atmosphere since the start of mass production. The need to measure performance has augmented since the creation and evolution of large companies [17]. As said by Neely et al. (1995) the performance measurement process implicates three central concepts: (1) performance measurement: it can be described as the process of quantifying the efficiency and the effectiveness of the actions; (2) performance measured: metric used to measure the efficiency and/or the effectiveness of the action and (3) system of performance measurement: it is the unity of metric used to measure effectiveness and efficiency of actions at the same time. The performance measurement is one of the essential elements in the management systems, and the performance involves the result of decisions taken [18].

By sweeping the literature on this topic of performance measurement in the automotive industry, it is clear that there are not many articles that deal with methods on this subject. Indeed, the most relevant articles will be quoted below with a summary of the results of each article, and this after filtering the search results in the following academic databases: Science Direct (www.sciencedirect.com), Scopus (www.scopus.com) and Springer (www.springer.com). An additional search was conducted using the Google Scholar search engine (scholar.google.com). The chain "performance measurement in automotive industries" was used to search for documents on databases, which followed the described criteria. Below, the most relevant works for our study are presented:

Dornhofer et al. (2016) have developed a modular performance measurement system for the automotive logistics chain. A process perspective is included in the performance measurement system to increase specificity and facilitate continuous improvement initiatives. Furthermore, by integrating the lean logistics principles, an alignment of their approach with the latest developments in logistics concepts in the industry is reached and lean implementation is maintained applying a consistent performance measurement system. To ensure standardization and, at the same time, the flexibility to adapt the performance measurement system to specific logistics processes within a specific company or even manufacturing site, they suggested a new, modular set-up of the performance measurement system, break down the logistics objectives identified at each stage of the process into the supply chain, from a supplier to the customer. The proposed modular performance measurement system can be designed in line with the processes, resulting in a consistent performance measurement system for the whole logistics chain and offering the potential for a supply chain-wide roll-out while, at the same time, covering site-specific processes in a standardized way. Additionally, the comparability of the performance measurement system in terms of benchmarking is ensured throughout the stages of the logistics process, as well as in the overarching performance measurement system [19].

Gaiardelli et al. (2007) proposed an integrated framework for measuring the performance of the aftermarket network and provided an empirical application to two automobile record companies and their official service network. The cases in this study show that performance measurement systems of different actors in the supply chain need to be aligned to achieve strategic coherence. Especially, the performance of different actors at the process level of the framework concurs in determining the after-sales service overall performance concerning the final customer. Furthermore, linkages at other levels (mainly the business and activity ones) may be needed or helpful in ensuring uniformity between strategic and operational objectives, inside the organizations and thus for the whole supply chain. The after-sales activities are currently acknowledged as a pertinent source of revenue, profit and competitive advantage in most manufacturing industries. Top and middle management, therefore, should concentrate on the definition of a structured business performance measurement system for the after-sales business. In addition, from the time when many actors are involved along the after-sale service supply chain, an integrated and multi-attribute set of measures requests to be designed consistently at every level of the supply chain [20].

Saad and Patel (2006) have studied the relevance of the concept of supply chain performance in developing countries. They have also tried to identify performance measure sets for supply chain performance in the context of a developing nation. Their research emphases on supply chain practices in the Indian automobile sector. It classifies and discusses the main motives and determinants for the adoption and implementation of supply chain management concepts. Their research is built on a combination of qualitative and quantitative methods. Primary data were collected by semi-structured interviews and an exploratory investigation [21].

Sanchez and Perez (2005) have explored the link between the dimensions of supply chain flexibility and firm performance in a sample of automotive suppliers. They have found in this research an optimistic relation between a superior performance in flexibility capabilities and firm performance, although flexibility dimensions are not equally important for firm performance. On the other hand, the results indicate that firms are developing more basic flexibility capabilities (at the shop floor level) than overall flexibility capabilities (at the client-supplier level). Nevertheless, aggregate flexibility capabilities are more positively connected to firm performance than basic flexibility capabilities. Therefore, companies might miss opportunities to increase competitiveness by underestimating customer-supplier flexibility capabilities. Finally, the results of this research show that flexibility capabilities are enhanced in supply chains with higher environmental doubt, technological complexity, and mutual understanding, but with lower interdependence among the agents involved in the supply chain [22].

Han et al. (2017) explored a performance measurement system for a dynamic supply chain partnership in an intercultural context. They built an initial framework by examining the existing literature, and then they concretize this framework through an in-depth case study in the Chinese auto industry. Afterward, a performance measurement system, including relational strategy and operational measurement criteria for a supply chain partnership, was developed. In fact, and according to them, the relational strategy contains elements of strategic orientation, management style, interdependence, mutual organizational characteristics and common objectives. Operational measurement criteria are a commitment, trust, communication behavior, information sharing, decision to participate, quality, production performance, delivery, cost, vendor strength, attitude, compromise, and loyalty. The last three operational measurement criteria are particularly relevant for the intercultural characteristic [23].

Velda and Dhiba (2017) have highlighted through the literature review the practices necessary for the implementation of supply chain management in the automotive sector in Morocco. They have chosen as the practice of the supply chain "the management of the supplier relationship"; they studied its impact on the performance of the automaker's suppliers. their results using a series of equation regressions have highlighted an object of supplier influence the performance of supplier relationship management practices of the automaker in Morocco [24].

On the basis of the review of the literature and the best knowledge of the authors, the following remarks were made:

- No article on the SCOR model in the automotive industry as a model for managing the performance of the automotive supply chain.
- No articles using an industrial application of the SCOR model in automotive industries
- The frameworks proposed in the existing articles do not cover the entire global automotive supply chain
- The proposed frameworks cannot be transversalized on the various typologies of the entries of the automobile sector (OEM, Tier 1, Tier 2 ...)

The purpose of this paper is to meet these needs by applying the SCOR model to an automotive company and taking advantage of the advice of industry professionals and practitioners to assess its benefits and limitations after deployment. In the following part, the case study will be developed with the different stages of the SCOR model, modeling, benchmarking, performance indicators, gaps and best practices for the company studied.

3 Methodology

This paper aims at providing for the automotive industries a more suitable, practical and concise path for SCOR® model. To fulfill this aim, it is well documented that case studies provide the very effective approach [29,30]. This is advocated by Yin (2009) who recommended that a case study was especially appropriate when trying to answer the "how" [27]. This is perfectly useful for our case, which focuses on an industrial application about the applicability of SCOR® model in automotive industry firm. Indeed, our methodology used elements of an explanatory case study focusing on the applicability of the SCOR® model. Furthermore, it could also be called an exploratory case study [28] when it came to understanding the metrics used presently within the company, their alignment with the proposed SCOR® model and the integration of the company into our framework during the implementation step. We consequently evaluated the framework using a case study at one manufacturing site of a Moroccan automotive industry. The company studied here is an international automotive company with more than 45 years of experience in the automotive sector, an annual turnover of approximately 3300 million EUR and 14,500 employees around the world. For evaluation purposes, we used a mixture of workshops across logistical functions and also based on interviews, documents made available and data analysis. The participants are dispersed across all levels of grading, from operative level up to logistics management of the different sites.

In the beginning, the first series of workshops was conducted with the aim of aligning the logistic objectives found in the literature with those used on the site. Founded on that, we proposed an open workshop format to collect the required performance indicators from logistics experts within the focus company. The focus was on the transparency of the performance indicators collected at the time, their relevance according to the experts' judgments, in addition to the supplementary performance indicators that were considered relevant for monitoring and improving logistics processes. Subsequently, the clustering and consolidation of all the performance indicators proposed by the SCOR® model and identified in the open workshop rounds were applied. We compared our proposed performance indicators with the ones identified in the experts' workshops. Performance indicators which were not named by the experts, but proposed in standard SCOR®, were highlighted and discussed in respect of their added value in a supplement meeting. The subsequent model was aligned in another round of workshops and additional interviews. After changing the detailed definition of the SCOR® model and each measure (alignment with data sources, adding responsibilities), the system was implemented and data evaluation started, as well as a dashboard allowing performance indicator evaluation launched. Based on that, a quantitative evaluation of each metric was started, which also emphasized initial improvement potentials.

After having implemented the model and finalized its deployment, two brainstorming sessions were conducted to compare the different experiences with this project and to capitalize on it. This allowed us to highlight the contributions of this model for the automotive industries and in return for the weaknesses that can be improved on future research.

4 Case Study

4.1 A brief description of the company

The company already present at Tunisia, chose to develop in Morocco in order to accompany its client and to have a base of production of plastic systems able to deliver the manufacturers based in Spain. The production site Metal understands about ten presses to be followed from 150 to 800 tons and transforms approximately 10000 tons of steel a year. It issues components for the body-in-white of the vehicles of a car maker. The plastic factory of production integrates approximately about ten press 350 and 2700 tons and transforms approximately 3500 tons of annual plastic. It issues components interior and external and engines for the two vehicles, but also of the components for the equipment suppliers also established with his of the free zone.

Supply and Production control: The trades of the Supply model and Production control are exerted in the factory where the role of the logistics consists in supplying the raw material, the components sub-contracted and the components necessary to the manufacturing of the components. The trades of handling interns (receipt, shipping, management of the warehouses...) are also part of this model. Administration of the sales: An administrator in charge of the receipt of the customer orders, imputation of the prices, the establishment of the articles client whose finality is the satisfaction of the client in terms of quantity, quality, and lead-times. Shipping, receipt, and transport: The trades of this model are exerted on upstream flows of supply raw material, components, and sub-contracted components

Currently, logistics is at the center of all the activities of the company, divided into 4 business families that cover all logistics activities: from the supply of raw material to the shipment of the final product. In addition, logistics is one of the services of the company that makes the most use of the computer tool: By following all the flows of the company, by anticipating these flows (forecasts, expression of needs, etc.) and overflowing beyond from the company to suppliers and customers.

For the case of our company, it uses as management tools: (1) SAP where the information is centralized (Systems, Applications, and Products, ERP), which integrates the different functions of the company (accounting, finance, production, supply, marketing, human resources, quality, maintenance, etc.). (2) EDI: for the exchange of various documents internally and externally of the company, it is indeed commercial documents or transport such as invoices. the planning process within the company that we are going to study in this paper, is managed by the planner who processes once a week his calculation of net need for finished products via SAP, checks the break dates of each part, and thanks to an Excel file it establishes the planning of all the week as well as the orders of manufacture of the parts to manufacture. These production orders are then issued to the production manager to be distributed to the production items, once the valid and closed planner has been completed and sent to the stock manager for the recorded on SAP.

4.2 Establishing performance metrics

The benchmarking of SCOR® model is done by comparing the indicators of level 1, 2 and 3, however, it is not necessary to apply the three performance levels in each axis. That is why it would be better to prioritize these axes in order to put the level indicators 1 for less important axes and up to level 3 for the most important. The prioritization of performance axes depending on the company's type of business and its management. SCOR standard set the notations for prioritizing strategic axes of the dashboard that will define for each one the objectives to be fixed during the benchmarking. Indeed, the notations S, A and P are used by the SCOR® model to prioritize the axes are:

- S: « Superior » corresponds to the 90th percentile: This means that it will line up with 10% of the best-performing companies.
- A: « Advantage » corresponds to the 70th percentile: The objective will be to reach the performance of 30% of the best companies.
- P: « Parity » corresponds to the 50th percentile: The objective is to exceed 50% of the best companies.

In order to evaluate the process performance, it will be very interesting to complete the indicators that were already in place by other indicators proposed by the SCOR® model, which correspond to its strategic axes.

According to the baseline with which to compare, there are three types of benchmarking:

- Comparison on a historical basis: monitoring indicator development.
- Internal Benchmarking: is the comparison between companies of the same group.
- External Benchmarking: The comparison is made by external organizations.

The SCOR® model proposes the external benchmarking because the more the reference database will be larger, the results will be significant. The objective here will be to line up with the best companies in the automotive sector basing on logistics activities. After defining the indicators on the dashboard, we preceded the collection of data to calculate their values. We worked depending on the importance:

- Reliability: basic information for these indicators was derived from the history of logistics rates and weekly reports.
- Management of assets: this data was collected from the Finance Department and management control. Most of them are confidential, so they will be expressed as a percentage.
- Logistics costs: this information was collected from the Accounts Department from history.
- Reactivity: This indicator corresponds to the lead-time or the time required to complete an order.
- Flexibility: only the part of the supply was held in account for this axis because it is the most important component in terms of time.

The next step is the measurement of the performance indicators for the dashboard. Indeed, the formulas for calculating the different KPIs are presented below:

$$\text{Service rate (SR)} = \frac{\text{Number of processed orders}}{\text{Number of total orders}} \tag{1}$$

$$\text{Order Fulfilment rate} = SR\ (\text{Quantity by reference}) \times SR\ (\text{deadline}) \times SR\ (\text{Accuracy of documents}) \tag{2}$$

$$\text{Fill rate} = \frac{\text{Delivered volume}}{\text{Capacity of the truck}} \tag{3}$$

$$\%\ \text{of sales costs (COGS)} = \text{sales} - \text{Profits} - \text{administrative costs} / \text{sales} \tag{4}$$

$$Working\ capital\ ratio = \frac{current\ assets}{current\ liabilities} = \frac{Receivables}{Payables} \quad (5)$$

$$Financial\ cycle\ delay = delay\ of\ customer\ debt + number\ of\ days\ of\ stock - payables\ delay \quad (6)$$

$$Payables\ delay = \frac{Payables\ x\ 365}{Annual\ sales} \quad (7)$$

$$Delay\ of\ customer\ debt = \frac{customer\ debt\ x\ 365}{Annual\ sales} \quad (8)$$

$$Number\ of\ days\ of\ available\ stock = \frac{stock\ value\ x\ 365}{Annual\ sales} \quad (9)$$

$$Forecast\ reliability\ rate = \frac{\sum_1^n Abs(sales\ forecast(i) - sales(i))}{\sum_1^n sales\ forecast\ (i)} \quad (10)$$

$$Coverage\ rate = \frac{\sum_1^n Warehouse\ stock(i) + Outstanding(i)}{Week\ average\ forecasts} \quad (11)$$

$$Internal\ service\ rate = \frac{\sum_1^n Abs(planned(i) - realized(i))}{\sum_1^n planned\ (i)} \quad (12)$$

$$Rate\ of\ stock\ variance = \frac{\sum_1^n Abs(SAP\ stock(i) - physical\ stock(i))}{\sum_1^n SAP\ stock\ (i)} \quad (13)$$

$$Rotation\ of\ fixed\ assets\ in\ the\ supply\ chain = \frac{Sales}{Net\ fixed\ assets} \quad (14)$$

$$Customers\ complaints\ rate = \frac{Number\ of\ written\ complaints}{Total\ number\ of\ affected\ deliveries} \quad (15)$$

$$Service\ suppliers\ rate = \frac{Number\ of\ complete\ delivered\ orders\ on\ time}{number\ of\ orders\ to\ be\ delivered\ by\ suppliers} \quad (16)$$

$$Cost\ of\ an\ out\ of\ stock = total\ costs\ related\ to\ a\ cessation\ of\ customer\ activity \quad (17)$$

$$Number\ of\ days\ of\ available\ stock = \frac{Total\ outstanding\ TTC\ x\ number\ of\ days}{Total\ turnover\ tax\ for\ the\ period} \quad (18)$$

$$Customer\ return\ rate = \frac{Number\ of\ returned\ deliveries}{Total\ number\ of\ deliveries} \quad (19)$$

$$Supplier\ return\ rate = \frac{Number\ of\ returned\ deliveries}{Total\ number\ of\ deliveries} \quad (20)$$

The established scorecard containing the benchmarking values provided by the SCOR standard as well as the improvement objectives for each metric are presented in Appendix1.

4.3 Best practices proposed by the SCOR® model

As presented in the dashboard set out in Appendix 1, the difference between the current situation of the company and the value of benchmarking differs from one indicator to another. It is clear that better gap analysis implies good performance improvements. So the first task to do is to know the root causes of these gaps and the processes involved.Regarding the processes mentioned above, the most performing companies have established certain practices that have reported to them tangible benefits. This is a kind of capitalization of experience that the SCOR® model offers through benchmarking. The following table lists best practices offered by SCOR® Framework, taking into account the processes previously diagnosed and the anomalies observed.

The following figure shows the best practices according to their difficulty of implementation and the impact they will have on business processes. Priority actions for this project will obviously be those who have the greatest impact and the least difficulty.

We will prioritize action plans to implement the best practices that have high impacts or benefits for the company and whose difficulty of implementation is low.

Table 2. The best practices proposed by SCOR® model

Processes	Best Practices
P2- supply planning	All Key Participants in the Supply Chain, Including Strategic Partners, Have Full Visibility of the Demand/Supply Plan
	VMI: Vendor-managed Inventory
	CPFR: Collaborative Planning, Forecasting, and Replenishment
P2.1 Identify prioritize product requirements	The Demand Plan is Updated Frequently to Reflect Actual Consumption or Customer Forecast Information
	Maximize Data Integrity and System Accuracy by Ensuring 99%+ Accuracy of BOM Configuration, Inventory Levels, and Schedule Requirements
P2.4 Establish procurement plans	Digital Linkage (EDI, XML, Etc.) is Used to Provide Real-Time Demand Information and Handle Routine Transactions
	Maintain Data and System Integrity by Ensuring Production Data, Inventory Levels, and Schedule Requirements Are 99+% Accurate A detailed production model that synchronizes PLAN and MAKE activities in real time
EP.7 Manages Planning Configuration	ABC Classification

S1.1 et S2.1 schedule deliveries	Electronic Kanban Pull Signals Are Used to Notify Suppliers of the Need to Deliver Product

5 Conclusion and prospects

Being part of a complex and highly competitive field, and in order to evaluate its position in the automotive industries, the company proposed to model its supply chain and to implement a Benchmarking to improve its performance. To meet this need, we opted for the SCOR® model as a diagnostic tool. The first step was to model its supply chain, according to the SCOR® model, this model aimed to standardize the structure of the process from the existing model. The next step is to develop the dashboard. We were able to compare the company studied with the leading companies in the automotive field through the service offered by the Supply Chain Council. Moreover, based on the results of this benchmarking, we collected the axes that represent an opportunity for improvement for the company; we selected a set of best practices that can rectify the weaknesses detected at certain axes. To do this, we developed an action plan that is likely to implement these best practices. The main prospects for this project are:

- For the company: Integration of proposals for improvements in future projects and Logistics Department of the annual performance of external benchmarking, in order to validate and renew the strategic direction of the company.
- For research projects: The proposal for a model of the SCOR® methodology adapted to the structure of Moroccan SMEs.

6 Bibliographies

1. J. T. Mentzer, J. S. Keebler, N. W. Nix, C. D. Smith, and Z. G. Zacharia, "Defining Supply Chain Management," J. Bus. Logist., **22**(2), 1–25 (2001)
2. H. Min and G. Zhou, "Supply chain modeling: past, present and future," Comput. Ind. Eng., 43, no. 1–2, 231–249 (2002)
3. Logistics Performance Index (LPI) Report (2016): The Gap Persists, WASHINGTON, March 20. Accessed Mai 21, 2018. URL: http://lpi.worldbank.org/international/global/2016
4. M. Sako and S. Helper, "The Information Requirements of Trust in Supplier Relations: Evidence from Japan, Europe and the United States," *Trust and Economic Learning*, 23–47 (1998)
5. P. Grosche, U. Mayrhofer, and S. Schmid, "La configuration et la coordination internationales de la chaîne de valeur dans l'industrie automobile allemande," *Financ. Contrôle Strat.*, 1–27 (2015)
6. H. Pierreval, R. Bruniaux, and C. Caux, "A continuous simulation approach for supply chains in the automotive industry," Simul. Model. Pract. Theory, **15** (2), 185–198 (2007)
7. A. Gunasekaran, C. Patel, and R. E. McGaughey, "A framework for supply chain performance measurement," International Journal of Production Economics, **87** (3), 333–347 (2004)
8. H. Garcia Reyes and R. Giachetti, "Using experts to develop a supply chain maturity model in Mexico," Supply Chain Manag. An Int. J., **15** (6), 415–424 (2010)
9. K. McCormack, M. Bronzo Ladeira, and M. Paulo Valadares de Oliveira, "Supply chain maturity and performance in Brazil," Supply Chain Manag. An Int. J., **13** (4), 272–282 (2008)
10. P. Bolstorff and R. Rosenbaum, *Supply Chain Excellence: A Handbook for Dramatic Improvement Using the SCOR Model, Third Edition* (2011)
11. K. Rotaru, C. Wilkin, and Ceglowski Andrzej, "Analysis of SCOR's approach to supply chain risk management," Meas. Bus. Excell., **21** (2), 191–206 (2014)
12. Supply Chain Council, *Supply Chain Operations Reference Model (SCOR)* (2012)
13. F. R. Lima-Junior and L. C. R. Carpinetti, "Combining SCOR® model and fuzzy TOPSIS for supplier evaluation and management," Int. J. Prod. Econ., **174**, 128–141 (2016)
14. G. M. D. Ganga and L. C. R. Carpinetti, "A fuzzy logic approach to supply chain performance management," Int. J. Prod. Econ., **134** (1), 177–187 (2011)
15. APICS Supply Chain Council, *SCOR Supply Chain Operations Reference Model ver. 12.0 quick reference guide* (2017)
16. M. J. Lebas, "Performance measurement and performance management," Int. J. Prod. Econ., **41** (1–3), 23–35 (1995)
17. M. S. Neto and S. R. I. Pires, "Performance Measurement in Supply Chains: A Study in the Automotive Industry," New Advances in Vehicular Technology and Automotive Engineering, **19** (4), 733-746 (2012)
18. A. Neely, M. Bourne, and M. Kennerley, "Performance measurement system design: Developing and testing a process based approach," Int. J. Oper. Prod. Manag., **20** (10), 1119–1145 (2000)
19. M. Dörnhöfer, F. Schröder, and W. A. Günthner, "Logistics performance measurement system for the automotive industry," Logist. Res., **9** (1), 1-26 (2016)
20. P. Gaiardelli, N. Saccani, and L. Songini, "Performance measurement of the after-sales service network-Evidence from the automotive industry," Comput. Ind., **58** (7), 698–708 (2007)
21. M. Saad and B. Patel, "An investigation of supply chain performance measurement in the Indian automotive sector," Benchmarking An Int. J., **13** (1/2), 36–53 (2006)

22. A. Martinez Sánchez and M. Pérez Pérez, "Supply chain flexibility and firm performance: a conceptual model and empirical study in the automotive industry," Int. J. Oper. Prod. Manag., **25** (7), 681–700 (2005)
23. W. Han, Y. Huang, and D. Macbeth, "Performance measurement of cross-culture supply chain partnership : a case study in the Chinese automotive industry," Int. J. Prod. Res., **56**, 2437-2451 (2017)
24. A. M. E. Velda and Y. Dhiba, *2017 International Colloquium on Logistics and Supply Chain Management: Competitiveness and Innovation in Automobile and Aeronautics Industries, LOGISTIQUA 2017*, 106–111 (2017)
25. K. M. Eisenhardt, "Building theories from case study research," Academy of Management Review, **14** (4), 532-550 (1989)
26. R. Yin, *Case study research: design and methods, Applied Social Research Methods Series, 1st edition* (1994)
27. R. Yin, *Case study research: design and methods, Applied Social Research Methods Series, 4th edition* (2009)
28. D. M. McCutcheon and J. R. Meredith, "Conducting case study research in operations management," J. Oper. Manag., **11** (3), 239–256 (1993)

Appendix

The scorecard containing performance metrics measured for the company:

Table 3. The scorecard of the company

Performance metrics	Criteria	Unit	Maximize / Minimize	Frequency	Actual value	Benchmark class			The company objective	GAP
						S	A	P		
Service rate (SR)	Reliability	%	Max.	Week	92%	100%	90%	80%	S	8%
Order Fulfilment rate	Reliability	%	Max.	Week	77%	100%	90%	80%	A	13%
Fill rate	Reliability	%	Max.	Week	76%	96%	82%	75%	A	6%
% of sales costs (COGS)	Logistics costs	%	Min.	Month	5%	3%	8%	13%	S	-3%
Working capital ratio	Management of assets	Ratio	Min.	Month	2.2	1,2	1,5	2	A	-0,7
Financial cycle delay	Management of assets	Day	Min.	Annual	90	13	42	80	A	-48
Payables delay	Management of assets	Day	Min.	Annual	30	-	-	-	-	
Delay of customer debt	Management of assets	Day	Max.	Annual	45	-	-	-	-	
Number of days of available stock	Flexibility	Day	Min.	Annual	85	13	34	78	A	-55
Forecast reliability rate	Reliability	%	Max.	Annual	91%	100%	90%	80%	S	9%
Coverage rate	Flexibility	Ratio	Min.	Month	2,5	1	1,4	2	S	-1,5
Internal service rate	Reliability	%	Max.	Week	95%	100%	95%	90%	S	5%
The rate of stock variance	Flexibility	%	Max.	Week	88%	100%	95%	90%	S	12%
Rotation of fixed assets in the supply chain	Management of assets	-	-	Annual	C[b]					
Customers complaints rate	Flexibility	%	Min.	Week	6%	0%	2%	5%	A	4%
Service suppliers rate	Flexibility	%	Max.	Week	95%	100%	95%	90%	S	5%
Cost of an out of stock	Logistics costs	K€	Min.	Month	12	0	1	2	A	-12
Number of days of available stock	Logistics costs	Day	Min.	Month	15	3	5	7	A	-10
Customer return rate	Logistics costs	%	Min.	Week	2%	0%	0,5%	1%	S	-2%
Supplier return rate	Flexibility	%	Min.	Week	1%	0%	0,5%	1%	S	-1%

[b] Confidential information

Multi-Objective Decision Aid Application for RFID Network Planning

Abdelkader Raghib[1] and *Badr* Abou El Majd[2]

[1]LIMSAD Lab, FSAC, Hassan II University of Casablanca, Casablanca, Morocco
[2]CeReMAR, LMSA Lab, FSR, Mohamed V University, Rabat, Morocco

Abstract. The Radio Frequency Identification system (RFID) is a generic term for technologies of a type of automatic identification system that use radio-frequency waves to automatically identify people or objects. Multiple RFID readers are required to detect and monitor the tags in the network area due to their limited interrogation range. Therefore, the cost of the RFID System, its efficiency in the communication process, and the number of tags covered directly depend on the number of deployed readers. In this work, we present a new many/multi-objective model with the aim to achieve the optimal solutions of the NP-hard problem of RFID Network Planning (RNP). The model involves many objectives and constraints as determining the minimum number of readers, finding their optimal $3D$ deployment and localization, guarantying a full coverage of all the tags distributed in the area, and minimizing the energy and the interferences caused by an overlap in the readers signals and the tags responds at the same time. Moreover, we provide a new decision aid application with the aim to access, manage, solve, and optimize the RNP problem involving the needs of researchers and decision makers.

1 Introduction

The term RFID refers to the Radio Frequency Identification system, which is an automated data collection technology that use radio-frequency waves to automatically identify and track objects.
The RFID system consists mainly of three components: tag attached to the object to be uniquely identified, reader to transmit and receive data through radio waves, and finally the RFID middleware [1] to manage the network, as well as to filter and formats data.
Due to the limitations of the reader's interrogation range, many sensors are required to cover all the tags in an area. Therefore, the efficiency of the RFID network depends mainly with the number of deployed readers and the coverage of the tags. In order to achieve the optimal deployment of the RFID Network Planning problem (RNP) [2, 3], it's becoming difficult to ignore the number of objectives to satisfy as determining the minimum number of deployed readers, guarantying a full coverage of all the tags distributed in the area, and minimizing the energy and the interferences caused by an overlap in the readers signals and the tags responds at the same time. Therefore, several attempts have been made to achieve the optimal RFID network planning, mostly by using the multi-objective optimization [4] based on operators and algorithms. There is a large volume of published studies describing the role of the RFID Network Planning (RNP) and presenting several keys in order to achieve the optimal deployment of RFID readers. As in [5], authors presented a tentative reader elimination operator (TRE) based on PSO algorithm, and with the aim to delete and recover the deployed readers during the search process. Detailed multi-objective optimization of RNP showed in [6, 7], in order to find the globally best Pareto-optimal solutions of the problem. In other major studies and with the aim to propose a new way to solve the RNP problem, [8–10] describe a new optimization algorithms called respectively, hierarchical artificial bee colony (HABC), the multi-colony bacteria foraging optimization (MC-BFO), and the cooperative multi-objective artificial colony (CMOABC). [11–13] provide in-depth analysis and results of the RNP problem, by using new models in order to obtain an optimal deployment. Moreover in other studies as in [14], the authors reached the robust optimal solutions of the deployment readers problem which are insensitive to the uncertainties and the uncontrollable variations on the optimization parameters using a robustness multi-objective optimization. Or in [15], where the research provides a probabilistic model of coverage for solving the RFID network planning.
In this work, we will describe a new robust many/multi-objective model for a $3D$ deployment and localization of the RNP problem involving many objectives and constraints. Moreover, we deal with the needs of the decision maker by providing a full decision aid platform with the aim to access, manage, solve, and optimize the RNP problem for academic and real-world applications of the RFID system.

2 Multi-objective Optimization

The optimization deals with the study of those kinds of problems with the aim to minimize or maximize one or more objectives simultaneously that are functions of some real or integer variables, and to obtain their optimal values

while respecting some constraints.
The process of modeling, solving and optimizing problems with two or more conflicting and non measurable goals is called Multi-objective optimization. Problems with multiple objectives can be found in many real-world applications and various fields [20, 21], and known as Multi-objective Optimization Problems (MOP). They generally have more than one solution in the feasible region, which satisfy the trade-offs among the objective functions. The set of those solutions is called the Pareto set [4].
A basic formulation of a multi-objective optimization problem can be defined as:

$$\begin{cases} Min_x\ F(x,p) = [f_1, f_2, ..., f_M] \\ St: \\ g_j(x,p) \leq 0 \quad (j = 1 \dots J) \\ h_k(x,p) = 0 \quad (k = 1 \dots K) \\ x_L \leq x \leq x_U \end{cases}$$

where $(f_i)_{i=1}^{M}$ are the objective functions, x is the design variable vector, x_L and x_U are respectively the lower and upper bounds of x, p is the parameter vector, and the functions $(g_j)_{j=1}^{J}$, $(h_k)_{k=1}^{K}$ are the constraints.

In the optimization's studies including multi-objective optimization, the main focus is placed in finding the global optimum or global Pareto-optimal solutions (called by economist Vilfredo Pareto), representing the best possible objectives values, and where the quality of each solution is incomparable without any prior knowledge of preference. In the optimization of problems with a single-objective, the superiority and the quality of a solution over others are easily determined by comparing their objective function values. On the other hand in the multi-objective optimization problems, the goodness of a solution is determined by the dominance. The multi-objective optimization techniques ideally generate a set of non-dominated solutions, and for this reason the concept of global optimum no longer makes sense. In general this set of solutions is known as the set of non-dominated solutions, or non-inferior or Pareto front. The set of Pareto optimal solutions is called the Pareto optimal set, and its image in the objective space is known as the Pareto front. Thus, it is important in the multi-objective optimization to find the solutions as close as possible to the Pareto front.

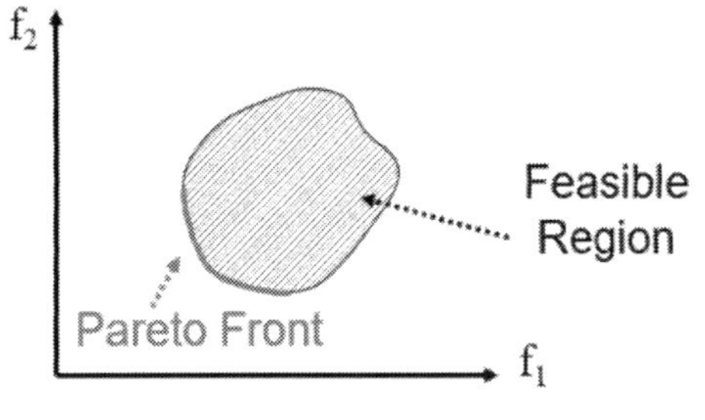

Figure 1. Illustrative example of The Pareto front of a set of solutions in a bi-objective space.

The figure 1 shows an example of the Pareto-optimal of a multi-objective problem, with the aim to minimize the two objectives f_1 and f_2. A solution is said to be dominated when it cannot be improved without worsen another solution of the set found so far. All Pareto-optimal solutions are non-dominated, which mean that no other solutions in the solution space (Feasible Region) are superior to them while taking into consideration all of the objectives.

3 OptRFID Solution for an optimal RNP Problem

OptRFID solution is a valuable tool for the decision makers, researchers, and industrial designed to solve and optimize the deployment of readers in the RFID network. It provides an interface to access, manage, solve, and optimize the RNP problem.
Our experiences using and maintaining of many academic and real-worlds of the RFID system, have helped us to develop and improve the OptRFID design and implementation by including some features as simplicity, facility of use, flexibility, extensibility and portability.

The traditional methods of the optimization management pose several problems of the incapacities such as the absence of an easy-to-use, flexible, extensible and portable software framework for solving the RNP problem. Indeed, the decision maker may be interested in consulting and identifying in real time all probable set of efficient solutions for different scenarios in the research or the decision making process. Therefore, it's very difficult to find a tool that gathers and provides all the above needs due to the complexity of the problem or to the large number and diversity of scenarios. Moreover, with the aim of allowing a minimum optimizing and programming effort, there is a lack of solutions which provide a set of ready-to-use algorithms implementations.
Our aim was to develop one of the world's leading optimization of the deployment of RFID readers. All functions of the problem deployment, used algorithms and optimization methods can be tracked and managed in real-time, through a cost effective and innovative solution 'OptRFID'.

3.1 OptRFID's Mathematical Modeling

In this section, we outline the different objectives and constraints that we considered and included in the OptRFID platform for reaching the optimal deployment of RFID readers. The goal is to find a set of the coordinates readers for covering all the tags deployed in the whole area. At the same time, the total number of readers and their interference must be minimized.
Figure 2 shows an example of RFID deployment: readers are presented by red cross and their interrogation range by red circles, tags are presented by blue diamond nodes.

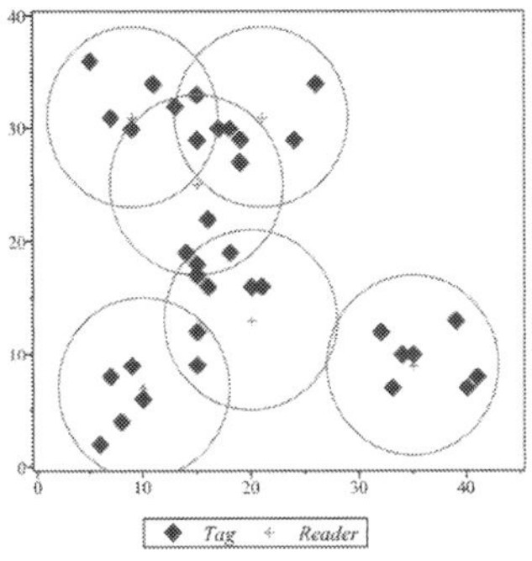

Figure 2. Example of RFID deployment.

Throughout this paper, we use the following notations:

- N_R : Number of the available RFID readers.
- N_T : Number of the tags deployed in the area.
- IR_j: Interrogation range of the jth reader.
- d_{ji} : Distance between the jth reader and the ith tag.

3.1.1 Number of deployed readers

The first objective function in OptRFID tool represents the number of RFID readers, which is an important index for evaluating the performance of the deployment. Minimizing the total number of readers aims to reduce the cost. The optimization algorithm will search for the minimum value of this objective function. It can be defined as:

$$f_1 = \sum_{j=1}^{N_R} r_j$$

Where,

$$r_j = \begin{cases} 1 \text{ if the } j^{th} \text{ reader is deployed.} \\ 0 \text{ otherwise.} \end{cases}$$

Moreover, OptRFID finds the optimal position of each deployed reader and ensures that we can only place each reader within a certain area. For a rectangular area with height H, width W, and depth D, this constraint is satisfied:

$$0 \leq x_j \leq H \text{ and } 0 \leq y_j \leq W \text{ and } 0 \leq z_j \leq D \quad \forall\, j = 1, \ldots, N_R$$

3.1.2 Full coverage

The main objective function for the deployment of RFID readers in OptRFID is to cover all the tags in the whole space. That means each tag must be in the interrogation range of at least one deployed reader. Therefore, we should reduce the number of non-covert tags, which can be formulated as follows:

$$f_2 = N_T - \sum_{i=1}^{N_T} y_{ji} \qquad (j = 1 \ldots N_R)$$

Where,

$$y_{ji} = \begin{cases} 1 \text{ if } \exists!\ j = 1 \ldots N_R \text{ such that } (d_{ji} \leq IR_j \text{ and } r_j = 1) \\ 0 \text{ otherwise.} \end{cases}$$

3.1.3 Tags Interference

The tags interference is the interrogation of a tag with several readers at the same time as we can see in figure 3. It decreases the efficiency of the RFID system. The interference mainly occurs when at least two deployed readers in the area cover a tag.

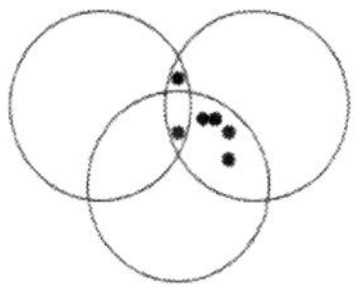

Figure 3. Example of the interference of six tags covered by three RFID readers.

The interference must be reduced by using the following expression:

$$f_3 = \sum_{i=1}^{N_T} c_i$$

Where,

$$c_i = \begin{cases} 1 & \text{if } \sum_{j=1}^{N_R} y_{ji} \geq 2 \\ 0 & \text{otherwise.} \end{cases}$$

3.1.4 Readers Interference

The readers interference is the interrogation of several readers at the same time as we can see in figure 3. It decreases the efficiency of the RFID system. The interference mainly occurs when at least two deployed readers in the area overlap. The interference must be reduced by using the following expression:

$$f_4 = \sum_{j=1}^{N_R} cr_j$$

Where,

$$cr_j = \begin{cases} 1 & \text{if} \sum_{k=1, j \neq k}^{N_R} IntR_{jk} \geq 2 \\ 0 & \text{otherwise.} \end{cases}$$

and,

$$IntR_{jk} = \begin{cases} 1 \text{ if } \exists!\ k = 1 \ldots N_R \,/\, (dr_{jk} \leq (IR_j + IR_k) \text{ and } r_j * r_k = 1 \\ 0 \text{ otherwise.} \end{cases}$$

3.1.5 Power of readers

In the RFID system, the identification of objects relates on the communication process between the reader and the tag. This process depends on correlated factors and parameters to find the ideal power received at an antenna using the Friis Transmission Equation:

$$\frac{P_r}{P_t} = G_r G_t \left(\frac{\lambda}{4\Pi R}\right)^2$$

Where, G_r and G_t are the gain of the transmitting and receiving antenna, respectively. P_r is the power at the receiving antenna, P_t is the output power of transmitting antenna, λ the wavelength, and R is the distance between the

antennas. Therefore, instead using the power of readers in the objective functions, we decide to work with the distance between the antennas, call it in the the manuscript d_{ji}: the distance between the jth reader and the ith tag, and using IR_j: the interrogation range of the jth reader.
Optimizing the RFID system related to minimizing the power of readers:

$$f_5 = \sum_{j=1}^{N_R} IR_j$$

Our OptRFID optimization deployment for the RNP problem is formulated as a multi-objective problem:

$$\begin{cases} Min\ F(x,y,z,ir,r) = (f_1(x,y,z,ir,r), f_2(x,y,z,ir,r), \\ \quad f_3(x,y,z,ir,r), f_4(x,y,z,ir,r), f_5(x,y,z,ir,r)) \\ (x,y,z,r) \in (\mathbb{R}_+^N, \mathbb{R}_+^N, \mathbb{R}_+^N, \{0,1\}^N) \text{ and } 0 \le ir \le IR \end{cases}$$

where N is the maximum number of RFID readers that can be deployed in the working area, (x_i^k, y_i^k) is the coordinate of the kth reader and r_i^k its availability, which equals 1 if its deployed and 0 otherwise.

3.2 OptRFID's Architecture, Design and Implementation

OptRFID is easy to understand and use, which respond to any specific needs of decision makers for the RFID deployment. It's allowing to automate the process of organization of the problem's optimization, and to facilitate the traceability. Furthermore, the framework includes the implementation of many algorithms, which can be used as drafts for developing and covering large scenarios. Also, it's flexible according to the decision needs by requiring minimum updates to solve the problem under different configurations, including algorithm parameters as well as those related to the problem, and the variable variation. OptRFID is indigenously interfaced, but can also be deployed separately in order to supports any needs from any user. Therefore, the solution allows to design new problems, to integrate new elements, or even to combine different optimizations methods and algorithms in order to be able to evaluate and to compare different scenarios fairly.
Moreover, OptRFID provides a single platform for managing and optimizing the RFID network planning problem, which means a single place for entering, managing, maintaining parameters, algorithms, etc. This drives naturally multiple benefits as being more efficient in decision-makers servicing; each user can have his own configuration matching with his own needs; A change of parameters is automatically spread across all the platform and optimization processes; and one system to install, maintain, upgrade, get trained on, etc.

OptRFID is divided into a set of modules that cover the entire spectrum of the optimal RNP deployment:

- RNP Module for defining the RNP problem parameters as tags deployments (tag's coordinates), readers coverage (reader interrogation range, variation), problem dimension (*2D* or *3D*), etc.
- Optimization module for manipulating and choosing the optimization setting as the robustness (acceptable variation for all the five objectives described in section 3.1), the solutions, the desired objectives (Choose the objectives to include in the optimization) and constraints, etc.
- Algorithm module which includes all provided algorithms (SMPSO, NSGA-II), Operators (Mutation, Crossover, etc), Quality indicators (Hyper Volume, GD, IGD, Spread, Epsilon, etc), etc. Through this option, OptRFID allows to the decision maker to choose specific settings about the desired algorithm for solving the RNP problem.

The following table 1 provide an example of the parameters allowed to the decision maker if using for SMPSO or NSGA-II algorithms.

Table 1. Allowed parameters if using SMPSO or NSGA-II algorithms.

SMPSO's Parameters	NSGA-II's Parameters
Max Iterations H	Max Evaluations
Swarm Size W	Population Size
Particle Length	Chromosome Length
Number of runs	Number of runs
	Mutation Probability
	Crossover Probability

OptRFID is a parametrizable system, which mean that the details of the RNP problem, tags, algorithms, optimizations are defined through a specific set of parameters that can be changed at any-time, and from any-where. Changing a parameter does not require any lines of codes, but at the opposite can be done by any user, and the new optimization can be started immediately.

OptRFID provides an easy to use interface that can be used by the users, and within two or three screens, a decision-maker puts all data required to define the parameters of all deployed components in the RFID network. Moreover, OptRFID platform provided with a database and an appropriate GUI. For this we opted for Oracle to create the database, and the languages PlSql, PHP, JavaScript, Json, Angular, etc for the development. It's designed for traceability, secure management, and adaptable the needs of the decision makers.
The figures 5 and 6 show some screens of OptRFID for managing the algorithms and readers parameters.

The figure 4 shows the Conceptual Data Model (CDM) by presenting the structure of the information system from the view of data (dependences or relations between the various data of the information system).

Our platform provides pre-formatted data that allows external system to reprocess and to build a proper statement out of data provided by OptRFID. Moreover, with the aim to get the expected data to the specific needs, OptRFID allows to the user choosing the results type by providing a set of data output available in multiple

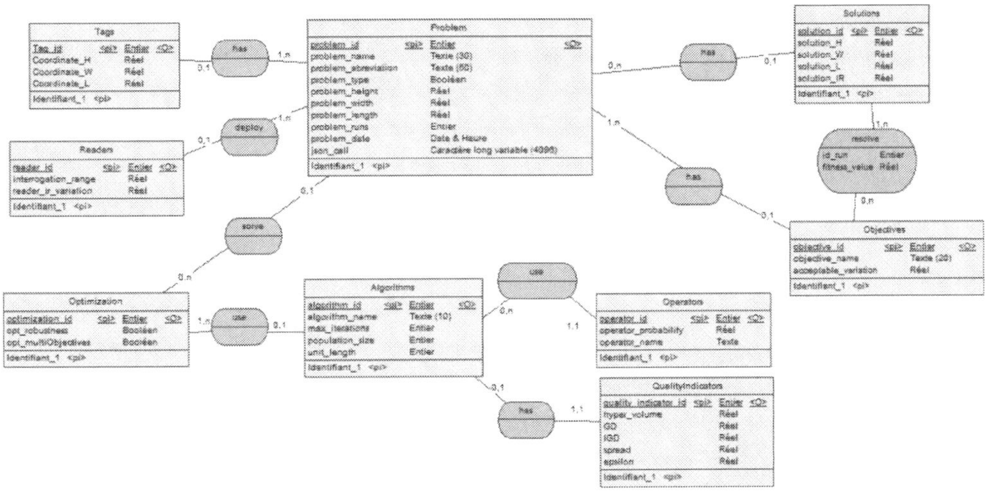

Figure 4. OptRfid's Conceptual Data Model.

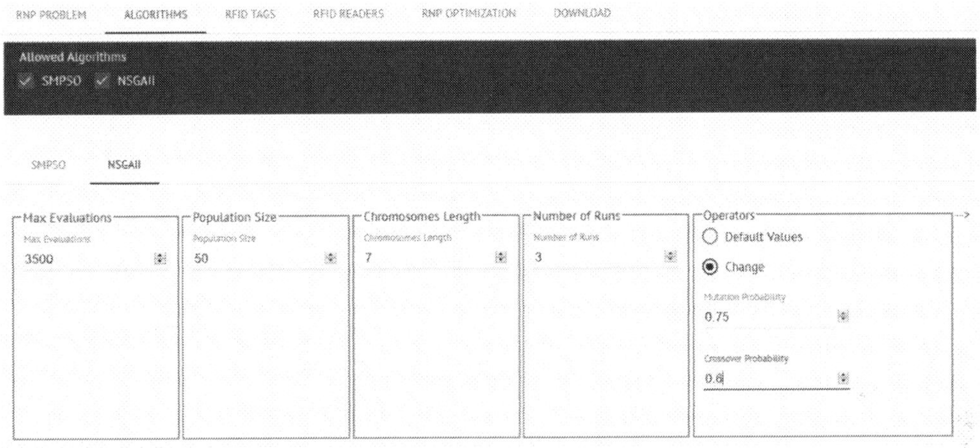

Figure 5. NSGA-II Algorithm Configuration Screen.

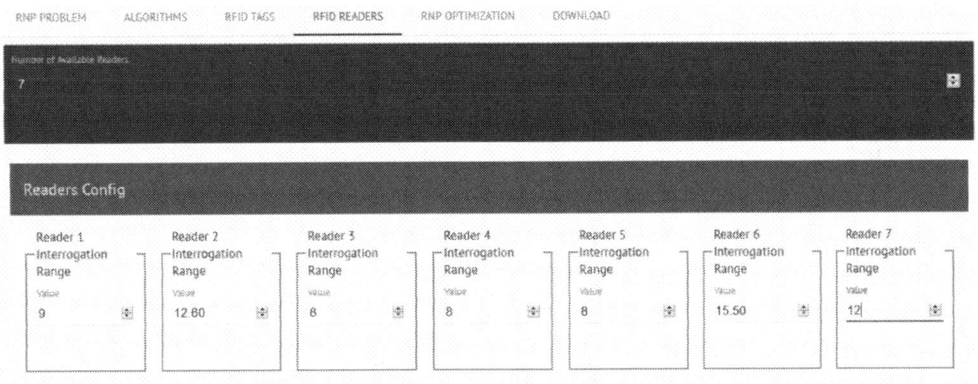

Figure 6. RFID Readers Deployment Screen.

formats (csv, matlab, maple, R, latex tables, json, oracle database, excels, matplotlib, scilab, etc), and reaching a clear and fast visibility on the optimal deployment of a specific problem.

The experiments were conducted to evaluate and validate the performance of OptRFID platform for solving a benchmark using the provided optimization and design solutions. The table 2 presents some used parameters for solving the benchmark using OptRFID.

Table 2. Properties of multi-objective optimization of RNP Problem using OptRFID.

Module	Settings	
	Parameter	**Value**
RNP	Dimension	$2D$
	Area	$50 * 50m^2$
	Dimension	$2D$
	Tags Number	100
Optimization	Type	Multi-Objectives
	Objectives	Nbr of Readers
		Full Coverage
		Tags Interference
Algorithm	Choice	SMPSO
	Maximum Iterations	20 000
	Swarm Size	20
	Length of Particle	10
	Number of runs	10

The results are graphically supported in Figures 7 and 8 by the Pareto optimal solution for the problem according to the desired objectives and the optimal deployment of RFID readers. The OptRFID tools provides the best overall performance by guarantying the quality of the non dominated solutions and yield higher accuracy and efficiency in a faster and cheaper way, which facilitates the decision aiding.

As perspectives, we opt opening the solutions to the external word by providing standard interfaces (mainly batch or web services) that can be used by any user in order to provide a great flexibility for the decision makers, and let them deploying OptRFID solution at the pace they wish. We plan also to integrate more algorithms and optimizations types to let more choices and visibility for the decision maker.

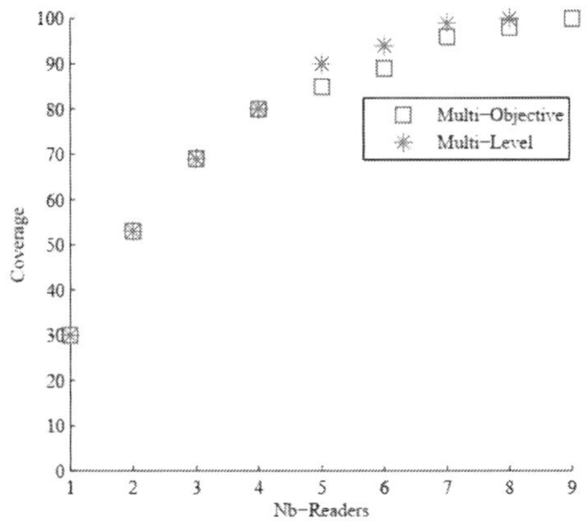

Figure 7. The Pareto front.

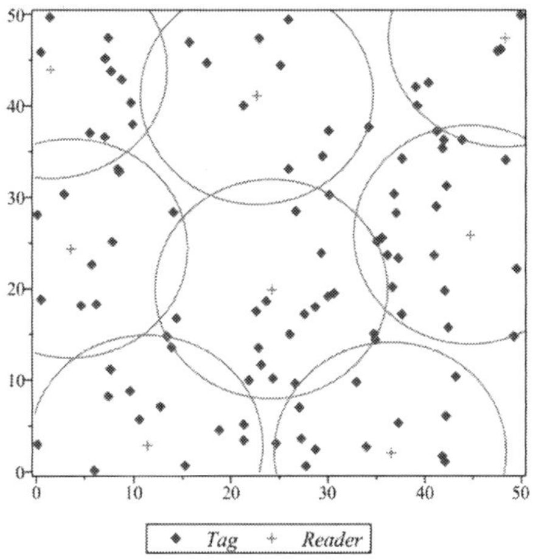

Figure 8. The optimal RNP Deployment.

4 Conclusion

The main goal of the current study was to develop a global decision aid solution of a complete RFID system application based on practical and theoretical propositions. Therefore, we proposed and developed a new mathematical model with the aim to achieve an optimal deployment for the $3D$ NP-HARD problem of RFID Network Planning (RNP). As perspectives, we plan to present a new approach in order to deal with the uncontrollable parameters and variables. We aim also to apply the game theory and hierarchical strategies [16–19] in the context of RFID network planning, in order to achieve a robust optimal deployment of RFID readers with lower computation times. Moreover, we plan to integrate our global solution for solving many problems in various fields such as logistics, air, military, security, and personal services.

References

[1] T. Bouhouche, A. Raghib, B. Abou El Majd, M. Bouya, M. Boulmalf, A Middleware Architecture for RFID-enabled traceability of air baggage, in *MATEC Web Conf.* 105, DOI:10.1051/matecconf/201710500008, (2017)

[2] A. Raghib, B. Abou El Majd, Hierarchical Multiobjective Approach for Optimizing RFID Reader Deploy-

ment, in *International Journal of Mathematical Modelling and Numerical Optimisation*, (To Appear 2018).

[3] A. Raghib, B. Abou El Majd, B. Aghezzaf, An Optimal Deployment of Readers for RFID Network Planning Using NSGA-II, in *Recent Developments in Metaheuristics (Springer, Cham, Operations Research/Computer Science Interfaces Series)*, pp. 463–476, ISBN 978-3-319-58252-8 978-3-319-58253-5, DOI: 10.1007/978-3-319-58253-5-27, (2018).

[4] K. Deb, Multi-Objective Optimization Using Evolutionary Algorithms, in *John Wiley & Sons*, Inc., New York, NY, USA, ISBN 978-0-471-87339-6, (2001)

[5] Y.J. Gong, M. Shen, J. Zhang, O. Kaynak, W.N.Chen, Z.H. Zhan, Optimizing RFID Network Planning by Using a Particle Swarm Optimization Algorithm With Redundant Reader Elimination, in *IEEE Transactions on Industrial Informatics*, Vol. 8, pp. 900–912, ISSN 1551-3203, DOI: 10.1109/TII.2012.2205390, (2012)

[6] H. Chen, Y. Zhu, L. Ma, B. Niu, Multiobjective RFID Network Optimization Using Multiobjective Evolutionary and Swarm Intelligence Approaches, in *Mathematical Problems in Engineering*, Vol. 2014, 13 pages, ISSN 1024-123X, DOI: 10.1155/2014/961412, (2014)

[7] O. Botero, H. Chaouchi, RFID network topology design based on Genetic Algorithms, in 2011 *IEEE International Conference on RFID-Technologies and Applications* (RFID-TA), pp. 300–305, DOI:10.1109/RFID-TA.2011.6068653, (2011)

[8] L. Ma, H. Chen, K. Hu, Y. Zhu, Hierarchical Artificial Bee Colony Algorithm for RFID Network Planning Optimization, in *The Scientific World Journal*, Vol. 2014, 21 pages, ISSN 2356-6140, DOI: 10.1155/2014/941532,10.1155/2014/941532, (2014)

[9] H. Chen, Y. Zhu, K. Hu, Multi-colony bacteria foraging optimization with cell-to-cell communication for RFID network planning, in *Applied Soft Computing*, Vol. 10, pp. 539–547, ISSN 1568-4946, (2010)

[10] L. Ma, K. Hu, Y. Zhu, H. Chen, Cooperative artificial bee colony algorithm for multi-objective RFID network planning, in *Journal of Network and Computer Applications*, Vol. 42, pp. 143–162, ISSN 1084-8045, DOI: 10.1016/j.jnca.2014.02.012, (2014)

[11] C.C. Hsu, P.C. Yuan, The design and implementation of an intelligent deployment system for RFID readers, in *Expert Systems with Applications*, Vol. 38, pp. 10506–10517, ISSN 0957-4174, DOI: 10.1016/j.eswa.2011.02.109, (2011)

[12] S. Lu, S. Yu, A fuzzy k-coverage approach for RFID network planning using plant growth simulation algorithm, in *Journal of Network and Computer Applications*, Vol. 39, pp. 280–291, ISSN 1084-8045, DOI: 10.1016/j.jnca.2013.07.015, (2014)

[13] H. Chen, Y. Zhu, K. Hu, RFID networks planning using a multi-swarm optimizer, in *Control and Decision Conference*, 2009. CCDC '09. Chinese, pp. 3548–3552, DOI: 10.1109/CCDC.2009.5192599, (2009)

[14] A. Raghib, B. Abou El Majd, O. Ouchetto, B. Aghezzaf, Robustness optimization for solving the deployment of RFID readers problem, in *5th International Conference on Multimedia Computing and Systems* (ICMCS), pp. 509–513, DOI: 10.1109/ICMCS.2016.7905627, (2016)

[15] V. Tuba, A. Alihodzic, M. Tuba, Multi-objective RFID network planning with probabilistic coverage model by guided fireworks algorithm, in *10th International Symposium on Advanced Topics in Electrical Engineering* (ATEE), pp. 882–887, DOI 10.1109/ATEE.2017.7905125, (2017)

[16] B. Abou El Majd, J.A. Desideri, A. Habbal, Aerodynamic and structural optimization of a business-jet wingshape by a Nash game and an adapted split of variables, in *Mécanique & Industries*, Vol. 11, pp. 209-214, (2010)

[17] R. Duvigneau, B. Abou El Majd, J. A. Désidéri, Towards a self-adaptive parameterization for aerodynamic shape optimization, in *ESAIM Proceedings*, EDP Sciences, (Vol. 22, pp. 169-174), (2008)

[18] J. A. Désidéri, R. Duvigneau, B. Abou El Majd, Z.Tang, Algorithms for efficient shape optimization in aerodynamics and coupled disciplines, in *42nd AAAF Congress on Applied Aerodynamics*, Sophia-Antipolis, France, (2007)

[19] B. Abou El Majd. *Optimisation de forme paramétrique: Stratégies hiérarchiques, adaptatives, et concourantes*. Éditions universitaires européennes, (2015).

[20] H. Rahil, B. Abou El Majd, M. Bouchoum, Optimized Air Routes Connections for Real Hub Schedule Using SMPSO Algorithm, in *Recent Developments in Metaheuristics*. Ed. by Lionel Amodeo, El-Ghazali Talbi, and Farouk Yalaoui. Cham: Springer International Publishing, pp. 369–384. ISBN: 978-3-319-58253-5. DOI: 10.1007/978-3-319-58253-5-21, (2018).

[21] R. Fadil, B. Abou El Majd, H. Rahil, H. El Ghazi, N. Kaabouch, Multi-objective optimization approach for air traffic flow management, in *MATEC Web Conf.* 105, p. 00005. . EDP Sciences, (2017)

Is Supply Chain a complex system?

Safaa Raaidi[1*], *Imane Bouhaddou*[1], and *Asmaa Benghabrit*[2]

[1]LM2I Laboratory, ENSAM, Meknes, Morocco
[2]LMAID Laboratory, ENSMR, Rabat, Morocco

Abstract. Nowadays, industries are continually looking to implement new subsidiaries in different continents, in order to better fulfill their customers' needs, generate the best products in the shortest time and cheaper than their competitors. Achieving these goals is no longer related to the company itself, but to all partners in the supply chain. This justifies the need for efficient and judicious management of the whole supply chain, through the collective intervention of all its actors. Needless to say, a supply chain is a system made up of a set of suppliers, producers, subcontractors, retailers, wholesalers and customers, between whom material, information and financial flows are exchanged. Management of these flows is becoming increasingly difficult and constitutes the main source of the supply chain complexity. In order to alleviate this problem and improve supply chain performance, it is necessary to model it, taking into consideration its characteristics, which make it a complex system. Hence, the scoop of this paper is to prove that supply chain is a complex system, by highlighting its most relevant characteristics that make it such a system. Complex means what is braided together or woven together. If we separate the elements, we get acquaintance elements, but we lose their interactions. Within this trend, our contribution subscribes with its ultimate purpose modelling supply chain as complex system.

1. Introduction

Open borders to international trade or what is so-called "free trade" proved to be one of the main companies' issues all over the world. Industries are continually looking to implement new subsidiaries in different continents, in order to better fulfill their customers' needs and requirements, generate the best products in the shortest time and cheaper than their competitors. Achieving these goals is no longer related to the functions, activities and processes of the company itself, but to all supply chain partners. This justifies the need for efficient and judicious management of the whole supply chain, through enhanced collaboration between all its actors.

A supply chain is a system made up of a set of suppliers, producers, subcontractors, wholesalers, retailers and customers, between whom material, information and financial flows are exchanged [1]. The management of these flows, particularly the material and information flows, is becoming increasingly difficult and constitutes the main source of the supply chain complexity. In fact, the factors of supply chain complexity have steadily increased in the last decades, namely: the uncertainty, the multiplicity, the variability, the size, the speed, the diversity, etc.

As a matter of fact, nobody can deny that supply chain management is becoming increasingly difficult, we can even say that the supply chain can be seen as a complex system. In fact, in order to propose an efficient and judicious model to reduce this complexity and improve the supply chain performance continually, our contribution consists on showing theoretically this axiom, in the perspective of the complex systems theory.

2. Supply chain

There is no universal definition that clarifies the global meaning of the term "supply chain". Some definitions adopt a product point of view and others an "enterprise" or "process" point of view [2]. For example, APICS dictionary defines supply chain as: "The global network used to deliver products and services from raw materials to end customers through engineered flows of information, physical distribution, and cash" [3].

According to Labarthe [4], "Supply chains constitute an economic and social system made up of a set of inter-acting enterprises, and implementing processes of cooperation, coordination and negotiation in order to ensure their efficiency and durability in satisfying the customer demand".

However, the literature shows that the majority of definitions take up a certain common ideas:

- A supply chain usually refers to finished good or a family of finished goods.
- It involves several companies, several actors.
- These companies are linked by three streams: the information flow, physical flow and financial flows.

* Corresponding author: raaidisafaa@gmail.com

• Each stakeholder performs the functions of supply, transformation / production, distribution or sale.
• A company is potentially involved in several supply chains. Indeed, a company generally seeks to multiply its client-companies and its products can be used for the development of several finished products [5].

Within each organization in a supply chain, a participant receives demands from the prior downstream stage and places orders with the next upstream stage to be able to supply the downstream customer demands [6].

Furthermore, some works put more emphasis on the purpose of a supply chain by introducing the notion of *performance*, this performance being mainly characterized by the satisfaction of the ultimate customer [2]. So as to measure this performance, *Michael H. Hugos* has pointed out in the fourth edition of his guide "*Essentials of Supply Chain Management*" [7], that companies in any supply chain must make decisions individually and collectively regarding their actions in five areas:

1. Production: this activity includes the creation of master production schedules that take into account plant capacities, workload balancing, quality control, and equipment maintenance.
2. Inventory: the primary purpose of inventory is to act as a buffer against uncertainty in the supply chain. However, holding inventory can be expensive, so what are the optimal inventory levels and recorder points?
3. Location: this activity aims to specify where facilities for production and inventory storage should be located, and choose then the most cost-efficient locations. In order to determine the possible paths available for product to flow through for delivery to the final consumer.
4. Transportation: Airfreight and track delivery are generally fast and reliable mode of inventory transportation from one supply chain to another, but they are expensive. However, shipping by sea or rail is much less expensive but usually involves longer transit times and more uncertainty.
5. Information: with good information, people can make effective decisions about what to produce and how much, about where to locate inventory, and how best to transport it.

In the balance of this paper, we adopt the following definition: "A supply chain is a system consisted of many participants (suppliers, producers, distributors, retailers, customers, etc), which collaborate, directly or indirectly, to fulfil customer demand along the supply chain, through three flows' types: materiel, informational and financial, all the while ensuring supply, transformation, production, distribution, and sale functions". Figure 1 shows an example of the defined supply chain.

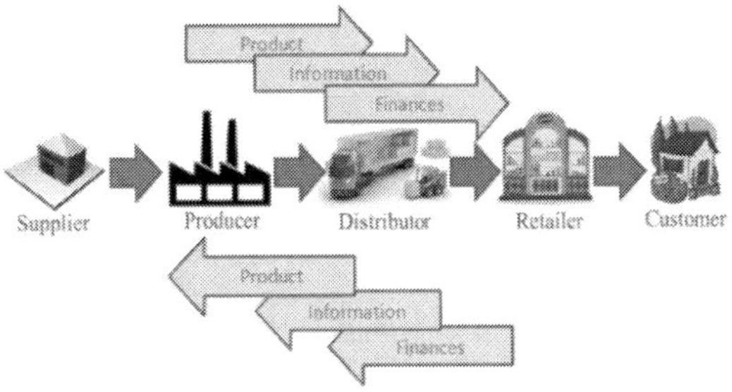

Fig. 1. Example of the defined supply chain.

Nowadays, information flows do not follow this linear form, but rather now look like a simultaneous exchange, as shown in the Figure 2, especially through advanced Information and Communication Techniques (ICT).

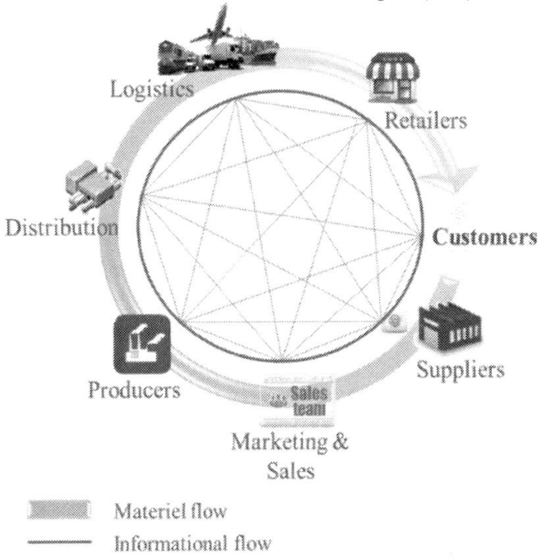

Fig. 2. The new flows distribution within supply chain (Inspired by [1]).

2. Complex systems

2.1. Definition

In fact, there is no agreement between Scientifics in the field of complex systems about these systems definition. However, there is a number of definitions that contain some similarities. According to E. Morin, who is considered as the father of complex thinking, "Complex comes from *complexus*, which means what is braided together or woven together" [8], we cannot deduce the overall system functioning by the addition of the subsystems or parts that compose it, because the whole is greater than the sum of its parts. We are therefore talking about the central property of complex systems, which is emergence. Mitchell defines complexity relying, in addition to emergence, on self-organisation: " ...system that exhibits nontrivial emergent and self-organizing behaviors" [9], that is refer to the ability of the

components of a complex system to create organised behaviour without an internal or external controller [10]. Of course, emergence and self-organization are not the only characteristics of complex systems. We will explain the remaining characteristics in Table 1.
Taking as an example of complex systems, a simple molecule of water. We often hear that "water is life", all living things are mostly composed of water, that is, it contains properties that ensure the living beings life. However, they are found neither in the hydrogen atom nor in the oxygen one.
Nevertheless, Ross and Ashby, both an engineer and a mathematician, defined complexity as "the degree of a system variety". A system is a unit composed of extremely diverse elements. The more varieties a system has, the more complex it is [8], and the more it becomes difficult even impossible to predict its behaviour.
Relying on the last example, to make this system even more complex, for example, we use the glucose molecule that dissolves easily in water. Considering the chemical reaction between these two molecules, we get a new system, with more components that are the atoms and more interactions, which generate a more complex system, as shown in the figure 3.
We then adopt the following definition: "A complex system is a system composed of a set of homogeneous or heterogeneous elements, interacting with each other in a non-linear way, thus implementing a dynamic allowing the whole system to exist as a whole [11].

2.2. Complex systems characteristics

Several authors (Filiz Isik, 2011), (Andreas Koch & Laurent Larsonneur, 2008), (Gezhi Weng et al., 1999), (Le Moigne J.L, 1999) highlight that Complex systems are characterized by multiple properties, (Abla Chaouni et al., 2017) grouped them in article [12], as showing in Table 1.

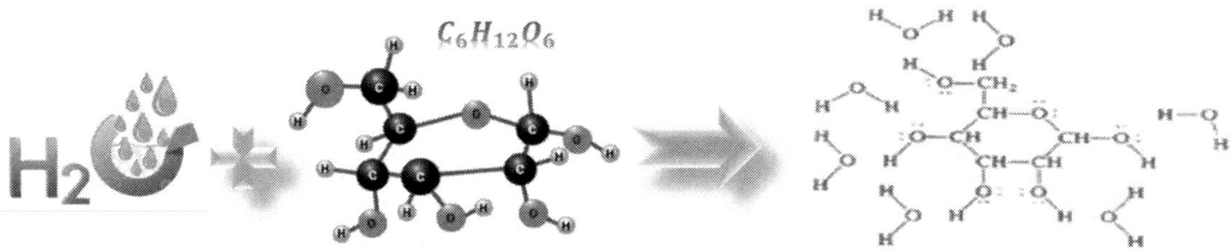

Fig. 3. Chemical reaction between water H_2O and glucose $C_6H_{12}O_6$ is a complex system.

Table 1. Complex system characteristics.

Characteristic	Definition
Emergence	We talk about emergent behavior when a property appears at the global level, without it being obvious or immediately predictable from the behavior of the elementary constituents of the system [13].
Connectivity	Complexity results from the inter-relationship, inter-action and inter connectivity of the elements within a system and between a system and its environment.
Co-evolution	Elements in a system can adapt based on their interactions with one another and with the environment. Additionally, patterns of behavior can adapt over time.
Distribution control	There is not centralized control mechanism that governs system behavior. Although the interrelationships between elements of the system produce coherence, the overall behavior usually cannot be explained merely as the sum of individual parts.
Non-linearity	Complex systems does not conform to the principle of adding, meaning that the output is not necessarily proportional to the input of the system
State of paradox	Complex systems have indicated dynamics combining both order and chaos. This reinforces the idea of bounded instability or the edge of chaos that is characterized by a state of paradox: stability and instability, competition and cooperation, order and disorder.
Far-from-equilibrium	This phenomenon illustrates how systems that are forced to explore their space of possibilities will create different structures and new patterns of relationships.

Heterogeneity	The system has different types of elements. The species of a rainforest, people living in large towns, vendors in malls, structure of the mammalian brain all exhibit extreme diversity.
Open	An open system is a system that has external interactions. Such interactions can take the form of information, energy, or material transfers into or out of the system boundary, depending on the discipline, which defines the concept.
Self-organization	Complex systems can change their internal structure in response to changes in their environment; this ability allows them to survive but also to learn [13].

3. Supply chain complexity

The management of material and information flows, along the supply chain system, is becoming increasingly difficult and constitutes the main source of the supply chain complexity. In fact, the supply chain complexity factors have been steadily increasing in recent decades, Table 2 summarizes the most influential factors in the supply chain.

Supply chain complexity can be defined as quantitative differences between predicted and actual states, which are associated with uncertainty and/or variety caused by internal and external drivers in a supply chain system [6].

Table 2. Supply chain complexity factors.

Uncertainty	Galbraith defines uncertainty in [14], as the difference between the amount of information needed to perform a task and the amount of information that is already available. The uncertainty sources within supply chains are multiple, namely: uncertainty about deadlines, customer demand, processes, markets, etc.
Multiplicity	This characteristic of the complexity covers the number of components such as items (raw, manufactured or end), products, processes, stakeholders, relationships, interactions, goals, locations, etc. A high number level of any components contributes increasingly complexity in a supply chain system [6].
Variability	Consists of elements or components, which are different from each other. It represents dynamical behaviour of a system [6].
Size	Industries are continually seeking to implement new subsidiaries in the various continents, in order to better meet their customers' requirements, to find new markets and new customers. As a result, supply chain becomes more extensive, its product lines become more diversified and the number of its processes multiplies.
Speed	The production speed is becoming higher, and product life cycles are becoming shorter, especially with advanced information and communication technologies, which constitute one of the main concerns for any company to survive and increase in the market.
Diversity	Related with the homogeneity or heterogeneity of a system. For example: A high level of diversity of any components (supplier, product, mean of transport) along the supply chain leads to system's heterogeneity and results a high level of complexity [15].

Complexity has many negative consequences on supply chains such as high operational costs, customer dissatisfaction, time delay in delivery, excess inventory or inventory shortage, lack of cooperation, collaboration and integration among supply chain participants, etc [6]. Confronted with such daunting complexity, supply chain executives reported that they face five major challenges, as shown in Figure 4 [16].

According to IBM Global Services, the most critical challenge that has the very significant impact on supply chain is "Supply chain visibility" with 70%, then "Risk management" with 60%, "Increasing customer demands" with 56%, "Cost containment" with 55%, and ultimately "Globalization" with 43%.

All are critically important, and must be addressed simultaneously. Together, they comprise what we call the Chief Supply Chain Officer agenda [16].

To overcome this problem of supply chain complexity and continuously improve its performance, it is necessary to model it. Indeed, researchers were proactive in the modeling of the supply chain (for example: by multi-agent systems, the SCOR model, mathematical models). However, our approach aims to model it in a complex perspective, taking into account its characteristics that make it a complex system. This correspondence will be the subject of the following section.

4. Supply chain as complex system

In the light of all these definitions and complex systems' characteristics, we will deduce that the supply chain is a complex system, as result of taking exactly these same characteristics. The detail of this correspondence is summarized in Table 3.

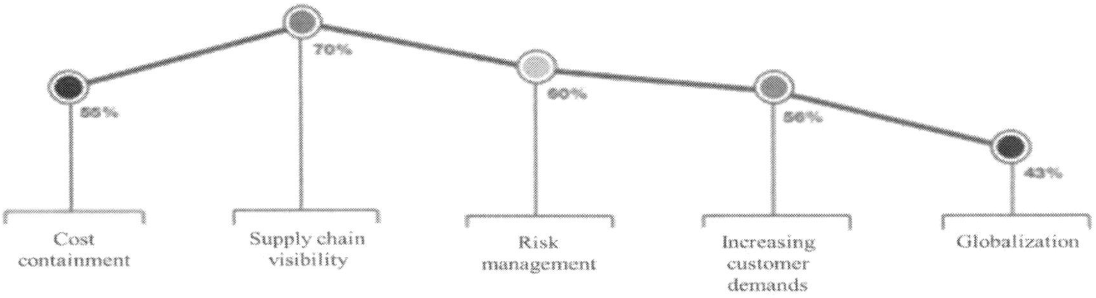

Fig. 4. Supply chain leaders wrestle with five major challenges [16].

Table 3. Supply chain as complex system.

Characteristic	Projection on supply chain
Emergence	The manufacture of a finished good, or even a semi-finished product, imperatively requires the collaboration of several supply chain entities. Therefore, the supply chain cannot be reduced to a single entity, but rather it is considered as a whole. For instance, when the company makes a product, it will be beneficial to deliver it to its customer right after its manufacture, in order to avoid the additional costs of storage. However, it is not always the case, because the delivery time represents a compromise between the supplier and each of its customers, who take, in turn, their customers in their decision-making process, and so on. Hence, each entity performs its function in collaboration with the other entities, which generates a collective behaviour.
Connectivity	Communication is one of the main keys to supply chain management. It allows obtaining from all its actors the information and the knowledge, which will favor the course of the current activities of the company. Especially with advanced information and communication techniques, which ensure faster and more timely information sharing.
Co-evolution	Each peer of (client, supplier) companies co-evolves in parallel, as the supplier company adapts its strategies and processes according to its customer requirements on the one hand, and on the other hand according to the evolution of its environment. In the field of smart phone manufacturing for example, it is remarkable that competition is fierce and life cycles are becoming even shorter, because of the speed of technological development. To face this challenge, supply chain partners must innovate continuously, and afterwards they accept and adapt with the innovative proposals of each of them, in order to co-evolve and survive in a hyper-competitive environment.
Distribution control	The actors of the current supply chains are geo-distributed. Therefore, supply chains exploit remote control systems, made up of several controllers that control the subsystems or units of the global installation. Hence, remote control systems ensure a reduction of uncertainty and improve the entire chain performance.
Non-linearity	The non-linearity within supply chains is revealed in the activities, processes, orders, flows and objectives of the different actors. For example, Bullwhip effect illustrates that a slight variation in the initial state of the system, generate a significant deviation from its usual progress.
State of paradox	To generate better quality products, the company has to invest even more and take the time needed to control the product quality. However, the high speed of global progress requires deliveries in the shortest possible time with optimal costs, compared to competitors. In addition, the companies are continually seeking to find a compromise between the piloting system by forecasts, and those by stock.
Far-from-equilibrium	Delays in delivery, stakeouts, unexpected breakdowns, fluctuations in demand, etc, are the main sources of imbalance in the supply chain.
Heterogeneity	The supply chain is composed of several participants (suppliers, factories, warehouses, distributors, wholesalers, etc.), concerns different types of products, relies on several functions, and uses different processes according to each actor activities.

* Corresponding author: raaidisafaa@gmail.com

Open	Supply chain is an open system that constantly interacts with its environment, it is often found that an actor is part of several supply chains, on the other hand, the supply chain involves several suppliers, producers, warehouses, distributors, wholesalers, etc.
Self-organization	Supply chain actors can modify their internal strategies in response to the market changes and its dynamism, to survive in the market and to ensure customer satisfaction. Sometimes suppliers fail to deliver the products at the right time, or the quantity delivered is not always of good quality. Then the company must manage this problem in order to ensure its customer's satisfaction, by providing a stock of risk management even if it can be expensive for her (hybrid strategy). Thus, the company self-organizes in front of the market dynamism.

5. Conclusion

The globalization of markets and the opening of borders have greatly complicated the supply chain management. Good management of supply chain complexity has become the major concern of any company, in order to remain competitive in the market and better meet its customer's needs.

In this context, our contribution objective is to integrate the two scientific disciplines: that of Supply chain management with the Complex systems field. In this article, we presented a literary overview on the supply chains, explaining its different actors, flows and the main sources of its complexity that hinder its improvement. Then, we have showed a state of art of complex systems, and we have explained all of its characteristics, in order to, theoretically, justify that the supply chain is indeed a complex system because it takes exactly the same characteristics.

As we have already mentioned, to manage the supply chain complexity, it is necessary to model it in a complex perspective, which will be the subject of our next contribution.

References

1. I. Bouhaddou. Vers une optimisation de la chaîne logistique : Proposition de modèles conceptuels basés sur le PLM (Product Lifecycle Management). Autre [cs.OH]. Université du Havre, 26,31 (2015). Français.
2. FA. Gruata La Forme-Chretien. Référentiel d'évaluation de la performance d'une chaîne logistique - Application à une entreprise de l'ameublement. Lyon : s.n., 28 (2007).
3. American Production and Inventory Control Society. [En ligne] 2001. http://www.apics.org.
4. O. Labarthe. Modélisation et simulation orientées agents de chaines logistiques dans un contexte de personnalisation de masse: Modèle et cadre méthodologique. Marseille : s.n., 26 (2006).
5. J. Francois. Planification des chaînes logistiques : modélisation du système décisionnel et performance. (2008).
6. F. Isik. Complexity in Supply Chains: A New Approach to Quantitative Measurement of the Supply-Chain-Complexity. Supply Chain Management, Dr. pengzhong Li (Ed.), ISBN: 978-953-307-184-8, InTech. (2011).
7. H. Michael. *Essentials of Supply Chain Management. s.l.* : Wiley, 5,6 (2018).
8. E. Morin. La complexité: un défi à la connaissance. s.l. : L'actualité Poitou-Charentes, N°68, (2004).
9. M. Mitchell. *Complexity: A Guided Tour.* s.l. : Oxford University Press, (2009).
10. B. Young. Complex Systems Engineering Applications for Future BMC2. Alexendria,VA : ICCRTS, **032** (2013).
11. S. Hassas. Systèmes Complexes à base de Multi-Agents Situés. Lyon 1 : s.n., 11,12 (2003).
12. A. C. Benabdellah, A. Benghabrit, I. Bouhaddou Supply Chain Challenges with Complex Adaptive System Perspective. Springer, Cham, 1085,1086 (2018).
13. H. Zwirn. *Systèmes complexes (Les): Mathématiques et biologie.* 224 (2006).
14. J. Galbraith. Designing complex organizations. Addison-Wesley, Massachusetts. (1973).
15. I. Bouhaddou, A. Benabdelhafid. PLM (Product Lifecycle Management): a key to manage supply chain complexity. CS-DC'15 World e-conference. (2016).
16. IBM Global Services. The smarter supply chain of the future. U.S.A. : s.n., (2009).

Benefits and limitations of the SCOR® model in Automotive Industries

Radouane LEMGHARI[1, a], *Chafik* OKAR[2], and *Driss* SARSRI[1]

[1]The Innovative Technologies Laboratory, National School of Applied Sciences, 90000, Tangier, Morocco
[2]The Higher School of Technology of Berrechid, Hassan the 1st university, 26000, Settat, Morocco

Abstract. The automotive industry's supply chain has been extensively researched but has tended to focus on the component production sections of the chain. Indeed, the automotive industry needs a structured model that encompasses the entire fragmented, uncertain and complicated industry. For this reason, the SCOR® model was chosen in this study because it contains well-defined and standardized processes and metrics for performance measurement and comparative analysis of the entire supply chain and not only internal processes. In this context, the objective of this paper is to concretize the application of the SCOR® model in an automotive company by following the steps and guidelines described by the supply chain council that designed this model. Then, several contributions and benefits of the application of this model in the automotive sector will be raised based on this case study and especially on the interviews conducted with the various practitioners and specialists in the supply chain. On the other hand, a list of limitations and difficulties encountered during and after the implementation of this standard will also be collected. Although this study is primarily oriented towards a university audience, it can also be interesting and useful for practitioners, who will be able to gain an understanding of the purpose of the existing research and to have access to a real-life example of the application of this model in the automotive sector.

1 Introduction

The automotive industry has complex supply chains because of the nature of the products they make up. A single car can have up to 15,000 individual pieces and in the event that it is not possible to deliver a single piece, it can slow down or stop a production line. Nowadays, many vehicles together in Europe and North America involve high-tech components manufactured in Asia-Pacific. In 2017, they represented 40% of the cost of a car compared to 10% in 2004 [1,2].

The automotive industry is a highly globalized sector, where for many actors there is competition from around the world. Faced with the increase in supply and the strong pressure on prices, reinforced by the power of Asian groups (Toyota, Hyundai, etc.) and the arrival of new players from emerging countries (Geely, Tata Group, etc.), many companies seek to optimize their value chain in order to remain competitive [3] In the supply chain of the automotive industry, many factories are working together to manufacture a product (car, motor, etc.). The automotive industry is relevant in terms of influence on the economy as well as technology. As Humphrey and Memedovic [4] have pointed out, the automotive sector is global and specified as a capital-intensive industry with vertical integration and economies of scale [5]. He was responsible for the development of technological innovation and management, as well as the original major change in industrial production processes [6,7].

On the other hand, to meet customer needs, they must be designed and organized appropriately [8]. Logistics can be defined as a key factor of competition in the automotive industry due to the increasing number of variants and options of the model. With the growing importance of logistics [9], the evaluation of logistics effectiveness and efficiency is gaining increased attention [10]. For this reason, the automotive sector characteristics and current performance measurement practice have been reviewed and analyzed to determine the main characteristics of an effective model for an automotive firm [9,10]. The main assumption is that automotive firm requires tools specifically designed and tailored to their characteristics and needs [10].

There are several models have been applied for business analysis. The SCOR® model is one of the most famous of them that has been widely applied by practitioners [11]. Thus, in order to improve business systems, the return on investment, to deal with competitiveness and optimize the efficiency of their supply chain, some global companies apply the SCOR® model. The SCOR® model develops a systems approach to identify, evaluate and monitor supply chain performance [12]. The model not only provides an opportunity to see how the business operates but also a common frame of reference and language across the supply chain [13,14].

Given its increasing use among professionals and academics who are directly involved in supply chains, the SCOR® model was used because of its process

[a] Corresponding author: lemghari.radouane@gmail.com

orientation and [15]. It has also become the common language for benchmarking and comparison of supply chains and supply chain management practices [15].

The purpose of this document is the application of the SCOR® model in a car company and to raise the different contributions and benefits of its application in this very dynamic sector. On the other hand, a list of limitations and difficulties encountered in the implementation of this standard will also be collected. At first, a literature review of the supply chain operations reference model and the performance measurement in automotive industries is performed in Section 2. In Section 3, the methodology followed in this research is presented, and in Section 4, the different stages of a practical case study of the automotive industry are developed. The benefits of the application of this model resulting from this case study and a roadmap for the success of this project are illustrated in Section 5. Finally, in Section 6, we conclude with a list of limitations and difficulties encountered during and after the implementation of this standard as well as some closing remarks.

2 The Supply Chain Operations Reference (SCOR) model

In SCOR®, the integrated processes of PLAN, SOURCE, MAKE, DELIVER, RETURN and ENABLE from the supplier's supplier to the customer's customer represent supply chain management. Elements of business process engineering, metrics, benchmarking, leading practices, and people skills into a single framework are combined with SCOR® [15]. The version 4 was the first to include return process in the supply chain. The Revisions of the model are made by the members of the Council when it is determined that modifications need to be made to facilitate the use of the model in practice [16]. The release of version 11 for access, as well as the incorporation of the new material into all SCOR® training and certification programs, are announced by the Supply-Chain Council (SCC), the global not-for-profit organization supporting supply chain professionals and educators [17]. This major update includes Enable Process, Best Practices, and Redesign Costs. These updates bring the current model with the way these processes, practices, and measures are needed to practitioners of the supply chain that implement the model [17]. SCOR® describes the business activities related to all phases of the satisfaction of a customer request. The model itself is structured around primary management processes, and on SCOR® version 11, ENABLE is advanced to one of these processes, so there are six: PLAN, MAKE, SOURCE, DELIVER, RETURN, and ENABLE. Using these process definition blocks, SCOR® can be used to model supply chains that are very simple or very complex using a common set of definitions in disparate industries. In fact, public and private organizations and companies around the world use the model as a basis for projects to improve the global supply chain [17].

These processes focus on performance management, information, politics, the inventory strategy, capital, transportation, physical logistics network, regulatory and other management processes to enable Planning and execution of supply chain activities. SCOR® covers all customer relationships, product and market surrounding customer orders, purchase orders, work orders, return authorizations, forecasting and replenishment orders. It also includes physical movement of raw materials, work in progress, finished goods and merchandise return [15].

The SCOR® model has three levels of process detail. In practice, Level 1 describes the number of supply chains, how their performance is measured, and necessary competitive requirements. Level 2 presents the configuration of the planning and execution strategies in the material flow, involving standard categories such as "make-to-stock", "make-to-order" and "engineer-to-order ". Level 3 considers the business processes and system functionality used to process sales orders, purchase orders, work orders, return authorizations, replenishment orders, and forecasts. Level 4 process details are not contained in SCOR® but must be defined to implement improvements and manage processes. Advanced users of the framework have defined process detail as far as Level 5, software configuration detail [15].

Organizations using the model SCOR® performance metrics can compare their performance levels compared to other organizations in the supply chain using a benchmarking tool called SCORmark. The SCORmark database contains historical data from over 1000 companies and 2000 supply chains. The benchmarking process using the SCORmark can be performed by the following steps: (1) defining the supply channels to be compared; (2) measuring the internal and external performances; (3) compares the performance to relevant industrial companies; (4) establish competitive demands; (5) calculate the opportunity value of improvement [18,19]. To facilitate the comparison, the SCORmark stratifies the process performance according to three points [19]:

- "Superior": is the performance (median value) on a specific indicator attained by 10% of the best classified SC's comparing to the total of the supply chains surveyed;
- "Advantage": is the performance (median value) among the top 10 companies and the median of all the supply chains considered;
- "Parity": is the performance (median value) of all the supply chains considered.

The APICS[1] association announced the launch of the SCOR model version 12.0. Developed by a panel of international supply chain experts, this latest version of SCOR incorporates Omni-channel, metadata, block chain and other emerging engines that supply chain professionals are using today. SCOR has been the global inter-sectorial standard for supply chain excellence over the last 20 years and, with this update, will continue to support ways to measure, improve and communicate the supply chain's business performance [20].

3 Experimental Method

The SCOR model is composed of several parts: a modeling of the processes of the SC with diagrams and

[1] APICS is the association for supply chain management and the leading provider of research, education and certification programs that elevate supply chain excellence, innovation and resilience.

for each process SCOR proposes metrics; based on these metrics the performance of the supply chain is evaluated. SCOR then provides a basis for comparison with the best global logistics chains, this is the benchmarking part. Lastly, SCOR proposes the best practices to overcome the differences in performance on the various processes.

Indeed, the processes are modeled with three standard levels and two levels specific to the studied society:

-Level 1 corresponds to the upper level of the supply chain. It is divided into 6 processes: Plan, Source, Make, Deliver, Return and Enable.

-Level 2 is a breakdown of Level 1 according to major production categories and business strategy. The choice of SCOR level 2 processes depends on the production strategy. Processes related to the "make-to-stock" category relate to production processes related to sales forecasts. There are two other categories: the "Make-to-Order" category for order-related production and the "Order-Engineered" category where the product is designed for a specific customer.

-Level 3 describes each process that makes up Level 2 processes with the process diagram ".

Metrics are classified with the same levels as processes and with performance attributes: reliability, responsiveness, agility, costs and assets.

The study was based on the analysis of 3 performance metrics that belong to level 1 of the SCOR model planning process, they are also used to measure the performance of the company's logistics chain, and calculated in based on the different balance sheets of the company.

The three Metrics of the level 1 chosen are:

- Inventory Days of Supply (IDS): This metric answers the question: how many days it takes to sell the entire inventory. It measures the average duration between the acquisition and the sale of goods. For a manufacturer, it covers the amount of time between the purchase of raw materials and the sale of the finished product.
- Days of sales outstanding (DSO) : This KPI measures the average number of days that a company takes to collect revenue after a sale has been made. It is often determined on a monthly, quarterly or annual basis and can be calculated by dividing the amount of accounts receivable in a given period by the total value of credit sales in the same period, and multiplying the result by the number of days in the measured period
- Days of payable outstanding (DPO): This performance metric indicates how long it takes for a company to pay its commercial creditors' bills, such as suppliers.

Regarding the second level of the SCOR model, the study was based on 5 performance metrics, which allow analyzing in a very precise way the current situation of the processes. These metrics of the level 2 are:

- Perfect order fulfilment: The percentage of orders complying with delivery performance with complete and accurate documentation and no delivery damage.
- Product rate with insufficient stock: This performance metric represents the percentage of products whose stock is less than five days.
- Product rate with overstock: This performance metric represents the percentage of products with a stock greater than one month.
- Unusual transportation usage rate: This performance metric represents the percentage of use of exceptional transportation.
- Delivery rate with complete documents: This performance metric represents the percentage of deliveries whose documents are complete and accurate, with no failures.

Above we represent the four performance metrics of the third level that is operational, these metrics study the execution of our process:

- Truck fill rate: The filling percentage of the delivery units.
- Late delivery rate of suppliers: This performance metric represents the rate of products received late by suppliers.
- Rate of products in rupture: This performance metric represents the number of products for which there are breaks in relation to the total number of products that exist.
- Late delivery rate of customers: This performance metric represents the rate of products delivered late to customers.

The applied experimental method comprises five stages:

Stage 1: Select a company in the automotive sector. this company is a supplier of metal parts to a car manufacturer. the optimization of the planning process for this company is very important given the criticality of this process in the automobile chain that has to release a vehicle per minute. The improvement of the logistics chain is one of the major interests of companies. This task, despite its complexity, remains indispensable for any evolving company. But before any improvement, a detailed analysis of the current state of performance indicators is required. These must be selected beforehand according to appropriate criteria.

Stage 2: Application of the SCOR model to the automotive industry;

The second stage is partially derived from the method proposed by the SCC[15]. The SCOR model used is SCOR version 11.0 [17]. The 3 sub-stages are as follows:

- Stage 2.1: scope determination: the determination of the practical case which is the most representative (number of activities, volume of orders and turnover)
- Stage 2.2: implementation of the SCORCARD of the performance metrics
- Stage 2.3: “AS IS” of the process: “process diagram”

Stage 3: Using Benchmarking to compare performance metrics for the planning process

Stage 4: Analysis and selection of Best Practices

Stage 5: Identification of the benefits and limitations of the SCOR model in automotive industries

4 Application to the case study

The following section describes the results of the application of the SCOR model to the automotive sector.

4.1 Stage 2.1: scope determination

The most critical process of the company's supply chain is highlighted in stage 1. All activities in the planning process will be detailed to highlight the different gaps penalising the overall performance of the supply chain. The study focuses in particular on this process, also seen as the counter-time (5 months average for each process), the other processes will be the subject of future projects.

3.2 Stage 2.2: SCORCARD performance metrics

Stage 2 describes the preparation of a SCORCARD composed of 12 metrics. Among these 12 metrics: one (Perfect order fulfilment) is already used by the planning department, and five are relevant in terms of application in the automotive sector. The rest of the metrics are essential to evaluate the profitability and the efficiency of the studied process. In the following, the performance metrics of the planning process are presented in detail (description, calculation formula, unit, improvement direction):

Table 1. KPI: Inventory Days of Supply (IDS)

Performance metric	Inventory Days of Supply (IDS)
Description	The number of days it takes to sell the entire inventory The average duration between the acquisition and the sale of goods
Formula	value of stock / average cost of products sold per day
Unit	Days
Improvement direction	Minimize

Table 2. KPI: Days of sales outstanding (DSO)

Performance metric	Days of sales outstanding (DSO)
Description	The average number of days a business takes to collect revenue after a sale has been made
Formula	(month-end total outstanding incl. taxes x number of days) / (Total turnover tax for the period)
Unit	Days
Improvement direction	Minimize

Table 3. KPI: Days of payable outstanding (DPO)

Performance metric	Days of payable outstanding (DPO)
Description	The number of days it takes a business to pay its commercial creditors bills
Formula	(Tax receivables / annual turnover incl. taxes) * 360 days
Unit	Days
Improvement direction	Minimize

Table 4. KPI: Perfect order fulfilment

Performance metric	Perfect order fulfilment
Description	The percentage of orders received by the customer with the right quantities ordered in terms of quantity and reference
Formula	number of orders fulfilled / total number of orders
Unit	%
Improvement direction	Maximize

Table 5. KPI: Product rate with insufficient stock

Performance metric	Product rate with insufficient stock
Description	The percentage of products whose stock is less than five days
Formula	number of products with a stock less than five days / total number of products
Unit	%
Improvement direction	Minimize

Table 6. KPI: Product rate with overstock

Performance metric	Product rate with overstock
Description	the percentage of products with a stock greater than one month
Formula	number of products with a stock greater than one month / total number of products
Unit	%
Improvement direction	Minimize

Table 7. KPI: Unusual transportation usage rate

Performance metric	Unusual transportation usage rate
Description	The percentage of use of exceptional transport
Formula	number of delivery with unusual transport / total number of deliveries
Unit	%
Improvement direction	Minimize

Table 8. KPI: Delivery rate with complete documents

Performance metric	Delivery rate with complete documents
Description	The percentage of deliveries whose documents are complete and accurate, with no failures
Formula	number of delivery with complete document / total number of deliveries
Unit	%
Improvement direction	Maximize

Table 9. KPI: Truck fill rate

Performance metric	Truck fill rate
Description	The filling percentage of the delivery units
Formula	number of units delivered / total capacity of trucks
Unit	%
Improvement direction	Maximize

Table 10. KPI: Late delivery rate of suppliers

Performance metric	Late delivery rate of suppliers
Description	The rate of products received by our suppliers late
Formula	number of units arriving late / total number of units delivered by our suppliers
Unit	%
Improvement direction	Minimize

Table 11. KPI: Rate of products in rupture

Performance metric	Rate of products in rupture
Description	The number of products for which there are breaks in relation to the total number of products that exist
Formula	number of products out of stock / total number of

	products
Unit	%
Improvement direction	Minimize

Table 12. KPI: Late delivery rate of customers

Performance metric	Late delivery rate of customers
Description	This performance indicator represents the rate of products received by our late suppliers
Formula	number of units delivered late / total number of units delivered to our customers
Unit	%
Improvement direction	Minimize

After defining the metrics on the dashboard, we preceded the collection of data to calculate their values; the result obtained is presented in the following table:

Table 13. Current values of performance metrics of the planning process

	KPI	Actual value
Level 1	Inventory days of Supply (IDS)	85
	Days of sales outstanding (DSO)	15
	Days of payable outstanding (DPO)	10
Level 2	Perfect order fulfilment	65,82%
	Product rate with insufficient stock	26%
	Product rate with overstock	32%
	Unusual transportation usage rate	34%
	Delivery rate with complete documents	67%
Level 3	Truck fill rate	88,46%
	Late delivery rate of suppliers	19,06%
	Rate of products in rupture	10%
	Late delivery rate of customers	10%

4.3 Stage 2.3: Process "AS IS"

Stage 3 describes the modelling of the process ("AS IS") and like presenting before the SCOR® model contains three levels of process detail. In practice, Level 1, the scope and content of a supply chain are defined. At Level 1, performance objectives based on supply chain competition are defined [17].

Business planning is a formalized process of decision-making which develops a representation of the future state of the enterprise and specifies the modalities of implementation of this will.

Level 2 presents the configuration of the planning and execution strategies in the material flow, involving standard categories such as "make-to-stock", "make-to-order" and "engineer-to-order " [17].

Each basic process of the SCOR® model is broken down into several sub-processes or also called a level 2 process.

Level 3 considers the business processes and system functionality used to process sales orders, purchase orders, work orders, return authorizations, replenishment orders, and forecasts [17]. Level 3 of the SCOR® model is provided for reference. It is therefore sufficient to choose a process of level 2 and to obtain the processes of the level 3 associated from the references, so it is easy to determine since it is not necessary to map the process ourselves.

Level 4 describes the activities performed within the supply chain, companies implement industry, and/or location specific processes and practices to achieve the required performance [17]. Level 4 can be achieved by detailing a Level 3 process. Although Level 4 is not detailed in the SCOR® model, SCC identifies Level 4 as the definition of firm-specific strategic practices to gain a competitive advantage [21].

Advanced users of the framework have defined process detail as far as Level 5, software configuration detail [17]

5 Benchmarking

The dashboard of the SCOR® model contains standard performance metrics that are prioritized in several levels so that the lower level metrics can be used to calculate the higher-level metrics. These KPIs are divided into five axes of performance, which are:

- Reliability: Delivery performance.
- Reactivity: Delay of command execution.
- Flexibility: Flexibility to command variations.
- Logistics costs: Costs of sales and logistics.
- Asset management: Financial flows and working capital.

The benchmarking of SCOR® model is done by comparing the indicators of level 1, 2 and 3, however, it is not necessary to apply the three performance levels in each axis. That is why it would be better to prioritize these axes in order to put the level indicators 1 for less important axes and up to level 3 for the most important.

The notations S, A and P are used by the SCOR® model to prioritize the axes:

- S: « Superior » corresponds to the 90th percentile: This means that it will line up with 10% of the best-performing companies.
- A: « Advantage » corresponds to the 70th percentile: The objective will be to reach the performance of 30% of the best companies.
- P: « Parity » corresponds to the 50th percentile: The objective is to exceed 50% of the best companies.

In order to evaluate the process performance, it will be very interesting to complete the indicators that were already in place by other indicators proposed by the SCOR® model, which correspond to its strategic axes.

According to the baseline with which to compare, there are three types of benchmarking:

- Comparison on a historical basis: monitoring indicator development.
- Internal Benchmarking: is the comparison between companies of the same group.
- External Benchmarking: The comparison is made by external organizations.

The SCOR® model proposes the external benchmarking because the more the reference database will be larger, the results will be significant. The objective here will be to line up with the best companies in the automotive sector basing on logistics activities.

These indicators have clear definitions and a link to process activities that provide a basis for effective benchmarking. This capability enables companies to benchmark their supply chain performance using the same methods currently used to compare the company's performance. Through benchmarking, managers can go beyond performance measurement and understand this performance against the backdrop of the performance of their industry peers [17]. The established scorecard containing the benchmarking values provided by the SCOR standard as well as the improvement objectives for each metric are presented in table 14.

Table 15. Benchmarking results

	KPI	Benchmarking			Objective
		Parity	Advantage	Superior	
Level 1	Inventory days of Supply (IDS)	80	42	13	A
	Days of sales outstanding (DSO)	15	10	4	A
	Days of payable outstanding (DPO)	11	5	2	A
Level 2	Perfect order fulfilment	82	94	97	S
	Product rate with insufficient stock	30	12	3	S
	Product rate with overstock	35	17	5	S
	Unusual transportation usage rate	15	9	1	S
	Delivery rate with complete documents	90	95	100	A
Level 3	Truck fill rate	90	95	100	S
	Late delivery rate of suppliers			0	S
	Rate of products in rupture			0	S
	Late delivery rate of customers			0	S

6 Analysis and selection of Best Practices

6.1 Results & discussion of the case study

After noting the various gaps noticed in indicators on the dashboard. We brought out the processes affected by these gaps. Each indicator of the dashboard is related to one or more processes. The preliminary step is to analyze the processes that affect the maximum weak indicators. This analysis will lead to determining the causes of this weakness and action variables on which it will take action to improve these indicators.

-Supply Planning Process P2: This process aims to establish a supply plan. This is the principle of planning required by the MRP method. It begins by identifying the raw material requirements based on customer orders and forecasts. The result of this part represents the gross requirement. Subsequently, we identify the available resources, whether in raw material stock or expected deliveries via SAP system (Treatment of CBN). Hence, we obtain net requirement. Once the net requirement is defined, we can establish the procurement plan and delivery programs taking into account the parameters of suppliers, in particular, procurement lead time, lot size and the minimum amount to be supplied. To achieve the best results in terms of the procurement planning process, it is necessary to have reliable data that will be used to establish an optimal supply plan. That means to supply the required amount at the right time with the lowest possible cost. Currently, this process represents several anomalies; the net requirement calculated by the MRP system is incorrect, therefore unreliable for 50% of the articles. This means that the actual consumption observed during a period of production is largely different from the calculated requirement.

-Procurement process S1 and S2: These two processes are execution processes. They ensure the implementation of procurement plans. We start by checking the daily calendar receptions. Once a truck arrives, the unloading is done in a temporary area without recourse or a qualitative and quantitative control procedure or a billing procedure. Among the anomalies noticed in these processes, there is the lack of enough space allocated for storage of PD, which forces to stack the pallets outdoor. Consequently, we often face the loss of the inaccuracy or the identification of certain products because of the wind; on the other hand, there is the risk of oxidation of pieces by moisture.

After the analysis of these processes and a brainstorming with the staff of the company, it proved that the various causes which lead to this erroneous computing of the requirements are:

With the level of process engineering used we noted that there is a work not standardized, errors of transaction and data capture in SAP, from the delays of updating of the master data, the non-observance of procedures of modification and the not enclosed production orders.

With regard to the labors available, we can define the failures continuations: Strong rotation, staffing less formed, weak rotation interns, Lack of communication interns

The materials which exist undergo also several defects, like under exploitation of SAP, of the not identified fields of obligatory input, of the lack of communication under SAP... Finally, side client, we notice delays in receipt of the orders, incorrect or incomplete data of modification as well as unreliable EDI forecasts.

6.2 Best practices proposed by the SCOR® model

Definitions of Best Practices that include lists of components that differ between authors is not satisfactory. Components should at least be organized by categorizing them in different ways, for example, for internal and external activities (customer and supplier relationships); direct and indirect activities; or to people and processes. In the industrial context, BP can be interpreted as a mechanism for defining and implementing a manufacturing strategy [22]. According to Dilworth (1993), the choice of a manufacturing strategy involves choosing the weights or weights to be given to profitability, flexibility, product quality and service (reliability and speed). Best practices should be chosen by a company to support the chosen strategy [23]. Motivated by the idea that the plant should be considered in a strategic context and the analogy with computer-integrated manufacturing [24].

Regarding the processes mentioned above, the most performing companies have established certain practices that have reported to them tangible benefits. This is a kind of capitalization of experience that the SCOR® model offers through benchmarking. The following table lists best practices offered by SCOR® Framework, taking into account the processes previously diagnosed and the anomalies observed. Priority actions for this project will obviously be those who have the greatest impact and the least difficulty.

We will prioritize action plans to implement the best practices that have high impacts or benefits for the company and whose difficulty of implementation is low.

Table 16. The best practices proposed by SCOR® model

Processes	Best Practices
P2- supply planning	All Key Participants in the Supply Chain, Including Strategic Partners, Have Full Visibility of the Demand/Supply Plan
	VMI: Vendor-managed Inventory
	CPFR: Collaborative Planning, Forecasting, and Replenishment
P2.1 Identify prioritize dproduct requirements	The Demand Plan is Updated Frequently to Reflect Actual Consumption or Customer Forecast Information
	Maximize Data Integrity and System Accuracy by Ensuring 99%+ Accuracy of BOM Configuration, Inventory Levels, and Schedule Requirements
P2.4 Establish procurement plans	Digital Linkage (EDI, XML, Etc.) is Used to Provide Real-Time Demand Information and Handle Routine Transactions
	Maintain Data and System Integrity by Ensuring Production Data, Inventory Levels, and Schedule Requirements Are 99+% Accurate
	A detailed production model that synchronizes PLAN and MAKE activities in real time
EP.7 Manages Planning Configuration	ABC Classification
S1.1 et S2.1 schedule deliveries	Electronic Kanban Pull Signals Are Used to Notify Suppliers of the Need to Deliver Product

7 SCOR Benefits and limitations in Automotive Industries

7.1 Benefits & Roadmap for success

The representation performed through the model allows the identification of potential sources of supply, organizational breaks, redundant processing, and information flows using tortuous circuits... The SCOR® model, therefore, leads to identifying the paths Critical Supply Chain (nodes and links). This is essential because it is necessary to focus its resources on the critical points. The metrics database proposed by the model can enable companies to select relevant key performance indicators to assess and monitor the level of risk in a previously defined framework. Once the points of failure are identified and evaluated, treatments can be offered through best practices identified by SCC. The SCOR® model makes the failure reduction actions consistent with each other so that risk management fits. Note that the Supply Chain can meet the strategic objectives through this model. It should be added that SCOR® modeling allows for quick configurations and reconfigurations of the supply chain, making it possible to set up economic models that respond to the changing environment. The SCOR® model, therefore, provides the Supply Chain with the capacity to be robust, because of the structure of the model, but also resilient, because of the possible reconfigurations, made indispensable. Recall that applied to the Supply Chain, resilience is the ability of a system to recover a stable state after a disruption (failure of a supplier, unforeseen increase in demand ...). The model adapts to most Supply Chain issues arising from strategic changes such as mergers/acquisitions of companies, analyzes prior to the introduction of new information systems. Of course, like any cross-cutting project, running a project on the SCOR® model and, in addition, if it is combined with other approaches linked, as well, to continuous improvement, requires the support of the general manager as well as the membership of all the actors involved in the project. In addition, it is clear that the model training phase is one of the keys to the success of the project.

The SCOR® process can get into a lot of process detail levels to help a company in the automotive sector to analyze its supply chain. It gives an idea of the progress of its supply chain and helps the company to understand and control its links with suppliers and customers. Each step in the process is a link in the supply chain that is critical to getting a product successfully along each level of the model. The SCOR® model provides benefits to companies that use it to identify supply chain issues. The model allows full leveraging of capital investment, a creation of a supply chain roadmap, alignment of business functions, and an average of two to six times the return on investment [25].

Indeed, for our case studied the benefits come immediately and are continuous in the complete process of applying the SCOR® model. We will detail these advantages following 5 steps of the successful roadmap as it is found after applying the SCOR® model to our case study. The first step is to educate using information and workshops from the Supply Chain Council. Every member of the team, including senior management, is educated in what Supply Chain Management is, how it works, and the benefits of this new business philosophy. Afterward, the roadmap uses the discovery phase of the process. In this phase, the business process that is currently in place as well as the flow of the "real" process, are identified, examined, measured, and compared to the automotive industry practices. It is essential to the success of the company, before any changes or improvement process implemented, that it knows how it really works. Not how the process is written in the policies and procedures manual, but how it works in reality. This knowledge can only come from the people who do the work, the real experts on how the business is executed. After the discovery phase, the model leads us through the analysis phase whose solutions are produced by developing a significant "gap analysis" and making a comparison of the current situation with that of the best companies in the world. Once the discovery phase is complete it is necessary to introduce the development phase which brings together all the changes and improvements proposed. This is where you need to prioritize process changes and validate that the desired changes and improvements are made. Once the process changes are validated you have to move to the fifth and final step in the roadmap, that's the implementation. It is in this phase that we choose a pilot program to implement the new modified business process. The pilot program must be chosen with great care and the planning must be representative of our current economic model, but it should not be of such magnitude that it could paralyze the current business. The pilot program not only validates the changes but also trains staff, modifies processes in real time and compares them to predetermined goals. This is an exciting and rewarding part of the SCOR® model process and rewards are generally far beyond the program's efforts and expenditures. Once the pilot program has validated workflow improvements, improvements to the new process can be disseminated to the rest of the business process in a methodical and managed process. The implementation of SCM in a company brings huge improvements and benefits to the business, but also it generates risks and costs. These last two elements can be managed and controlled with the appropriate tools and training. The SCOR® model of the supply chain council provides the step-by-step process, the training required and the support needed to achieve this goal successfully [26].

7.2 Limitations & Weakness

In this part, we will discuss the different difficulties encountered during the implementation of the model in our society, and these remarks were reported by the practitioners and managers involved in the selected processes during the interviews organized with them after the implementation phase. Indeed, the first difficulty reported is the fact that there were 159 metrics proposed by the SCOR® standard. This has made the task of selecting and monitoring all these metrics long and tedious. Each measure involves measuring, monitoring and defining action plans. Help in identifying the most important metrics was required. This help has helped reduce cost overruns in terms of time spent using the model. An in-depth understanding of model definitions was needed to optimize the use of this model. Modeling via SCOR® requires a thorough understanding of how activities actually work and the completion of a field study. Acquiring and using the business model is a long and tedious process. This generates additional costs and time. Computerization is a key future development to take full advantage of the model acquired [27].

The SCOR® model has limitations that lead to gaps in the need for performance evaluation in a car industry company. All processes and sub-processes proposed by SCOR® in level 3 are not chosen for the application case. The method recommends making choices according to the present activities. However, the choice was made using the definition describing the content of the processes in the model and an adaptation of the activities proposed by the model to be inadequacy with the specificity of the automotive sector, in addition, some activities present may belong to the different process. The choice of affiliation of these activities was delicate. SCOR brings names to processes. Since affiliation is not obvious for some processes, the names do not characterize the content of the process. Level 3 process names are changed to more common business terms. Navigating the structure provided by the SCOR model requires its complete representation on paper, taking in our case the dimensions of an A0 only for a single process. On the one hand, if the study is extended to all processes, the complete drawing is difficult to obtain due to its size and it makes the navigation in the complete system long and arduous. On the other hand, Correlations between categories of indicators are not observable. For example, it is not possible to show that time has an impact on cost. A good knowledge of the model definitions is necessary for its use. Modeling by SCOR requires a good knowledge of the actual operation of the activities and conducting the study in the field. In summary, obtaining the business model and its exploitation is long and tedious. Computerization seems to be an inevitable future development to exploit the wealth of the model obtained [27].

Lockamay III and McCormack (2004) argue that the importance of supply chain management planning is clearly revealed in the literature is crucial. There is a lack of research linking supply chain planning practices with supply chain performance. They also reveal the need to use information technology in the supply chain to improve information sharing, supply chain competitiveness and the use of ERP systems, advanced planning, and Internet technologies [28]. (Wang et al 2005), also argues that the SCOR® model is not a complete solution for supply chain development. There are limitations in the SCOR® model. This is why they proposed a new framework for global logistics systems. However, they mentioned that this framework is not the only way to solve all supply chain management issues [29][30].

8 Conclusion and prospects

The SCOR® model provides a framework of the common supply chain, standard terminology, common measures with associated benchmarks and best practices. It can be used as a common template for the evaluation, positioning, and implementation of supply chain application software. It is in its growing life cycle phase and is leveraging to become a standard in the automotive industry that has complex supply chains due to the nature of the products manufactured.

Indeed, based on the opinion of practitioners in the field model and specialists during the implementation of the model, several benefits were collected such as securing the logistics chain since the SCOR® model provides the supply chain with the ability to be robust, because of the structure of the model, but also resilient, because of the possible reconfigurations, made indispensable. Also, it gives an idea of the progress of its supply chain and helps the company to understand and control its links with suppliers and customers. Each stage of the process is a link in the supply chain essential to the success of a product at each level of the model. The

SCOR® model provides benefits to companies that use it to identify supply chain issues. On the other hand, the model has several limitations during its implementation in the automotive sector, there was, in fact, a large number of metrics proposed by the SCOR® standard. This made the task of selecting and tracking all these metrics long and tedious. In addition, some activities in the sub-processes may belong to a different process. The choice of affiliation of these activities was delicate. SCOR® provides names for processes that do not characterize the content of these processes. Also, there is a lack of research linking supply chain planning practices with supply chain performance. In addition, the best practices proposed by the model SCOR® to mitigate the various raised gaps remain not sufficient to improve the situation in the company. Therefore, it will be very interesting for future research to supplement the best practices by those proposed by the literature for automotive industries.

Although this study is primarily oriented towards a university audience, it can also be interesting and useful for practitioners, who will be able to gain an understanding of the purpose of the existing research and to have access to a real-life example of the application of this model in the automotive sector. The number of cases is too limited to make a broad generalization. An extension should consider the planning and scheduling of more automobile manufacturers and also include tier 1 and 2 of automotive subcontractors. It is therefore appropriate, in the context of future research, to continue this study by extending the sample to a larger number of companies. Future work will also focus on adapting the SCOR® model to the context of the automotive sector to facilitate its implementation and use.

9 Bibliographies

1. E. Karpova, *Challenges of Supply Chain Management in Brazil Degree Programme in International Business* (2017)
2. J. Manners-Bell, *Supply Chain Risk - Understanding Emerging Threats to Global Supply Chains - Knovel* (2014)
3. P. Grosche, U. Mayrhofer, and S. Schmid, "La configuration et la coordination internationales de la chaîne de valeur dans l'industrie automobile allemande," Financ. Contrôle Strat., 1–27 (2015)
4. J. Humphrey and O. Memedovic, "The Global Automotive Industry Value Chain: What Prospects for Upgrading by Developing Countries," SSRN Electron. J. (2003)
5. A. Schulze, J. Paul MacDuffie, and F. A. Taube, "Introduction: knowledge generation and innovation diffusion in the global automotive industry--change and stability during turbulent times," Ind. Corp. Chang., **24**(3), 603–611 (2015)
6. M. Goldenstein and B. P. Casotti, "Panorama Do Setor Automotivo : As Mudanças Estruturais Da Indústria E As Perspectivas Para O Brasil," BNDES Setorial, **28**, 147–188 (2008)
7. C. R. Vaz, T. R. Shoeninger Rauen, and álvaro G. Rojas Lezana, "Sustainability and innovation in the automotive sector: A structured content analysis," Sustainability (Switzerland), **9**(6), 1-23 (2017)
8. H. Pierreval, R. Bruniaux, and C. Caux, "A continuous simulation approach for supply chains in the automotive industry," Simul. Model. Pract. Theory, **15** (2), 185–198 (2007)
9. A. Gunasekaran, C. Patel, and R. E. McGaughey, "A framework for supply chain performance measurement," International Journal of Production Economics, **87** (3), 333–347 (2004)
10. M. Sako and S. Helper, "The Information Requirements of Trust in Supplier Relations: Evidence from Japan, Europe and the United States," *Trust and Economic Learning*, 23–47 (1998)
11. T. Cavusoglu, T. Gulledge, and T. Kessler, "Aligning the supply chain operations reference (SCOR) model with enterprise applications," *Int. Conf. Manag. Eng. Technol. Proc.*, **1** (2001)
12. S. Stephens, "Supply Chain Operations Reference Model Version 5.0: A New Tool to Improve Supply Chain Efficiency and Achieve Best Practice," *Inf. Syst. Front.* (2001)
13. S. Holmberg, *A Systems Perspective on Supply Chain Measurements*, **30**(10) (2000)
14. B. G. Jamehshooran, A. M. Shaharoun, and H. N. Haron, "Assessing Supply Chain Performance through Applying the SCOR Model," Int. J. Supply Chain Manag., **4**(1), 1-11 (2015)
15. P. Bolstorff and R. Rosenbaum, *Supply Chain Excellence: A Handbook for Dramatic Improvement Using the SCOR Model, Third Edition* (2011)
16. K. Rotaru, C. Wilkin, and Ceglowski Andrzej, "Analysis of SCOR's approach to supply chain risk management," *Meas. Bus. Excell.*, **21**(2), 191–206 (2014)
17. Supply Chain Council, *Supply Chain Operations Reference Model (SCOR)* (2012)
18. F. R. Lima-Junior and L. C. R. Carpinetti, "Combining SCOR® model and fuzzy TOPSIS for supplier evaluation and management," Int. J. Prod. Econ., **174**, 128–141 (2016)
19. G. M. D. Ganga and L. C. R. Carpinetti, "A fuzzy logic approach to supply chain performance management," Int. J. Prod. Econ., **134**(1), 177–187 (2011)
20. APICS Supply Chain Council, *SCOR Supply Chain Operations Reference Model ver. 12.0 quick reference quide* (2017)
21. J. Richey, *Maximizing Supply Chain Visibility Solutions with SCOR® How to Maximize Supply Chain Visibility Solutions with SCOR® and Create Long-Term, Measurable Results,* 1–7 (2013)
22. M. Ketokivi and R. Schroeder, "Manufacturing practices, strategic fit and performance: A routine-based view," Int. J. Oper. Prod. Manag., **24**(2), 171–191 (2004)

23. J. B. Dilworth, *Production and Operations Management: Manufacturing and Services, fifth edition, McGraw-Hill, New York*, Fifth edit. (1993)
24. N. Beaumont, "Best practice in Australian manufacturing sites," Technovation, **25**(11), 1291–1297 (2005)
25. N. Bauhof, "SCOR Model: Supply Chain Operations Reference Model," *Beverage Ind.*, **95**(8), 78 (2004)
26. L. M. Ellram and W. Tate, *Annual International Supply Management Conference*, May (2006)
27. E. Lepori, D. Damand, and B. Barth, "Benefits and limitations of the SCOR model in warehousing," *IFAC Proceedings Volumes*, 424–429 (2013)
28. A. Lockamy and K. McCormack, "Linking SCOR planning practices to supply chain performance," Int. J. Oper. Prod. Manag., **24**(12), 1192–1218 (2004)
29. W. Y. C. Wang, M. S. C. Ho, and P. Y. K. Chau, *A Process Oriented Methodology For The Supply Chain Analysis Of Implementing Global Logistics Information Systems*, 1–11 (2005)
30. G. E. Delipinar and B. Kocaoglu, "Using SCOR Model to Gain Competitive Advantage: A Literature Review," Procedia - Soc. Behav. Sci., **229**, 398–406 (2016)

Supply Chain Management based on Blockchain: A Systematic Mapping Study

Youness Tribis[1,*], *Abdelali* El Bouchti[2,3], and *Houssine* Bouayad[1]

[1]Faculty of Technical Sciences, Hassan 1st University, Settat, Morocco
[2]National School of Business and Management, Hassan 1st University, Settat, Morocco
[3]Institute for Forecasting and Futuristics

Abstract. Groundbreakingly, blockchain technology (BCT) has gained widespread acceptance and importance in the last few years. Implemented in different areas of applications such as social and legal industries, finance, smart property, and supply chain networks. This technology assures immutability and integrity of data without the need of a third trusted party. Furthermore, BCT could guarantee a transparent and decentralized transaction system in businesses and industries. Even though general research has been done in the BCT, however, there is a lack of systematic analysis on current research challenges regarding how BCT is effectively applicable in supply chain management (SCM). A systematic literature review (SLR) of SCM based on blockchain does not exist yet. This work aims to explore and analyse the state-of-the-art on the BCT applications for SCM. We synthesize existing evidence, and identify gaps, available in the literature. The survey uses a systematic mapping study (SMS) method to examine 40 extracted primary studies from scientific databases.

Keywords: blockchain, supply chain, supply chain management, systematic mapping study,

1 Introduction

To improve the supply chains performance for many industries and provide new supply chain management (SCM) solutions, some systematic literature review (SLR) on SCM have, recently, demonstrated the importance of information and communication technology (ICT) ([1], [2], [3], [4] and [5]), but without focusing on blockchain technology (BCT).

The BCT has been generating interesting research areas because of its innovative characteristics that provide efficient solutions for the current gaps related to several supply chains industries. Indeed, existing SLR on blockchain conducted until now did not address such topic linked to the blockchain application for SCM. For example, a SLR performed by BOJANA KOTESKA et al. [6], Stefan Seebacher and Ronny Schüritz [7], and SMS established by Yli-Huumo J, et al. [8], did not cover blockchain application for SCM that should be possible to map out the current status of research area.

Blockchain is a decentralized transaction and data management technology firstly designed, by Satoshi Nakamoto, for Bitcoin cryptocurrency [9]. Based on the literature, the blockchain definition was synthesized as a distributed database, which is shared among and agreed upon a peer-to-peer network. It consists of a linked sequence of blocks, holding timestamped transactions that are secured by public-key cryptography and verified by the network community. Once an element is appended to the blockchain, it cannot be altered, turning a blockchain into an immutable record of past activity [7].

Trust and decentralization were identified as the two key BCT characteristics, which are closely connected and interrelated [7].

The remaining parts of the paper are structured as follows: In Section 2, we describe the applied research methodology and the process of collecting relevant research papers. Section 3 presents the results of the collected papers and the classification mechanism extracted from data. Section 4 discusses the study and answers the research questions. Section 6 presents the limitations of this mapping study, and Section 7 concludes the paper.

2 Research methodology

In this study, we follow the SMS process, described by Petersen et al. [10], to explore the existing studies related to BCT applied for SCM. We also use guidelines for a SLR described by Kitchenham and Charters [11] to search for relevant papers which covered the literature from all years to 2018. The results of the SMS would help us to identify and map research areas related to BCT based SCM and possible research gaps. The SMS process

* Corresponding author: y.tribis@uhp.ac.ma

is presented in Figure 1 and consists of five process steps and outcomes.

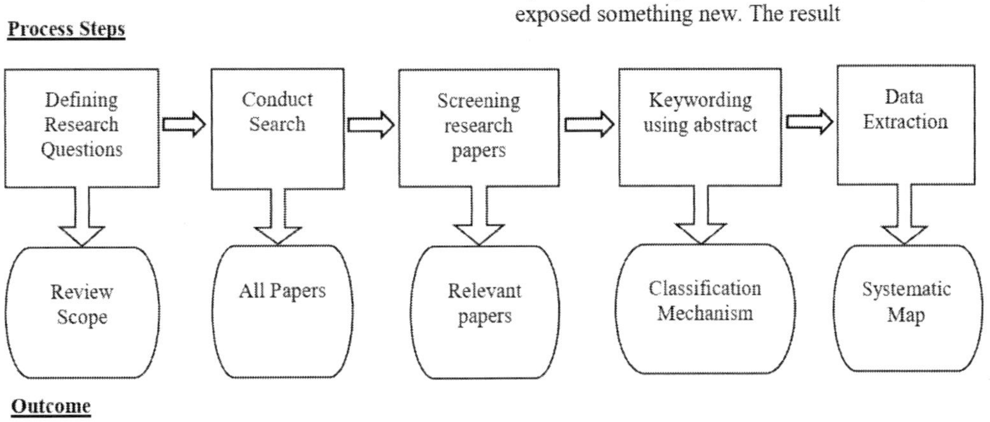

Fig. 1. The systematic mapping process

We defined three research questions that form the basis for the review:

- RQ1: What research topics have been addressed in existing research on SCM based on BCT?
- RQ2: What are the benefits would BCT bring to resolve the current problems in the supply chain industries?
- RQ3: What are the current research gaps in SCM related to BCT?

To search the relevant papers on the research topic, we decided to use every time the terms "supply chain" and "blockchain" in the search string. After designing and testing the search protocol, we chose the scientific databases, for the searches. The search was conducted on January 2018 and updated on May 2018.

A screening process was composed of four filters, in order to select the most suitable results. Firstly, we screened all papers based on their types and titles similarity to exclude duplicates, technical reports and notes. Secondly, we screened the papers based on their titles and excluded studies that were not relevant to the scope of this mapping study. Thirdly, the authors read the abstracts of every paper that passed the previous filter. Finally, the authors read the full text of all included papers in the third filter.

To screen each paper, we used inclusion and exclusion criteria. We exclude the following types of papers: (1) papers that were duplicates, (2) papers that were technical reports and thesis, (3) papers without full text availability (4) papers where the main language was not English, (5) papers that had some other meaning than blockchain used in SCM, and (6) the phrases of at least one of the search terms didn't be included in the heading, keywords or abstract of the paper.

The next stage in a SMS is keywording as defined by Petersen et al. [10]. Firstly, we read the abstract and identified keywords and concepts that represent the contribution of the paper. Next, we develop a higher level of understanding based on these keywords. We used the keywords to group and form classes for the mapping the studies. After that, we read all the selected papers and updated the categories or created new ones, if the paper exposed something new. The result was a systematic map of collected categories designed from all the relevant papers.

The data extraction form should be designed to collect all the information needed to address the research questions of this mapping study. The following information, was extracted from each study; Title of the paper, Name of the author(s), Country of authors, Type of the publication (e.g. conference/journal), Abstract, Methodology, Research questions and Major study findings.

3 Results

Based on the extracted data, the search and selection results are presented in Figure 2. As a result, 40 primary papers have been included in this study.

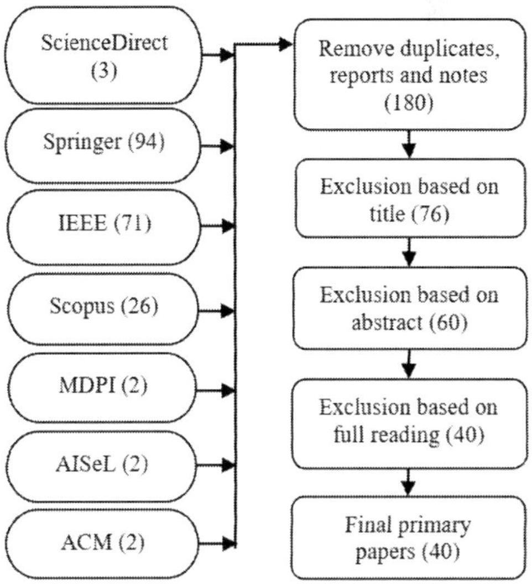

Fig. 2. Search and selection process of the papers.

Considering the publication year, the distribution of the primary studies was shown in Figure 3. Remarkably, all the selected papers were published after the year 2015. This demonstrates that this research area is a modern and novel one. When zoom in at the publication year distribution, out of all the selected papers, 2 papers (5.71%) were published in 2016, 22 papers (62.86%) in 2017 and 16 papers (45.71%) in April 2018. Thus, the number of publications increases each year, which reflects the rise of interest in BCT.

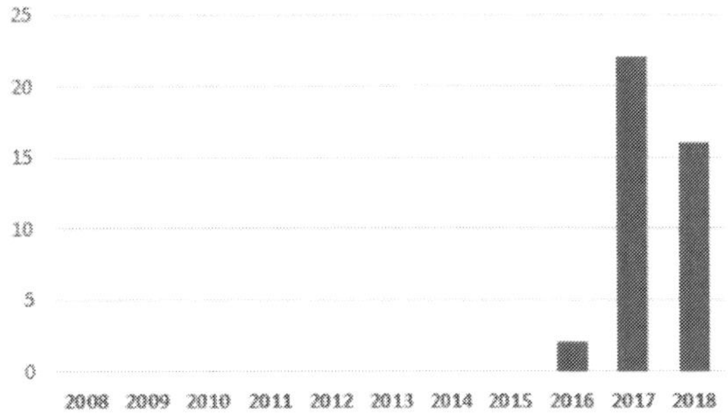

Fig. 3. Publication year of the selected primary papers.

Figure 4 illustrates the geographical distribution of the primary papers. Dispersed at 16 countries, this geographical distribution shows that BCT has gained research attention around the world. The major number of papers 12 (34.29%) were published by universities or companies in the USA. After this, the second position was occupied by Switzerland and China with 5 papers (14.29%). The rest of the countries had two or less papers published.

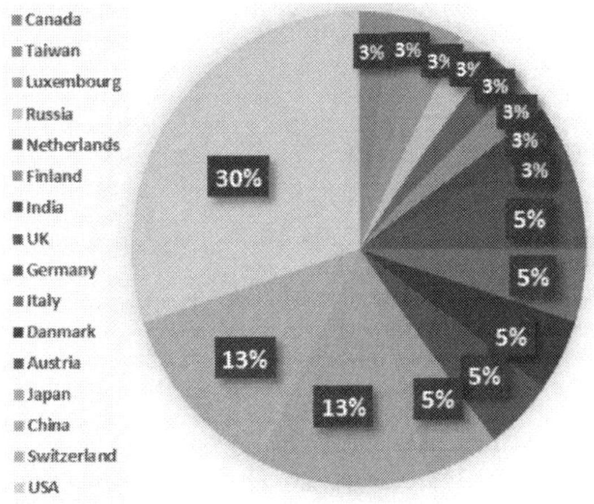

Fig. 4. Geographic distribution of the selected primary papers.

The publication channel distribution was presented in Figure 5. Most of the papers were published in journals (19) (45.29%) and conference (12) (34.29%). The rest of the papers were published in book section (5) (14.29%), workshop (3) (8.57%), or in symposium (1) (2.86%).

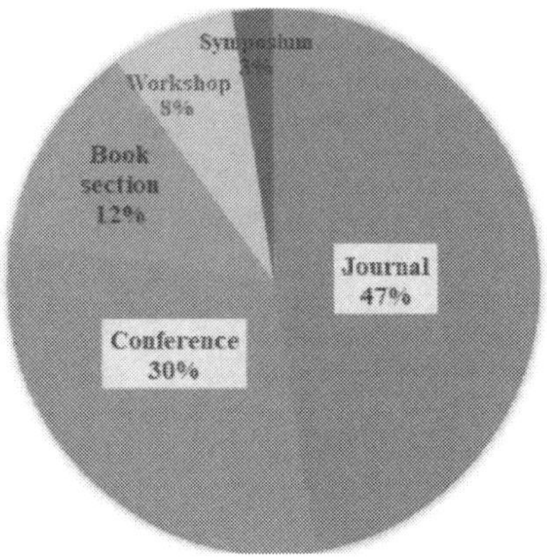

Fig. 5. Publication type.

In addition, Table 1 displays the global list of publication channel of the selected papers.

The classification mechanism, presented in a bubble plot (Figure 6), assembled two facets. One facet structured the topic in terms of the research thematic. The other considered the research type as described by Wieringa et al. [12].

Within the research topic facet, we summarize the primary studies on 10 main research topics. We identified one thematic, supply chain traceability, that represent the majority of topics (17 papers, 42.5%) and focus on some special supply chain industries such as food supply chain (3 papers) and pharma supply chain (3 papers). It has really been the overall interest of researchers.

Furthermore, some topics present a relevant amount of entries in this analysis, such as information security of SCM system (7 papers, 17.5%) and supply chain finance (7 papers) were the second important topics. In addition, we have pointed out topics in which new solutions are required; it is the case of supply chain quality management, product ownership management, additive manufacturing and agricultural supply chain.

According to the research category, the majority of these papers (45.71 %) were solution proposal papers (16 papers) which aim to propose a new blockchain based frameworks related to several supply chain industries. We realize that currently, validation and evaluation research are weakly addressed in SCM based on blockchain papers. We also appreciate that researchers are not concerned about expressing opinions and experience reports on their personal experience.

4 Discussion

Besides the identified research topics, the SMS showed that a majority of research was focused in three main themes: physical traceability of supply chains (17 papers), information security of SCM system (7 papers), and supply chain finance (7 papers). The rest of the identified topics were gained a few interests of the scholars.

Table 1. Publication channels

Publication channels		Reference
IJIM	International Journal of Information Management	[13]
ICEBE	International Conference on e-Business Engineering	[14]
AISC	Advances in Intelligent Systems and Computing	[15]
Information (Switzerland)		[16]
IEEE Access		[17]
EODS	Expert Opinion on Drug Safety	[18]
ICSSSM	International Conference on Services Systems and Services Management	[19] [20]
IEEE Software		[21]
ICIEEM	IEEE International Conference on Industrial Engineering and Engineering Management	[22]
WiMob	IEEE International Conference on Wireless and Mobile Computing, Networking and Communications	[23]
Telecommunications Policy		[24]
ISINSM	IFIP/IEEE International Symposium on Integrated Network and Service Management	[25]
JCST	Journal of Computer Science and Technology	[26]
ICCCN	International Conference on Computer Communication and Networks	[27]
Springer, Cham		[28] [29] [30] [31]
ISAFM	Intelligent Systems in Accounting, Finance and Management	[32]
JIIS	Journal of Intelligent Information Systems	[33]
HICSS-50	Hawaii International Conference on System Sciences	[34]
JMCC	Journal of Materials Chemistry C	[35]
(AISeL)	AMCIS Workshops	[36]
CBI	IEEE Conference on Business Informatics	[37]
ICDCSW	IEEE International Conference on Distributed Computing Systems Workshops	[38]
JPSM	Journal of Purchasing and Supply Management	[39]
FGCS	Future Generation Computer Systems	[40]
FAS*W	IEEE International Workshops on Foundations and Applications of Self* Systems	[41]
Palgrave Macmillan, Cham		[42]
NTMS	IFIP International Conference on New Technologies, Mobility and Security	[43]
Logistics (MDPI)		[44]
ISAFM	Intelligent Systems in Accounting, Finance and Management	[45]
LNICST	Lecture Notes of the Institute for Computer Sciences, Social-Informatics and Telecommunications Engineering	[46]
JEFC	Journal of Excipients and Food Chemicals	[47]
ICHNAS	IEEE International Conference on e-Health Networking, Applications and Services (Healthcom)	[48]
BISE	Business & Information Systems Engineering	[49]
ACM CSG	ACM Conference on Supporting Groupwork	[50]
RCR	Resources, Conservation and Recycling	[51]
ACM CENSS	ACM Conference on Embedded Network Sensor Systems	[52]

Regarding the traceability thematic, there are two important fields that gain a more researcher's interest; the pharma supply chain ([41], [18], [34] and [36]) and food supply chain ([19], [20] and [22]). In addition, some specific subjects were introduced, such as, traceability of transportation dangerous good [31], aircraft's Part SCM [52], supply chain information system & traceability [40], shipping domain [38] and cyber-physical systems for material flow analysis and life cycle assessment [39]. The others studies ([16], [33], [32], [21], and [26]) treat the traceability subject in general without focusing in particular supply chain domain.

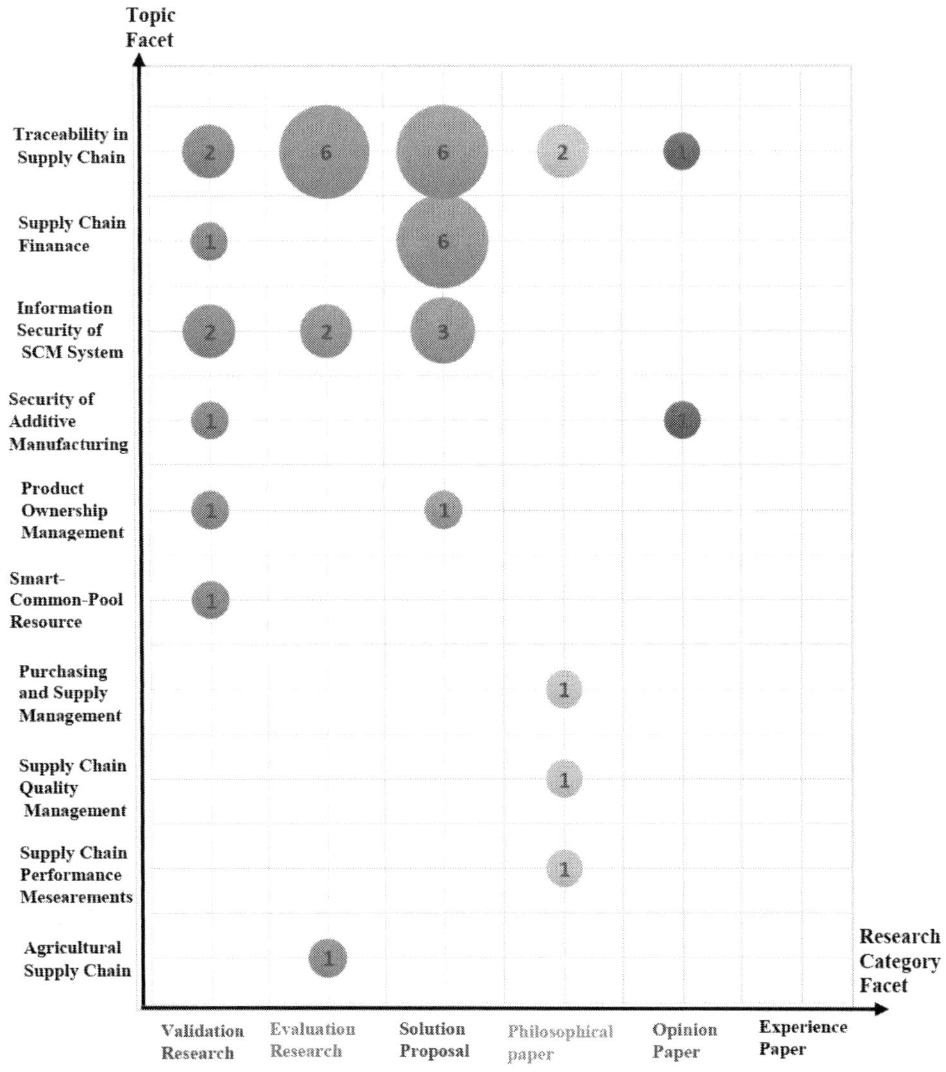

Fig. 6. A systematic map in the form of a bubble plot

We summarize the main benefits providing by this technology. In physical supply chain, the main BCT benefits for businesses are ensuring real-time traceability with trusted information and high level of reliability to all supply chain members with openness, transparency, reliability and security and providing pseudo real-time status of the shipment. BCT can increase the transparency of physical distribution process and eliminating the ability to cheat and provides each participant end-to-end visibility based on their level of permission. In addition, this technology could reduce operational costs by removing intermediaries' fees. For government, BCT could help to track, monitor and audit the supply chain and serving manufacturers to record the transactions in authenticity. For customers, BCT would enhancing the products quality and safety.

In the context of SCM information, the key BCT benefits for businesses are managing supply chain information more efficiently without being overly concerned about properly securing them, privacy immutability of data, public accessibility of records and access for diverse populations and locations. The government could receive more trusted information for better and focused inspections. For customers, BCT would guarantee public security of personal data.

In supply chain finance, BCT benefits for businesses are simplifying and making more agile and secure the trade finance commercial transactions among companies and financial institutions. Claims and contracts could be recorded onto a blockchain and validated by the network which allows all the stakeholders to identify the movements of funds between contractors, customers, and financial institutions. For government, auditing digital currency transactions help to raise the tax collection. For customers, BCT would bring more transparency and lower fees of transactions.

The majority of the identified gaps were related to external factors such as regulation and technical aspects. The first gap is the regulation conformance and legal barriers that limit the application of BCT. There is the lack of common standards for completing transactions. The second gap is the lack of adaptability and adoption. Many people are short of understanding of how it works that generate the large difficulty of bringing all the relevant parties together and convince the supply chain actors to adapt their old systems to the new blockchain based systems. The third gap is scalability and size. The majority of the proposed blockchain based frameworks were only tested on a limited scale in a laboratory environment; some challenges may emerge in scaling blockchains network with a large number of nodes. Furthermore, the sizes of the current blockchain applications are practically small. The fourth gap is high degree of computerization requirement. Indeed, many supply chain actors' in developing countries are not prepared to implement blockchain. The fifth gap is complexity and uncertainty. The latency of transactions that spend some hours to finish until all participants update their ledgers and the smart contract can be publicly accessible, but the data required for verification may or may not be accessible to all. The sixth gap is security and data integrity. The latency of transactions opens a window for cyber-attacks. Also, limited set of participants accessing to blockchain platform offering blockchain based solution, e.g. IBM and Ethereum, can cause more vulnerability to these blockchain based system. Thus, the security of these systems is impacted by that of these platforms. The seventh gap is the cost of implementation. For blockchain operation, the virtual network depended on a physical system for electricity, infrastructure, and hardware computing power. The eighth gaps are that the majority of the proposed frameworks were not evaluated to design systems for a real-world application, thus researchers should considerate the feasibility of the blockchain based solutions and verifying their applicability to the industry sectors.

5. Limitations

Publication bias, selection bias, imprecision in data extraction, and misclassification are the main limitations for this study. Publication bias is associated to the problem that is possible that some relevant studies were not chosen throughout the searching process. In fact, the search string included only the terms "blockchain" and "supply chain", there is a probability that not all the relevant research papers were found. Selection bias is assigned to the misinterpretation of to the criteria used to select the publications. Thus, we cannot guarantee that all relevant primary studies were selected. Imprecision in data extraction and misclassification are related to the risk that reviewers diverged when extracted the information.

6. Conclusion

This paper presents a systematic mapping study in order to map out all relevant research on SCM based on BCT. Guided by a set of research questions, the main motivation for this work was to identify research trends, open subjects, and gaps for improvement in this discipline. The results demonstrate that 42.5% of the papers focus on supply chain traceability based on blockchain, 17% of the papers address SCM information and 17% of the studies was conducted on the blockchain driven supply chain finance. The rest of the papers (23.5%) treat other blockchain applications in SCM that need additional investigation, such as agricultural supply chain, security of additive manufacturing, product ownership management, common-pool resource management, purchasing and supply management, supply chain quality management, supply chain performance measurements. The large part of selected studies (45.71%) is concentrating on proposing solutions of currents SCM challenges by designing new blockchain based frameworks. Nevertheless, many of the proposed frameworks-based solutions lack real performance evaluation on the industrial context.

References

[1] V. Maestrini, D. Luzzini, P. Maccarrone, and F. Caniato, "Supply chain performance measurement systems: A systematic review and research agenda," Int. J. Prod. Econ., vol. **183**, pp. 299–315, Jan. (2017).

[2] N. Zaheer and P. Trkman, "An information sharing theory perspective on willingness to share information in supply chains," Int. J. Logist. Manag., vol. **28**, no. 2, pp. 417–443, May (2017).

[3] Q. Shi, X. Ding, J. Zuo, and G. Zillante, "Mobile Internet based construction supply chain management: A critical review," Autom. Constr., vol. **72**, pp. 143–154, Dec. (2016).

[4] J. Volland, A. Fügener, J. Schoenfelder, and J. O. Brunner, "Material logistics in hospitals: A literature review," Omega, vol. **69**, pp. 82–101, Jun. (2017).

[5] J. Barata, P. Rupino Da Cunha, and J. Stal, "Mobile supply chain management in the Industry 4.0 era: An annotated bibliography and guide for future research," J. Enterp. Inf. Manag., vol. **31**, no. 1, pp. 173–192, Feb. (2018).

[6] B. KOTESKA, E. KARAFILOSKI, and A. MISHEV, "Blockchain Implementation Quality Challenges: A Literature." in *SQAMIA 2017: 6th Workshop of Software Quality, Analysis, Monitoring, Improvement, and Applications.* Belgrade, Serbia, September 11-13, (2017), vol. **1938**, pp. 8:8

[7] S. Seebacher and R. Schüritz, "Blockchain Technology as an Enabler of Service Systems: A Structured Literature Review," in *Za S., Drăgoicea M., Cavallari M. (eds) Exploring Services Science.*

IESS 2017. Lecture Notes in Business Information Processing, vol **279**. Springer, Cham, (2017), pp. 12–23.

[8] J. Yli-Huumo, D. Ko, S. Choi, S. Park, and K. Smolander, "Where Is Current Research on Blockchain Technology?—A Systematic Review," PloS One, vol. **11**, no. 10, p. e0163477, (2016).

[9] S. Nakamoto, "Bitcoin: A Peer-to-Peer Electronic Cash System." [Online]. Available: https://bitcoin.org/bitcoin.pdf.

[10] K. Petersen, R. Feldt, S. Mujtaba, and M. Mattsson, "Systematic Mapping Studies in Software Engineering," in *Proceedings of the 12th International Conference on Evaluation and Assessment in Software Engineering. EASE'08.* Swinton, UK, UK: British Computer Society; (2008). p. 68–77.

[11] Kitchenham B (2007) Guidelines for performing Systematic Literature Reviews in Software Engineering, Version 2.3, *EBSE Technical Report EBSE-2007-01*, Keele University and University of Durham

[12] R. Wieringa, N. Maiden, N. Mead, and C. Rolland, "Requirements engineering paper classification and evaluation criteria: a proposal and a discussion," Requir. Eng., vol. **11**, no. 1, pp. 102–107, Mar. (2006).

[13] N. Kshetri, "1 Blockchain's roles in meeting key supply chain management objectives," Int. J. Inf. Manag., vol. **39**, pp. 80–89, (2018).

[14] S. Chen, R. Shi, Z. Ren, J. Yan, Y. Shi, and J. Zhang, "A Blockchain-Based Supply Chain Quality Management Framework," in *2017 IEEE 14th International Conference on e-Business Engineering (ICEBE)*, (2017), pp. 172–176.

[15] P.-Y. Chang, M.-S. Hwang, and C.-C. Yang, "A blockchain-based traceable certification system," Adv. Intell. Syst. Comput., vol. **733**, pp. 363–369, (2018).

[16] H. Wu, Z. Li, B. King, Z. B. Miled, J. Wassick, and J. Tazelaar, "A distributed ledger for supply chain physical distribution visibility," Inf. Switz., vol. **8**, no. 4, (2017).

[17] K. Toyoda, P. T. Mathiopoulos, I. Sasase, and T. Ohtsuki, "A Novel Blockchain-Based Product Ownership Management System (POMS) for Anti-Counterfeits in The Post Supply Chain," IEEE Access, vol. **5**, pp. 17465–17477, (2017).

[18] T. K. Mackey and G. Nayyar, "A review of existing and emerging digital technologies to combat the global trade in fake medicines," Expert Opin. Drug Saf., vol. **16**, no. 5, pp. 587–602, (2017).

[19] F. Tian, "A supply chain traceability system for food safety based on HACCP, blockchain & Internet of things," in *the 14th International Conference on Services Systems and Services Management, ICSSSM 2017 - Proceedings*, (2017).

[20] F. Tian, "An agri-food supply chain traceability system for China based on RFID & blockchain technology," in *the 2016 13th International Conference on Service Systems and Service Management, ICSSSM 2016*, (2016).

[21] Q. Lu and X. Xu, "Adaptable Blockchain-Based Systems: A Case Study for Product Traceability," IEEE Softw., vol. **34**, no. 6, pp. 21–27, (2017).

[22] D. Tse, B. Zhang, Y. Yang, C. Cheng, and H. Mu, "Blockchain application in food supply information security," in *the IEEE International Conference on Industrial Engineering and Engineering Management*, (2018), 2017-December, pp. 1357–1361.

[23] P. W. Chen, B. S. Jiang, and C. H. Wang, "Blockchain-based payment collection supervision system using pervasive Bitcoin digital wallet," in *2017 IEEE 13th International Conference on Wireless and Mobile Computing, Networking and Communications (WiMob)*, (2017), pp. 139–146.

[24] N. Kshetri, "Blockchain's roles in strengthening cybersecurity and protecting privacy," Telecommun. Policy, vol. **41**, no. 10, pp. 1027–1038, (2017).

[25] M. Nakasumi, "Information sharing for supply chain management based on blockchain technology," in *2017 IEEE 19th Conference on Business Informatics, CBI 2017*, (2017), vol. **1**, pp. 140–149.

[26] Z. Li, H. Wu, B. King, Z. B. Miled, J. Wassick, and J. Tazelaar, "On the Integration of Event-Based and Transaction-Based Architectures for Supply Chains," in *the Proceedings - IEEE 37th International Conference on Distributed Computing Systems Workshops, ICDCSW 2017*, (2017), pp. 376–382.

[27] K. Foerstl, M. C. Schleper, and M. Henke, "Purchasing and supply management: From efficiency to effectiveness in an integrated supply chain," J. Purch. Supply Manag., vol. **23**, no. 4, pp. 223–228, Oct. (2017).

[28] K. Leng, Y. Bi, L. Jing, H.-C. Fu, and I. Van Nieuwenhuyse, "Research on agricultural supply chain system with double chain architecture based on blockchain technology," Future Gener. Comput. Syst. Vol. **86**, (2018), Pages 641-649.

[29] D. B. Kurka and J. Pitt, "Smart-CPR: Self-Organisation and Self-Governance in the Sharing Economy," in *2017 IEEE 2nd International Workshops on Foundations and Applications of Self* Systems (FAS*W),* (2017), pp. 85–90.

[30] B. Nicoletti, "The Future: Procurement 4.0," in *Agile Procurement*, Palgrave Macmillan, Cham, (2018), pp. 189–230.

[31] A. Imeri and D. Khadraoui, "The Security and Traceability of Shared Information in the Process of Transportation of Dangerous Goods," in *2018 9th IFIP International Conference on New Technologies, Mobility and Security (NTMS)*, (2018), pp. 1–5.

[32] K. Francisco and D. Swanson, "The Supply Chain Has No Clothes: Technology Adoption of Blockchain for Supply Chain Transparency," Logistics, vol. **2**, no. 1, p. 2, Jan. (2018).

[33] H. M. Kim and M. Laskowski, "Toward an ontology-driven blockchain design for supply-chain provenance," Intell. Syst. Account. Finance Manag., vol. 25, no. 1, pp. 18–27, (2018).

[34] Archa, B. Alangot, and K. Achuthan, "Trace and track: Enhanced pharma supply chain infrastructure to prevent fraud," Lect. Notes Inst. Comput. Sci. Soc.-Inform. Telecommun. Eng. LNICST, vol. **218**, pp. 189–195, (2018).

[35] S. Apte and N. Petrovsky, "Will blockchain technology revolutionize excipient supply chain management?," J. Excip. Food Chem., vol. **7**, no. 3, pp. 76–78, (2016).

[36] W. Liu, S. S. Zhu, T. Mundie, and U. Krieger, "Advanced block-chain architecture for e-health systems," in *2017 IEEE 19th International Conference on e-Health Networking, Applications and Services (Healthcom)*, (2017), pp. 1–6.

[37] R. Beck, M. Avital, M. Rossi, and J. B. Thatcher, "Blockchain Technology in Business and Information Systems Research," Bus. Inf. Syst. Eng., vol. **59**, no. 6, pp. 381–384, Dec. (2017).

[38] K. Jabbar and P. Bjørn, "Infrastructural Grind: Introducing Blockchain Technology in the Shipping Domain," in *Proceedings of the 2018 ACM Conference on Supporting Groupwork*, New York, NY, USA, (2018), pp. 297–308.

[39] S. Smetana, C. Seebold, and V. Heinz, "Neural network, blockchain, and modular complex system: The evolution of cyber-physical systems for material flow analysis and life cycle assessment," Resour. Conserv. Recycl., vol. **133**, pp. 229–230, Jun. (2018).

[40] M. Ruta, F. Scioscia, S. Ieva, G. Capurso, and E. Di Sciascio, "Supply Chain Object Discovery with Semantic-enhanced Blockchain," in *Proceedings of the 15th ACM Conference on Embedded Network Sensor Systems*, New York, NY, USA, (2017), pp. 60:1–60:2.

[41] T. Bocek, B. B. Rodrigues, T. Strasser, and B. Stiller, "Blockchains everywhere - A use-case of blockchains in the pharma supply-chain," presented in *the Proceedings of the IM 2017 - 2017 IFIP/IEEE International Symposium on Integrated Network and Service Management*, (2017), pp. 772–777.

[42] Z. Gao, L. Xu, L. Chen, X. Zhao, Y. Lu, and W. Shi, "CoC: A Unified Distributed Ledger Based Supply Chain Management System," J. Comput. Sci. Technol., vol. **33**, no. 2, pp. 237–248, (2018).

[43] L. Xu, L. Chen, Z. Gao, Y. Lu, and W. Shi, "CoC: Secure Supply Chain Management System Based on Public Ledger," in *2017 26th International Conference on Computer Communication and Networks (ICCCN)*, (2017), pp. 1–6.

[44] E. Hofmann, U. M. Strewe, and N. Bosia, "Concept—Where Are the Opportunities of Blockchain-Driven Supply Chain Finance?," in *Supply Chain Finance and Blockchain Technology*, Springer, Cham, (2018), pp. 51–75.

[45] E. Hofmann, U. M. Strewe, and N. Bosia, "Conclusion—What Can We Learn from Blockchain-Driven Supply Chain Finance?," in *Supply Chain Finance and Blockchain Technology*, Springer, Cham, (2018), pp. 89–91.

[46] E. Hofmann, U. M. Strewe, and N. Bosia, "Discussion—How Does the Full Potential of Blockchain Technology in Supply Chain Finance Look Like?," in *Supply Chain Finance and Blockchain Technology*, Springer, Cham, (2018), pp. 77–87.

[47] E. Hofmann, U. M. Strewe, and N. Bosia, "Introduction—Why to Pay Attention on Blockchain-Driven Supply Chain Finance?," in *Supply Chain Finance and Blockchain Technology*, Springer, Cham, (2018), pp. 1–6.

[48] D. E. O'Leary, "Configuring blockchain architectures for transaction information in blockchain consortiums: The case of accounting and supply chain systems," Intell. Syst. Account. Finance Manag., vol. **24**, no. 4, pp. 138–147, (2017).

[49] S. van Engelenburg, M. Janssen, and B. Klievink, "Design of a software architecture supporting business-to-government information sharing to improve public safety and security," J. Intell. Inf. Syst., pp. 1–24, Jul. (2017).

[50] K. Korpela, J. Hallikas, and T. Dahlberg, "Digital Supply Chain Transformation toward Blockchain Integration," *Hawaii Int. Conf. Syst. Sci. 2017 HICSS-50*, Jan. (2017).

[51] Z. C. Kennedy *et al.*, "Enhanced anti-counterfeiting measures for additive manufacturing: Coupling lanthanide nanomaterial chemical signatures with blockchain technology," J. Mater. Chem. *C*, vol. **5**, no. 37, pp. 9570–9578, (2017).

[52] Y. Madhwal and P. B. Panfilov, "Industrial Case: Blockchain on Aircraft's Parts Supply Chain Management," in *American Conference on Information Systems 2017 Workshop on Smart Manufacturing Proceedings. 6.* (2017).

Measuring urban public transport performance on route level: A literature review

Zehmed Karim [1], *Jawab* Fouad [2]

[1,2] Research Laboratory in International Management, Decision Making and Logistics (IMDLOG), High School of Technology (HST), Sidi Mohamed Ben Abdellah University (SMBAU), PB 2427 Imouzzer Road 30000, Fez, Morocco

Abstract. In this paper, we conducted a systematic literature review of empirical studies that focus on measuring urban public transport performance on route level using frontier methods. 22 papers were identified and included for full text review and were classified according to paper category, year of publication, country context, performance dimensions discussed, methodological approach used, sample size, data type and inputs, outputs and other variables used in each study.

1 Introduction

The measurement of public transport performance is a crucial tool for transport operators. It generally allows them to verify whether the service is provided efficiently and effectively, to identify areas where performance improvement may be needed, to ensure that community and users are satisfied; and to support decision-making bodies, such as transport authorities and funding institutions (etc...), to make decision about where, when, and how service should be provided.

The performance measurement in public transport is commonly carried out from four different perspectives. The first perspective is based on the user's perception and satisfaction, or aspects of service, such as reliability, frequency, fares, comfort, cleanliness, etc, are noted by users through surveys satisfaction (See reviews of [1]; [2]). The second perspective consists to measure the objectives achievement expected by the community in which public transport is served. In this sense, the objectives of the community are mainly social and environmental nature such as mobility of aged and disabled people , accessibility of precarious people to employment, reduction of air pollution , reduction of the congestion ... etc. The third perspective is about measuring and comparing the public transport performance from a service provider point of view. A fourth emerging perspective is one that tries to combine the aforementioned perspectives into a single analytical framework (See [3], [4], [5]).

From the third perspective, substantial efforts have been made to develop various methods and models to evaluate and compare public transport performance from system-level, operator-level and route-level's. At the system level, index measures are normally employed to produce a single value to reflect the combined and weighted result covering various kinds of transit activities [6]. At the operator level, there is a large and extensive literature on measuring and comparing the performance of urban public transport firms using, namely, frontier methods (See literature reviews of [7], [8], [9]).

In their comprehensive literature review about efficiency and effectiveness of public transport, [8] have found that 76.7%, 20.9% and 1.6% of studies treat performance measurement from Operator-level, System-level and route-level respectively. The measurement of public transport performance on route level is an emerging research field where researchers consider transit routes as production lines. Therefore, the aim of this paper is to review empirical studies that focus on measuring urban public transport (UPT) performance on route level using frontier methods.

The remainder of this paper is organized as follows. Firstly, a brief overview about performance concepts and measurement methodologies is presented. Secondly, the literature review procedure used is described. Thirdly, the results of literature review are presented and finally some concluding remarks are outlined.

2 Concepts and methodologies

The public transport literature generally distinguishes two performance dimensions, namely efficiency and effectiveness [10]. Efficiency is the relationship between the inputs (resources) and the outputs (production) of what is called "productive" or "technical" efficiency in the economic literature. On the other hand, effectiveness refers to the use of products to achieve goals or the consumption of services [11]. Due to the non-storable

[1] Corresponding author: zehmed.karim@gmail.com

characteristics of public transport services, these two measures should be considered separately in the evaluation of public transport systems [10-11].

To accommodate non-storable characteristics, [12] have proposed three performance measures that reflect efficiency and effectiveness dimensions. The Cost-efficiency is defined as the ratio of service inputs (Labor, Capital, and Fuel) to service outputs (Bus-Hours, Bus-km, Seat-km). The Service-effectiveness is defined as the ratio of service consumption (passengers, passenger-km, operating revenue...) to service outputs. The cost-effectiveness is defined as the ratio of service consumption to service inputs. However, if input factor prices are not available, it would be more appropriate to use the terms of production efficiency, service effectiveness, and operational effectiveness instead of cost-efficiency, service effectiveness and cost-effectiveness respectively [13].

In term of measurement methodologies, the most used methods in the literature over the past decades are frontier methods. These methods are classified into two groups: Parametric frontiers and Non-Parametric frontiers. The Stochastic Frontier Analysis (SFA) and Data envelopment analysis (DEA) are two most important methods used respectively in this regard. Due to lack of space for our article, we will present these two methods without mathematical details. For more details, the reader may refer to [14], [15], [16], [17] and [18].

The Stochastic Frontier Analysis (SFA), which is proposed independently by [19] and [20], is a parametric approach and is based on regression. SFA requires the specification of functional forms (production, cost or profit functions) for the technology. This approach takes into account, in addition to technical inefficiency, another random term that encompasses any errors in the observation or output measurement.

The most used version of Non parametric approach is the DEA (Data Envelopment Analysis) method which does not require an assumption of a functional form and it can handle multiple inputs and multiple outputs. DEA is firstly developed by [21] and used to evaluate the relative efficiency of organizational units that transform resources (inputs) into services (outputs). These units are called Decisions Making Unites (DMUs). The DEA technique involves the use of linear programming methods to construct a non-parametric piecewise surface (or frontier) over the data. The efficiency measures are then calculated relative to this surface [17]. DMUs located at the frontier have a score of 1 (or 100%) while those below the frontier score have less than 1 (or 100%) and therefore have a scope for improvement of their performance.

Note that no DMU can be greater than the efficiency frontier because it is not possible to obtain a score greater than 1 (100%). DMUs at the frontier serve as peers (or benchmarks) for inefficient DMUs. These peers are associated with observable best practices.

The two basic models of the DEA method most used in the literature are CCR and BCC. The CCR model developed by [21] assumed constant returns to scale, (a model also named Constant Returns to Scale-CRS), whereas the BCC model, developed by [22], assumed variables returns to scale (model also named variable returns to scale -VRS-)

Finally, a DEA model (CCR or BCC) can be oriented towards inputs or outputs. In an inputs-oriented approach, the DEA model minimizes inputs for a given level of outputs. In an outputs-oriented approach, the DEA model maximizes outputs for a given level of inputs.

3 Literature review procedure

The aim of our paper is to present a systematic literature review about empirical studies that focuses on measuring urban public transport (UPT) performance on route level using frontier methods (namely DEA and SFA methods). This literature review is devoted solely to papers published in specialized scientific journals and conference proceedings (from electronic sources).

To find these papers, a search was carried out on the databases like, ScienceDirect, SpringLink, Scopus, JSTORE ... etc, using expressions like " bus route performance", "efficiency of bus routes", "transit route performance"....

After executing the procedure, we found 29 papers, 6 of them were excluded because of their methodologies which does not full within frontier methods and one which is a project report.

In the end, 22 empirical papers were included for full text review. They are classified according to their category, year of publication and country context, performance dimensions discussed, methodological approach used, sample size, data type and inputs, outputs and other variables used in each study.

4 Results

4.1 General considerations

Among 22 examined papers, 18 are published in scientific journals and only four are published in conference proceedings. Moreover, the majority of examined studies are published after 2010 (figure 1) which means that the measurement of performance of urban public transport on route level is an emerging field in transit performance searches.

According to the examined papers, we observe that the majority of studies come from developed countries (figure 2). These may be explained by the fact that the intelligent public transportation has reached a higher level of maturity and so route-level data collection has been made easy with GPS and AVL (automatic vehicle location) technologies.

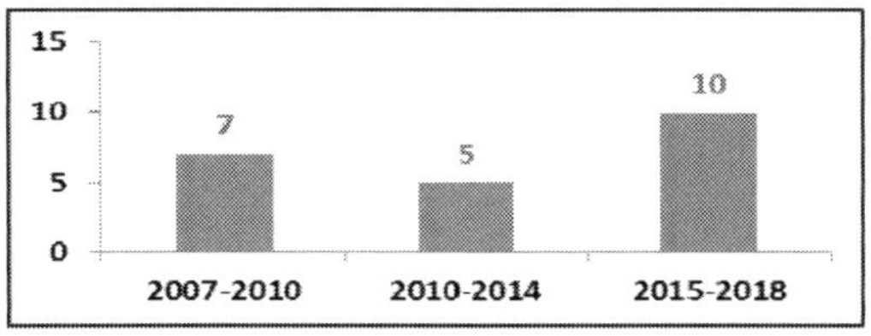

Fig.1. Number (Nb) of papers per year

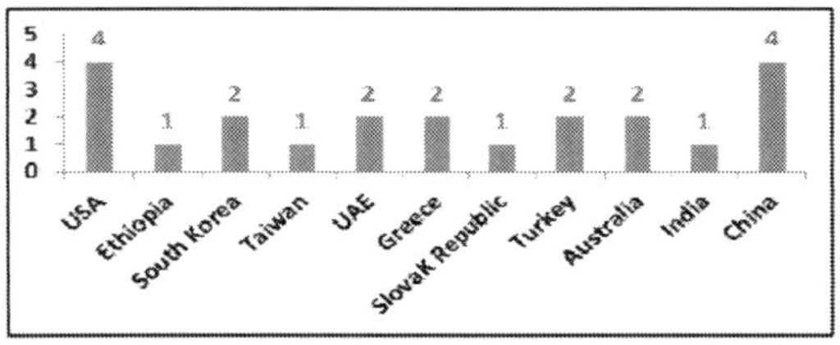

Fig.2. Nb of papers per country

4.2 Performance dimensions:

In addition to the traditional dimensions of performance that relates to efficiency and effectiveness, namely technical or operational efficiency, cost efficiency, operational effectiveness, technical efficiency or service effectiveness, some authors have explored other new dimensions of performance, namely quality efficiency, planning efficiency and financial efficiency. Furthermore, these dimensions have been measured separately or jointly. The most studied performance dimension in the examined studies is efficiency (figure 3).

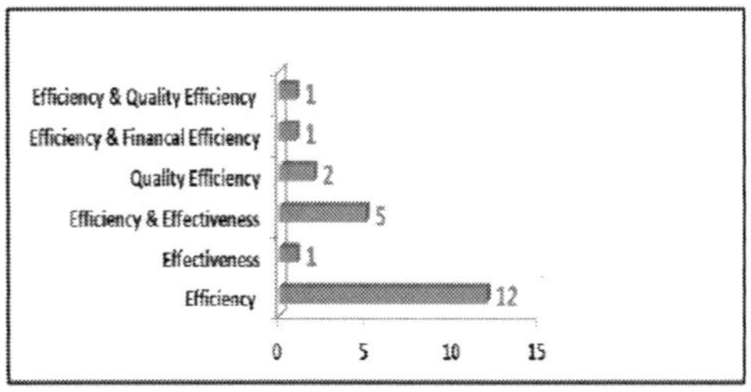

Fig.3. Studied performance dimensions

4.3 Methodological approach adopted:

From the figure 4, we notice the dominance of studies based on the non-parametric DEA method. The basic or the modified conventional DEA Models are most used. In addition, we notice the emerging trend of using DEA method with other techniques (GIS, Panel data analysis, multi-objective spatial optimization techniques, AHP…).

In contrast, only one study has adopted the parametric SFA method. Finally, we observe that some studies have adopted a mixed approach (use of DEA and SFA methods).

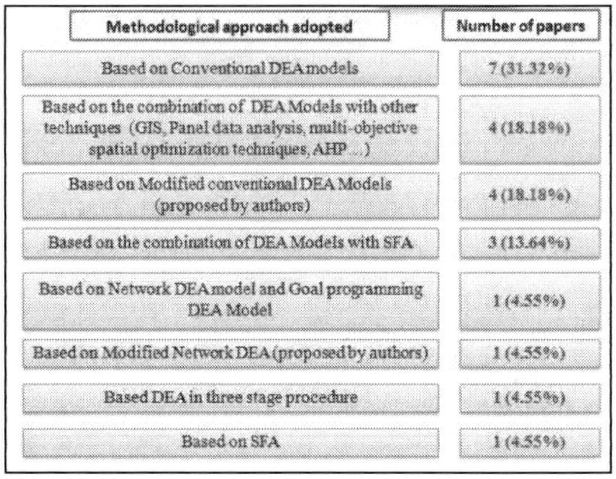

Fig.4. Methodological approach adopted

4.4 Sample size and data type:

Concerning the sample size (figure 5), the majority of studies are based on a sample size that ranges from 10 to 100 bus route, a few studies are based on a sample size less of 10 bus route or sample size more than 10 bus routes.

When it comes to data type (figure 6), the majority of studies are based on yearly or weekday data. A very few studies are based on monthly data and the remained studies did not specify the type or their data.

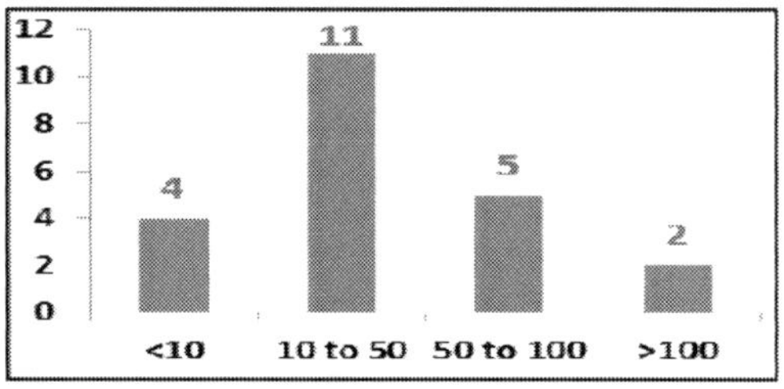

Fig.5. Sample size

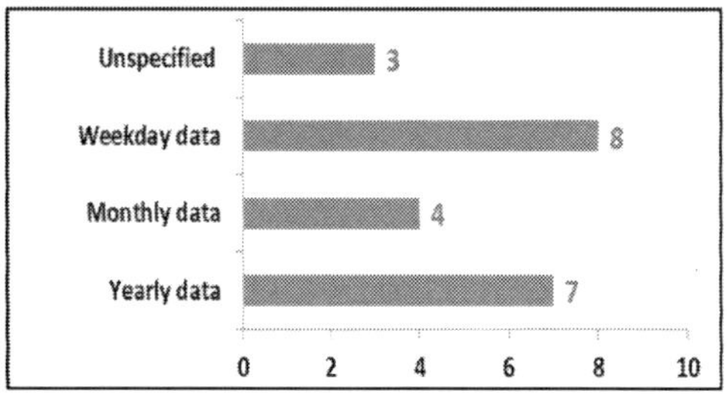

Fig.6. Data Type

4.3 Inputs, outputs and other variables

The three most used inputs are related to geographic characteristics of route namely route length and number of stops per route, and those related to service supply on route namely service duration or operation time and

Table 1. Most used inputs/outputs and other variables

Most used Inputs	Nb of papers	%
Geographic Characteristics of the route		
Intersections/transfers between routes	6	27.27
Route lengths	7	31.82
Nb of stops per route	9	40.91
Quality of the service on the route		
Service Reliability	6	27.27
Waiting time	4	18.18
Travel Time	5	22.73
Physical measures		
Nb of Buses/vehicles	10	45.45
Operating expenses		
Operating costs	6	27.27
Fuel cost	3	13.64
Service supply		
Virtual input	1	4.55

Most used Outputs	Nb of papers	%
Service supply		
Vehicle-km (miles)	5	22.73
Service consumption		
Passengers-km (miles)/Unlinked Passengers trip	5	22.73
Nb of passengers/Ridership	10	45.45
Quality of the service on the route		
Service Reliability	8	36.36
Vehicle Speed	6	27.27
Revenue		
Operating revenues/income	3	13.64
Virtual output	1	4.55
Variables that are considered as neither input nor output		
Environmental factors (population density, Parking, Nb of vehicles, Intersections, lanes, traffic condition, ...)	6	27.27
Externalities (Accidents, Emissions, pollution...)	2	9.09

finally inputs related to physical capital namely the number of buses or vehicles per route.

The most used output are those related to service consumption namely number of passengers/ridership and passengers-km (miles) and those related to quality of service namely indicators related to service reliability (on-time performance, running adherence, regularity adherence) and finally vehicle speed.

Finally, a very few authors have used some variables that are considered as neither inputs nor outputs. Some of them have used some variables that reflect environmental factors of routes as population density, parking, number of vehicles, intersections, lanes, traffic condition...and others have used variables that reflect externalities on the route as accidents, emission, and pollution.

5 Conclusions:

In this paper, we have carried out a systematic literature review of empirical studies that focus on measuring urban public transport performance on route level using frontier methods. 22 papers were identified and included for full review and were classified according to paper category, year of publication and country context, discussed performance dimensions, methodological approach used, sample size and data type and inputs, outputs and other variables used in the study. A summary of examined studies, following the aforementioned classification criteria, is given in appendix 1 (At the end of this paper).

Finally, it is important to outline the following remarks:

-An important set of studies have explored a new performance dimensions in urban public transport (quality efficiency, financial efficiency, spatial effectiveness....) rather than the traditional ones discussed on operator-level or system-level performance measurement studies (technical efficiency, cost efficiency, scale efficiency...).

- A varied use of inputs and outputs reflecting new aspects (geographical characteristics of the route, quality and characteristics of service, virtual input/output...) instead of traditional inputs/outputs ((Labour, capital, fuel)/ (passenger-km, vehicle-km) used at system-level or company-level performance measurement studies.

- An important set of studies have used daily and monthly data instead of annual data mostly used when evaluating the operators and systems performance.

- The majority of studies are based on DEA than SFA approach. Only one study has adopted SFA and limited studies that adopted a comparative perspective, or mixed for two DEA and SFA methods.

- The absence of other urban transport modes (Rail, metro, tram....) (All identified papers focuses on Bus transport mode).

- Very few studies that have incorporated variables reflecting user and community perspective's (3 among 22).

- The majority of studies come from developed countries (17 among 22).

- Few studies that have taken into consideration the impact of the external environment in performance measurement (6 among 22).

References

1. J. Oña, R. Oña, Quality of service in public transport based on customer satisfaction surveys: A review and assessment of methodological approaches, Transportation Science, **49 (3)**, pp. 605-622. (2014).
2. T.k. Ojo, Quality of public transport service: An integrative review and research agenda, Transport Letters,pp.1-14, (2017).
3. C. Sheth, K. Triantis, D. Teodorovic, Performance evaluation of bus routes: A provider and passenger perspective, Transportation Research Part E: Logistics and Transportation Review, 43 (4), pp. 453–478, (2007).
4. D.A. Abreha, Analysing Public Transport Performance Using Efficiency Measures and Spatial Analysis: the case of Adis Ababa-Ethiopia, Ms Thesis. International Institute for Geo-information Science and Earth Observation, Enschede, (2007).
5. C. Zhang, Z. Juan, Q. Luo, G. Xiao, Performance evaluation of public transit systems using a combined evaluation method, Transport Policy, **45,** pp 156-167, (2016).
6. L. Xin, Multi-Dimensional Assessment of Transit System Efficiency and Incentive-based Subsidy Allocation, Theses and Dissertations. Paper 1064, (2015).
7. B. Borger, K. Kerstens, A. Costa, Public transit performance: What does one learn from frontier studies?, Transport Reviews, **22(1)**, pp.1– 38, (2002).
8. S. Jarboui, P. Forget, Y. Boujelbene, Public road transport efficiency: a literature review via the classification scheme, Public Transport, **4(2)**, pp.101-128. (2012).
9. C. Daraio, M. Diana, F.D. Costa, C. Leporelli ,G. Matteucci, A. Nastasi, Efficiency and effectiveness in the urban public transport sector: A critical review with directions for future research, European Journal of Operational Research, **248 (1)**, pp.1–20. (2016).
10. W. K. Talley. P.P. Anderson, Effectiveness and efficiency in transit performance: A theoretical perspective, Transportation Research Part A: General, **15 (6)**, pp. 431–436. (1981).
11. X. Chu, G.J. Fielding, B.W. Lamar, Measuring transit performance using data envelopment analysis, Transportation Research Part A: Policy and Practice, **26 (3)**, pp. 223–230, (1992).
12. G.J. Fielding, T.T. Babitsky, M.E. Brenner, Performance evaluation for bus transit, Transportation Research Part A: General, **19 (1)**, pp. 73–82. (1985(a)).
13. M.M. Yu, L.H. Chen, *Assessment of Transportation Performance: A Network Structure. In: Zhu J. (eds) Data Envelopment Analysis. International Series in Operations Research & Management Science*, **Vol 238**. Springer, Boston, MA. (2016).
14. P.W. Bauer, Recent developments in the econometric estimation of frontiers, Journal of Econometrics, **46 (1/2)**, pp. 39-56, (1990).
15. M.E. Chaffai, Estimation de frontières d'efficience: un survol des développements récents de la littérature, Revue d'économie du développement, **5 (3)**, pp. 33-67, (1997).
16. L.R. Murillo-Zamorano, Economic efficiency and frontier techniques, Journal of Economic Surveys, 18 (1), pp. 33–77, (2004).
17. T.J. Coelli, D.S.P. Rao, C.J. O'Donnell, G.E. Battese, *A introduction to efficiency and productivity analysis*, Springer-Verlag New York Inc, **2nd ed.** (2005).
18. J. Zhu, *Quantitative models for performance evaluation and benchmarking: Data envelopment analysis with Spreadsheet, International Series in Operations Research & Management Science*, **213**. Springer International Publishing, Edition N° 3. (2014).
19. D.J. Aigner, C.A.K. Lovell, P.J. Schmidt, Formulation and estimation of stochastic frontier production function models, Journal of Econometrics, **6(1)**, pp.21-37, (1977).
20. W. Meeusen, J. Van den Broeck, Efficiency Estimation from Cobb-Douglas Production Functions with Composed Error, International economic review, 18, pp. 435- 444. (1977).
21. A. Charnes, W.W Cooper, F. Rhodes, Measuring the efficiency of decision making units, European Journal of Operational Research, **2(6)**, pp.29-44, (1978).
22. R.D. Banker, A. Charnes, W.W. Cooper, Some Models for Estimating Technical and Scale Inefficiencies in Data Envelopment Analysis, Management Science, **30 (9)**, pp. 1078–1092, (1984).
23. D.T. Barnum, S. Tandon, S .McNeil, comparing the Performance of Bus Routes after Adjusting for the Environment Using Data Envelopment Analysis, Journal of Transportation Engineering, **134(2)**, pp. 77–85, (2008).
24. J. Lin, P. Wang, D.T. Barnum, A quality control framework for bus schedule reliability, Transportation Research Part E: Logistics and Transportation Review, **44 (6)**, pp. 1086-108, (2008).
25. J. Hahn, H. Kim, S. Kho, A DEA Approach for Evaluating the Efficiency of Exclusive Bus Routes, Proceedings of the Eastern Asia Society for Transportation Studies,**7**, pp.227, (2009).
26. Y. Lao, L. Liu, Performance evaluation of bus lines with data envelopment analysis and geographic information systems, Computers, Environment and Urban Systems, **33(4)**, pp. 247–255, (2009).
27. G.P. Michaelides, A. Belegri-Roboli, T. Marinos, Evaluating the Technical Efficiency of Trolley Buses in Athens, Greece, Journal of Public Transportation, **13(4)**, pp. 93-109, (2010).
28. M. Gebeyehu, S. Takano, evaluation of bus routes performance in the city of addis ababa using stochastic frontier model, Infrastructure Planning Review, **24 (3)**, pp.447-457, (2007).

29. J. Hahn, H. Kim, S. Kho, Analysis of the efficiency of Seoul Arterial Bus routes and its determinant factors, KSCE Journal of Civil Engineering, **15(6)**, pp. 1115-1123, (2011).
30. Y. Chiou, L.W. Lan, T.H.Y. Barbara, Route-based data envelopment analysis models, Transportation Research Part E: Logistics and Transportation Review, **48(2)**, pp. 415-425, (2012).
31. H. Yaser, B. Nandita, Evaluating and Enhancing the Operational Performance of Public Bus Systems Using GIS-based Data Envelopment Analysis, Journal of Public Transportation, **15(2)**, pp.19-44, (2012).
32. J. Li, X. Chen, X. Li, X. Guo, Evaluation of Public Transportation Operation based on Data Envelopment Analysis, Procedia-Social and Behavioral Sciences, 96, pp. 148-155, (2013).
33. G. Georgios, P. Ioannis, P. Panagiotis, Measuring and improving the efficiency and effectiveness of bus public transport systems, Research in Transportation Economics, 48, pp.84-91, (2014).
34. V. Roháčová, A DEA based approach for optimization of urban public transport system Central European Journal of Operations Research, **23 (1)**, pp.215–233, (2015).
35. S. Güner, E. Coşkun, Determining the best performing benchmarks for transit routes with a multi-objective model: the implementation and a critique of the two-model approach, Public Transport, **8(2)**, pp 205–224, (2016).
36. K. Tran, A. Bhaskar, J. Bunker, B. Lee, Data envelopment analysis (DEA) based transit route temporal performance assessment: a pilot study proceedings of 38th Australasian Transport Research Forum, 16-18 November, (2016).
37. A. Bandyopadhayay, A. Banerjee, N. Abidi, Measuring Routes Efficiency of Kolkata Bus Transport: A Modified DEA Approach, Annals of Data Science, **3(3)**, (2016).
38. D.J Sun, S. Chen, C.Zhang, S. Shen., A bus route evaluation model based on GIS and super-efficient data envelopment analysis. Transportation Planning and Technology, **39 (4)**, pp.407-423, (2016).
39. W.Zhu, X. Yang, J. Preston, Efficiency measurement of bus routes and exogenous operating environment effects on efficiency, Transportation Planning and Technology, **39 (5)**, pp. 464-483, (2016).
40. K. Tran, A. Bhaskar, J. Bunker, B. Lee, Data Envelopment Analysis (DEA) based transit routes performance evaluation, In Transportation Research Board 96th Annual Meeting (TRB), Washington D.C, 8-12 January, (2017).
41. K.Zhang, Y. Xu, J. Sun, A mixed frontier model for urban bus performance evaluation, Proceedings of the Institution of Civil Engineers – Transport, **171 (2)**, pp. 65-74, (2018).
42. R. Wei, X. Liu, Y. Mu, L. Wang, A. Golub, S. Farber, Evaluating public transit services for operational efficiency and access equity, Journal of Transport Geography, **65**, pp. 70-79, (2017).
43. G. Samet, E. Coskun, Estimating the operational and service efficiency of bus transit routes using a non-radial DEA approach, EURO Journal on Transportation and Logistics, pp. 1-20, (2018).

Appendix 1: Summary of reviewed studies

References	Category	Performance dimensions	Methodological approach	Sample size	Geographic context	Period time
[3]	Journal paper	Efficiency of services	Based on Network DEA model and Goal programming DEA	60 bus routes	USA	Unspecified period
[23]	Journal paper	Efficiency	Based on Conventional DEA models	46 bus routes	USA	Spring 2005 weekday trips.
[24]	Journal paper	Quality efficiency	Based on the Combination of Conventional DEA Models with other techniques	48 routes directions	Chicago and forty surrounding suburbs, USA	weekday morning peak hours (6:30 to 9) 29 weeks (between January and June 2006)
[25]	Conference paper	Technical (operational) efficiency	Based on Conventional DEA models	18 bus lines	Seoul, South Korea	first half of 2008
[26]	Journal paper	Operational efficiency and spatial effectiveness	Based on the Combination of Conventional DEA Models with other techniques	24 fixed bus routes	Monterey County-California, USA	2004
[27]	Journal paper	Technical efficiency	Based on the combination of DEA models with SFA	20 Trolley bus lines	Athens, Greece	2003
[28]	Journal paper	Cost efficiency	Based on SFA	93 Bus routes	Addis Ababa	2004
[29]	Journal paper	Efficiency	Based on the Modified Network DEA	113 Bus Routes separated in 7 district	Seoul, South Korea	2009
[30]	Journal paper	Efficiency	Based on DEA in three stage procedure	1035 Bus routes currently operated by 37 intercity bus companies	Inter-city routes in Taiwan	2005
[31]	Journal paper	Efficiency & Effectiveness	Based on Conventional DEA models	8 Bus routes	Al Ain, UEA	Three different peak periods (7–9 AM, 12–2 PM, and 5 to 9 PM) daily, over one-week duration.
[32]	Journal paper	Quality efficiency	Based on Conventional DEA models	3 Bus routes	Beijing, China	8 march to 22 march 2012 (each Tuesday, peak and off peak period)
[33]	Journal paper	Efficiency & operational Effectiveness & combined effectiveness	Based on Conventional DEA models	50 fixed bus lines	Thessaloniki, Greece	2009-2011
[34]	Journal paper	Efficiency	Based on Modified conventional DEA Models	28 lines (7 trolleybus+3 minibus+18 bus)	Banská Bystrica (Slovak Republic)	41st week of 2008
[35]	Journal paper	Operational efficiency and	Based on Conventional DEA models	30 Bus Routes	Sakarya urban area,	2012

		service effectiveness			Turkey	
[36]	Conference paper	Service effectiveness	Based on Conventional DEA models	1 Bus Route	Brisbane, Australia.	19th August 2013
[37]	Journal paper	Operational and financial efficiency	Based on Modified conventional DEA Models	18 Bus routes	Kolkata, India	Unspecified period
[38]	Journal paper	Planning efficiency & operation efficiency & service efficiency	Based on the Combination of Conventional DEA Models with other techniques	18 Bus lines	Shenzhen, China	Unspecified period
[39]	Journal paper	Operating efficiency & quality efficiency	Based on the combination of DEA Models with SFA	39 Bus routes (14 inner city routes, 15 radial routes, 10 village routes)	Jiangyin City, China	April 2011
[40]	Conference paper	Technical efficiency & Service effectiveness	Based on Conventional DEA models	42 Bus routes	Brisbane, Australia.	21th August 2013 (morning and afternoon peak hours and off-peak hours)
[41]	Conference paper	Technical efficiency	Based on the combination of DEA Models with SFA	1 Bus route	Shanghai, China	8 July to 12 July 2013 and from 05:00 to 24:00 h.
[42]	Journal paper	Operational efficiency & Access equity	Based on the Combination of Conventional DEA Models with other techniques	92 fixed Bus routes	Wasatch Front, Utah, USA	August 2016
[43]	Journal paper	Operational and service efficiency	Based on Modified conventional DEA Models	30 Bus Routes	Sakarya City, Turkey	2012

References	**Inputs**	**Outputs**
[3]	*The provider node of the network :* Headway +Service duration (hours)+Costs (dollars)+Number of intersections +Number of Priority Lanes/*The passenger node of the network :* Vehicle-miles+ Schedule Reliability (ratio)+Average Traveling Time (hours)/	*The outputs for the provider node :* Vehicle-miles+ Schedule Reliability (ratio)+Average Traveling Time (hours)/*The output of the passenger node :* Passenger-miles
[23]	Seat kilometers + seat hours	Ridership+ Span of service+ Average frequency+ Maximum frequency+ On-time performance.
[24]	% longer running time +% shorter running time+ % longer headway+% shorter headway	Virtual output =1
[25]	Number of buses, distance of routes, Number of stops, length of bus intervals, and operating costs	Number of passengers+ Profit
[26]	*Operational efficiency inputs :* (Operation time + round-trip distance+ The number of bus stops) - *Spatial effectiveness inputs :* (Population 65 years and over+ Workers 16 years and over+ means of transportation to work+ Public transportation+ bus or trolley bus+ Population with disabilities)	Annual ridership (number of passengers)
[27]	Total labor expanded+ Total available vehicles+ Total energy expanded (electricity) by the fleet of the vehicles of each lines	Total vehicles-kilometers
[28]	Passenger per km+ Number of trips+ subsidy+ fare+ route length + N bus stops per km+ employee pop at route origin+ pop density at route origin+ waiting time+ N of buses+ trip length in minute+ linkage (n of transfers); bus capacity	Revenue per km
[29]	Possession costs (million won)+ Operating costs (million won)+ Headway (second)+ Number of bus stops+ Ratio of median bus stops (%)+ Overlapped route lengths (km)	*Desirable outputs :* Total riders (ten thousand)+service satisfaction score (point)/ *Undesirable outputs :* Emissions (ten million g)/ *Middle Outputs* : Vehicle-km (veh-km)+Travel

		Speed (km/h)
[30]	Fuel cost+ Labor+ Bus	Operating revenue+ Passenger-km
[31]	Average travel time per round trip (hr)+ total nb of vehicles; total nb of operators +Total nb of stops (round trip)	Total avg. Nb of passengers per day (effectiveness measure) + Vehicle-km (per day) (efficiency measure)
[32]	Virtual input =1	Passenger load rate+ Service reliability + Average dwell time (s)+ Average running speed (km/h)
[33]	*Efficiency + Operational effectiveness:* Span of service + line length+ Vehicles/*Combined effectiveness :* Revenue vehicle-kms +Vehicles	*Efficiency* : Revenue seat-kms /*Operational effectiveness and combined effectiveness* : Passengers
[34]	Distance+ capacity+ costs	Passengers+ sales
[35]	*Operational efficiency :* the number of busses (in numbers)+ the fuel consumption (in Turkish Lira), route distance (in km) / *Service effectiveness* : frequency+ service minutes+ stops per km+ deviation from shortest distance and travel time	unlinked passenger trip
[36]	*CCR DEA Model 1* : Number of service/*CCR DEA Model 2* : Nb of service + Total travel time/ *CCR DEA Model 3* : Nb of service + total travel time	*CCR DEA Model 1 and 2* : Transit work/*CCR DEA Model 3* : Transit work+ On time performance
[37]	Labor cost per bus per day + Fuel cost per bus per day+ Other operating expenses (related to servicing, maintenance, and stationery) per bus per day+ Total number of buses operating in the route	Average passenger per bus per day+ Profit per bus per day
[38]	*Planning efficiency:* dispatching frequency + total number of stops + route demand detour coefficient + the minimal average transfer time/ *Operation efficiency :* dispatching frequency + the total number of stops + bus arrival stability rate /*Service efficiency :* dispatching frequency + the total number of stops +peak hour maximum standing rate + peak hour delay rate	*Planning efficiency :* transit line 500-m travel demand coverage+500-m transfer rate/ *Operation efficiency :* average route operational speed + average on-road operational speed + arriving stop on time rate/*Service efficiency :* number of passengers by the line studied
[39]	Labor (total number of full-time drivers)+fuel (total amount liters of diesel consumed)+vehicle (total number of operating buses)	*Quantity outputs :* vehicle-km+ operating revenues /*Quality outputs :* average operating speed+ average punctuality rate
[40]	*Technical efficiency (service availability) :* Route length+ Service duration+ Number of services+ Busway length /*Technical efficiency (quality of service) :* Signalized intersection spacing+ Stop spacing+ Priority lane /*Service effectiveness :* Space-km+ Average vehicle speed	*Technical efficiency (service availability) :* Space-km/ *Technical efficiency (quality of service)* : Average vehicle speed /*Service effectiveness :* Transit work+ On time performance
[41]	Corrected fares+ Waiting time cost + In-vehicle time cost+ Reliability penalty cost	Total income+ Passenger-kilometers
[42]	total operation time per day+ number of operating buses per day+ total mileage per day	the average number of passengers per day
[43]	*Operational efficiency :* capacity and number of stops /*Service efficiency* : frequency+ stops per km+ service hours	average unlinked passenger trips per day for both models

References	**Other variables**
[3]	*Environmental factors:* parking space availability factor+ connectivity factor+ accessibility factor+ population density factor + comfort standard factor/*Undesirables variables (Externalities)* : Number of accidents + Emissions (parts per million-vehicles second)+ Noise Pollution (decibels per min per vehicle)+Resources degraded (dollars)
[25]	Interval between buses
[27]	Two dummy variables: influence of the Athens Metro+ the distance of the areas that each line serves.
[28]	Origin-destination of the routes+ Number of buses allocated per route+ Route length (km)+ Waiting time/headway (minutes) of each route.
[29]	Population density (ten thousand) + the number of registration vehicle (ten thousand).
[33]	Traffic conditions+ population density

The influence of organizational culture on performance: Case study of the Moroccan ATM system

Âta Ghalem[1,*], *Chafik* Okar[2], *Razane* Chroqui[2] and *El Alami* SEMMA[1]

[1]Laboratory IMII, Faculty of Sciences and Techniques, University Hassan 1st, Settat, Morocco
[2]LAMSAD, Higher School of Technology, Berrechid, Morocco

Abstract. According to Ferreira and Otley (2009) contextual factors and organizational culture are considered as variable factors that have a significant impact on the performance management system framework, which comprises 12 questions/issues. The intention of this work is to define how organizational culture influences the question/issue of performance information use, through the example of Air Traffic Management (ATM). In fact, based on Schein's model we determine how two of the levels of organizational culture influence the performance information use within the Moroccan ATM system.

1 Introduction

Performance is a concept newly introduced to the aviation sector. Yet, it quickly became a requirement for the Air Navigation Service Providers (ANSP) organizations in order to improve and prosper in the future. However, most of the focus has been oriented towards the performance measurement, while overlooking the performance management. Many frameworks have been developed for the performance measurement systems, such as the balanced scorecard, the performance measurement matrix, the SMART pyramid and the performance prism. While, fewer frameworks were developed to study the performance management systems (PMS), which were mostly based on the management accounting and management control systems, such as the extended PMS framework of [1]. Some studies, which focused on the performance management systems, were interested in the elements that impact this system. Within the literature, some scholars admit the existence of a relationship between performance and organizational culture ([2]; [3]; [4]; [5]). In fact, [1] consider that culture is one of other factors that have an impact on control systems design and use, and considering that the extended framework is seen by the authors to be including all aspects of organizational control, culture accordingly has an impact on the PMS framework. Reference [1] argues that the consideration of the influencing factors on the PMS, of which culture is part, even if only implicitly, is a requirement in the study of the operations of the PMS.

This paper is organized through three sections. The first section discusses the PMS framework based on the work of [1]. The second section presents the existing Air Traffic Management (ATM) performance frameworks, specifically the one proposed by the International Civil Aviation Organization (ICAO). The third section studies the influence of corporate culture on the performance of the Moroccan ATM system.

2 Performance Management System framework

Reference [6] proposed a framework for studying the operations of Management Control Systems (MCS) and completing the traditional framework suggested by [7]. Reference [6]'s framework focused on the management of organizational performance, and was structured around 5 issues that are believed to help developing a coherent structure for PMS and are represented as a set of questions:

1 Objectives: What are the key objectives that are central to the organization's overall future success, and how does it go about evaluating its achievement for each of these objectives?
2 Strategies and plans: What strategies and plans has the organization adopted and what are the processes and activities that it has decided will be required for it to successfully implement these? How does it assess and measure the performance of these activities?
3 Targets settings: What level of performance does the organization need to achieve in each of the areas defined in the above two questions) and how does it go about setting appropriate performance targets for them?

* Corresponding author: ata.ghalem@gmail.com

4 Incentive rewards: What rewards will managers (and other employees) gain by achieving these performance targets (or, conversely, what penalties will they suffer by failing to achieve them)?
5 Information feedback loops: What are the information flows (feedback and feed-forward loops) that are necessary to enable the organization to learn from its experience) and to adapt its current behaviour in the light of that experience?

Later in 2009, [1] proposed an extension of these five questions as an extended framework named performance management systems framework. The authors specify that the naming of the framework, as 'performance management system', aims to reflect a shift from the traditional compartmentalized approaches to control in organizations to a broader perspective of the role of control in the managing of organizational performance.

The extended framework of the twelve questions proposed by [1] is represented as followed:

Q1. Vision and Mission: What is the vision and mission of the organization and how is this brought to the attention of managers and employees? What mechanisms, processes, and networks are used to convey the organization's overarching purposes and objectives to its members?
Q2. Key Success Factors: What are the key factors that are believed to be central to the organization's overall future success and how are they brought to the attention of managers and employees?
Q3. Organizational Structure: What is the organization structure and what impact does it have on the design and use of performance management systems (PMSs)? How does it influence and how is it influenced by the strategic management process?
Q4. Strategies and Plans: What strategies and plans has the organization adopted and what are the processes and activities that it has decided will be required for it to ensure its success? How are strategies and plans adapted, generated and communicated to managers and employees?
Q5. Key Performance Measures: What are the organization's key performance measures deriving from its objectives, key success factors, and strategies and plans? How these are specified and communicated and what role do they play in performance evaluation? Are there significant omissions?
Q6. Targets Setting: What level of performance does the organization need to achieve for each of its key performance measures (identified in the above question), how does it go about setting appropriate performance targets for them, and how challenging are those performance targets?
Q7. Performance Evaluation: What processes, if any, does the organization follow for evaluating individual, group, and organizational performance? Are performance evaluations primarily objective, subjective or mixed and how important are formal and informal information and controls in these processes?
Q8. Reward System: What rewards (financial and/or non-financial) will managers and other employees gain by achieving performance targets or other assessed aspects of performance (or, conversely, what penalties will they suffer by failing to achieve them)?
Q9. Information Flows, Systems and Networks: What specific information flows (feedback and feed-forward), systems and networks has the organization in place to support the operation of its PMSs?
Q10. PMSs Use: What type of use is made of information and of the various control mechanisms in place? Can these uses be characterized in terms of various typologies in the literature? How do controls and their uses differ at different hierarchical levels?
Q11. PMSs Change: How have the PMSs altered in the light of the change dynamics of the organization and its environment? Have the changes in PMSs design or use been made in a proactive or reactive manner?
Q12. Strength and Coherence: How strong and coherent are the links between the components of PMSs and the ways in which they are used (as denoted by the above 11 questions)?

In fact, according to [1], the PMS is the wider context of MCS, as it is used to capture a holistic approach to the management and control of organizational performance. Thus, [1] define the PMS as the evolving formal and informal mechanisms, processes, systems, and networks used by organizations for conveying the key objectives and goals elicited by management, for assisting the strategic process and on-going management through analysis, planning, measurement, control, rewarding, and broadly managing performance, and for supporting and facilitating organizational learning and change.

Reference [1] points out that contextual factors and organizational culture are two aspects that permeate the PMSs and are not explicitly addressed by the twelve questions (see Fig 1 below). They add that contextual factors are shown by the literature to be influencing the control systems design and use, so their consideration is required even if they didn't include them in the framework. Those factors include: external environment, strategy, culture, organizational structure, size, technology, and ownership structure. [1] believe that strategy and organizational structure are already explicitly built into their framework and the remaining factors are considered as external influences, separating 'Culture' as a notable contextual variable as shown in Fig 1. Thus, the other contextual factors include: external environment, size, technology and ownership structure.

The PMS framework was discussed in the study of [8] and [9], who developed a conceptual model of performance management systems (Alternative models of rationality) building on the work and insights of [6] and [10] and [1], to give a 'middle range' theoretical framework that can provide a language for analysing any PMS ([9]).

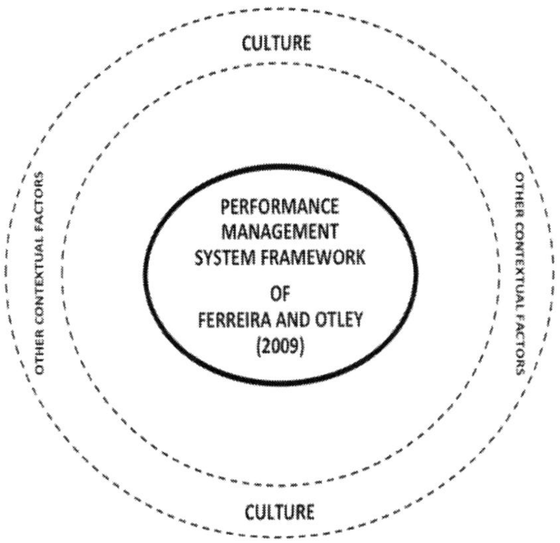

Fig 1. External influences on the PMS framework ([1])

References [8] and [9] consider that the first eight questions, which are related to the PMS design, would provide a conceptual nature of any PMS. Yet the remaining four questions, which are related to culture and context, will capture more underlying factors influencing the nature of the PMS. This creates a disagreement with [1]'s point of view, since they don't consider any longer these four questions as outlining cultural and contextual factors, as they initially did in 2005 ([10]). In fact, they put the contextual factors and culture outside the framework composed of all the twelve questions (bearing in mind that strategy and organizational structure are already explicitly built into the framework).

From both discussions above of [9] and [1], it is clear that the authors emphasis the importance of the influence of the contextual factors on the PMS. Reference [1] states that organizational culture is not explicitly discussed in their paper, they argue 'it is seen more as a contingent variable that may influence the PMS design rather than an organizational characteristic that can be manipulated... However, it is recognized that culture ... may be a highly influential feature of PMSs use' ([1]). Our mission is to study the influence of this organizational culture on the PMS through the case study of ATM; therefore, the next section discusses the performance frameworks that already exist the ATM sector.

3 ATM performance frameworks

ATM performance has known a great evolution in the last years, together with the evolution of the Air Navigation Systems. The International Civil Aviation Organization's (ICAO) performance framework has grown to be a Performance Based Global Air Navigation System with ASBU Methodology. The evolution started in the early 90's. Before 1992, the system was known as the Ground-based Air Navigation system. Then, in order to answer to the civil aviation growth, it evolved to be the Future Air Navigation Systems (FANS). Later in 1994, ICAO transitioned to the communications, navigation and surveillance/air traffic management (CNS/ATM) systems. In 2006, the system became mature enough to be considered as a Global ATM system. And then, in 2008, it was ready to be the Performance-based Global Air Navigation System. Since the civil aviation sector is always on continued and non-stopping roads of developments, the system was enhanced by the addition, in 2012, of the Aviation System Block Upgrade (ASBU) methodology.

ICAO has developed a Performance-Based Approach (PBA), which is a decision-making method for organizing the performance management process. This process is explained step-by-step in [11] in a set of six steps. The first step of the process defines the scope, the context and the general ambitions/expectations. These expectations of the ATM community [11] are categorized into 11 Key Performance Areas (KPAs): safety, security, environmental impact, cost effectiveness, capacity, flight efficiency, flexibility, predictability, access and equity, participation and collaboration, interoperability. The second step sets the objectives. The third step describes the quantification of these objectives, which should be SMART (Specific, Measurable, Achievable, Relevant and Time-bound). The forth step 4 is destined to the selection of solutions to exploit opportunities and resolve issues. The fifth step explains the implementation of solutions. The sixth step explains the assessment of achievement of objectives.

ICAO invited the member states and the regions to adopt the PBA methodology and the 11 KPAs. However, it recommended, as a starting point, to focus on a particular set of KPAs, or even a particular focus area within a KPA. Consequently, the EUR region adopted only 5 KPAs (safety, capacity, efficiency & environment, cost-effectiveness and participation by the ATM community) ([12]), while the European Union (EU) focused on only 4 KPAs (safety, environment, capacity and cost-efficiency). In fact, the EU proposed its own framework that is the Single European Sky (SES) performance scheme ([13]). Moreover, other than the SES performance scheme other performance frameworks exist in Europe, such as: the Single European Sky ATM Research (SESAR) performance framework and EUROCONTROL's performance monitoring.

Morocco is member state of ICAO, a member of the EUR Region and is engaged with the EU by the agreement of 2006 and with EUROCONTROL by the agreement of 2016, which gives him the possibility of adopting any of the frameworks mentioned above. In fact, the performance measurement has only been introduced recently to the Moroccan ATM system. Firstly, following the steps of ICAO based on [11], secondly following the steps of EUR Region based on [12] and then considering the European Union framework described in [13] along with EUROCONTROL's performance monitoring. In other words, since 2015, the Air Navigation Pole (PNA) - Which is the body of the Moroccan Airports Authority (ONDA) responsible for delivering Air Navigation Services in Morocco - adopted the same EUR Region approach and started with the subset of indicators in the five KPAs: Safety, Capacity, Efficiency and Environment, Cost-effectiveness and Participation by the ATM community. However, after the signature of the comprehensive agreement, in 2016, with EUROCONTROL, the ONDA/PNA altered its framework and adopted the one proposed by EUROCONTROL. This new agreement will help Morocco benefit from all the services and expertise offered by the agency, moreover, the framework will help the ONDA/PNA measure its performance and smooth the process of measurement. According to the interviews we had with the ONDA/PNA and based on the observations we made while being part of the organization, the management of performance is currently an on-going process that hasn't reached a maturity level yet. Today, the ONDA/PNA hasn't established, yet, a performance management system, instead it is focused on measuring its performance, based on the EUROCONTROL framework, in the four KPAs proposed by the EU, which are: Safety, Capacity, Environment and Cost-effectiveness. Note that, the ONDA/PNA is remarkably advanced in the measurement of its safety Key Performance Indicators (KPI) compared to capacity and environment KPIs, while it is still in the very first stage when it comes to the Cost-effectiveness KPA.

4 Methodology

Many studies were based on Schein's model such as [14]. According to Schein's model, there are three levels of organizational culture: artefacts, espoused values and beliefs and basic assumptions ([14]). So, in order to study the influence of culture on the PMS, which comprises twelve issues, we have to study the influence on each issue throughout the three different levels. However, the artefacts do not seem to have any type of influence on the PMS, thus this study will focus on the espoused values and some of the basic assumptions of the organization. On the other hand, due to the current state of the performance management process in the ONDA/PNA and in the absence of a complete PMS, it is difficult to study the influence of all the 12 questions of the framework of [1]. For this reason, instead of all the issues we will focus on Q10, which deals with the PMS's use since this aspect exist in the organization despite the absence of a mature PMS. In this case, the objective is to define how culture influences the use of information. Since the performance information received from the measurement system adopted by the ONDA/PNA is a very important decision-making aid.

5 Influence of culture on performance

In order to study the influence of culture on the PMS it is essential to define what does culture stand for. In the literature there are several definitions of corporate culture. Reference [4] chooses to use the definition of [15], they define culture as "a set of norms and values that are widely shared and strongly held throughout the organization". Reference [4] argues that this definition is the most suitable for their empirical study because the value component of culture is easier to measure. We believe that this definition presents, indeed, in a very precise way the direct meaning of corporate culture as shared values if the intention is to measure culture. However, in our case the intention is to study the influence of culture on the PMS. Thus, we need more meticulous definition that provides more aspects of culture. Reference [3] uses the definition of [16] who defines culture as "a pattern of basic assumptions—invented, discovered, or developed by a given group as it learns to cope with its problems of external adaptation and internal integration—that has worked well enough to be considered ... the correct way to perceive, think, and feel in relation to those problems" [16]. According to [14], Schein's model presents three levels of organizational culture, the artefacts (the visible products or physical manifestations of an organization's culture), the espoused values and beliefs (or the documented norms, ideals, goals, and aspirations, of the organizational group), level 3: the basic underlying assumptions (the unconscious, taken-for-granted, and non-negotiable beliefs and values that influence how group members think, view, and feel about things and guide their behaviour).

According to the strategic plan of 2011-2016 of the ONDA/PNA the espoused values shared in the organisation are the following:

- Respect: Act with consideration, courtesy, honesty, listening and respect towards others.
- Team spirit: Work together for a collective success in all transparency and with a particular attention to the respect of the dignity and the personal life of the collaborators.
- Transparency: Avoid any form of opacity in our daily actions.
- Loyalty: Behave at all times towards ONDA in a loyal and diligent manner that protects its interests.
- Integrity: Behave honestly and fairly and treat your employees and partners equally.
- Responsibility: Perform our duties with loyalty, professionalism, effectiveness, efficiency and objectivity.
- Excellence: Always be part of the quest for performance and enjoy exemplary management.

On the other hand, one of the main basic assumptions of the ONDA/PNA is:

- Safety: safety of air traffic comes first regardless of any other condition.

Reference [2] argues that performance information use is defined as consistent when the desired behaviour is realized in the actual behaviour. The author mentions some of the uses of performance information: the three uses of [17]: (1) to coordinate; (2) to monitor; and (3) to diagnose; and the eight purposes of [18]: (1) to evaluate,

(2) to control, (3) to budget, (4) to motivate, (5) to promote, (6) to celebrate, (7) to learn, and (8) to improve. Noting that that improvement is the main purpose of performance information.
Reference [2] states that the three levels of culture developed in Schein's model ([16]) influence the performance information. To show the impact of culture we suggest to consider the following hypothesizes. So hypothetically, in the Absence of respect, team spirit, transparency, loyalty, integrity, responsibility and excellence the information received from the measurement of performance will have one of the following possibilities: it will not exist or it will be incomplete in the lack of team spirit, it won't be correct in the lack of transparency, it won't follow its normal circuit between employees and top management in the lack of respect or it can even be late and not available when needed in the lack of responsibility. Through these hypothesises we can touch the aspects of impact on the information use. Moreover, safety of air traffic is the most basic and common assumption in all the ONDA/PNA. Every act, decision or suggestion is made based on the idea of safety. It is in itself a purpose of the whole performance measurement system adopted by the ONDA/PNA. According to EUROCONTROL Safety Culture is "the way safety is perceived, valued and prioritised in an organisation. It reflects the real commitment to safety at all levels in the organisation. Safety Culture is not something you get or buy; it is something an organisation has. Safety Culture can therefore be positive, negative or neutral. Its essence is in what people believe about the importance of safety, including what they think their peers, superiors and leaders really believe about safety's priority" [19]. If this culture is not taken into consideration when measuring performance a simple error in the indicators will give false information, which will lead to wrong decisions regarding the safety measures, consequently, the monitoring of performance, its evaluation, its improvement and our learning about it.

6 Conclusions

Through these hypothesizes we were able to visualize the important role of organizational culture in the ATM system, and the direct impact on the information use for instance, which can be easily manipulated by the different values and assumptions of the organization. However in order to generalize the results of the influence we believe that other ANSPs should be involved to test our hypothesis. The case study of Morocco isn't sufficient to study the influence of a vast concept such as culture, especially that, no two organizations use the same cultures. On the other hand, the immature performance management process of ONDA/PNA made it difficult to study the influence of culture on all the twelve issues of the PMS of [1]. With the possibility of studying the culture of different organizations we will be able to study only the mature issues of the PMS of every organization and combine the different results to obtain the final analysis of the impact of organizational culture on the twelve issues of the PMS. Moreover, this work focused on the influence of culture on performance, whilst the influence is plausible to be bidirectional or at least instead of an influencing effect other types of relationships might exist. Some researchers such as [20] argue about the existence of a relationship between both concepts, [20] state that 'here exists the problem of a dearth of research focusing exclusively on the effects of organizational culture on firm performance and more research is needed in understanding and exploring organizational culture-performance relationship'. Therefore, future studies should be oriented to discuss the nature of the connection between culture and performance.

References

1. Ferreira, Aldónio, and David Otley. "The design and use of performance management systems: An extended framework for analysis." *Management accounting research* 20.4 (2009): 263-282.
2. Taylor, Jeannette. "Organizational culture and the paradox of performance management." *Public Performance & Management Review* 38.1 (2014): 7-22.
3. Martinez, Elizabeth A., et al. "Organizational culture and performance." *American Economic Review* 105.5 (2015): 331-35.
4. Guiso, Luigi, Paola Sapienza, and Luigi Zingales. "The value of corporate culture." *Journal of Financial Economics* 117.1 (2015): 60-76.
5. Youngbantao, Unchan, and Nopadol Rompho. "The Uses of Measures in Performance Prism in Different Organizational Cultures." *Journal of Accounting and Finance* 15.6 (2015): 122.
6. Otley, David. "Performance management: a framework for management control systems research." *Management accounting research* 10.4 (1999): 363-382.
7. Anthony, R. N. "Planning and Control Systems: A Framework for Analysis, Cambridge, MA, Harvard University Graduate School of Business Administration." *Studies in Management Control* (1965).
8. Broadbent, Jane, and Richard Laughlin. "Performance management systems: a conceptual model and analysis of the development and intensification of 'new public management' in the UK." *APIRA Conference, Auckland.* 2007.
9. Broadbent, Jane, and Richard Laughlin. "Performance management systems: A conceptual model." *Management Accounting Research* 20.4 (2009): 283-295.
10. Ferreira, Aldónio, and David Otley. "The design and use of management control systems: an extended framework for analysis." *AAA Management Accounting Section 2006 Meeting Paper* (2005).
11. ICAO. "Manual on Global Performance of Air Navigation System (ICAO Doc 9883)." *1st ed.* (2009).

12. ICAO. "EUR Region Performance Framework Document (ICAO EUR Doc 030) " (2013).
13. European Union. "COMMISSION IMPLEMENTING REGULATION (EU) N° 390/2013, laying down a performance scheme for air navigation services and network functions." *Official Journal of the European Union L128* (2013): 1.
14. Hogan, Suellen J., and Leonard V. Coote. "Organizational culture, innovation, and performance: A test of Schein's model." *Journal of Business Research* 67.8 (2014): 1609-1621.
15. O'Reilly, Charles A., and Jennifer A. Chatman. "Culture as social control: Corporations, cults, and commitment." In: B.M. Staw & L.L. Cummings (Eds.), *Research in Organizational Behavior, JAI Press* 18 (1996): 157-200.
16. Schein, Edgar H. "Organizational Culture and Leadership." San Francisco: Jossey-Bass (1985).
17. Atkinson, Anthony A., John H. Waterhouse, and Robert B. Wells. "A stakeholder approach to strategic performance measurement." *MIT Sloan Management Review* 38.3 (1997): 25.
18. Behn, Robert D. "Why measure performance? Different purposes require different measures." *Public administration review* 63.5 (2003): 586-606.
19. EUROCONTROL. "Safety culture: an overview." (2008).
20. Mwaura, N. E., and H. Bula. "Organizational Culture and Firm Performance: A case of the Institute of Certified Public Accountants of Kenya." *International Journal of Business, Social Sciences and Education* (2017): 74-105.

Analysis of the transport - precariousness relationship in the peripheral areas of Casablanca

Ishak Hbiak[1,2,a], Abdelaziz Adidi[1] , El Hassan El Brirch[2] and Jean-Pierre Nicolas[3]

[1]*Institut National d'Aménagement et d'Urbanisme (INAU), Madinat Al Irfane, Rabat, Maroc*
[2]*Ecole Hassania des Travaux Publics (EHTP), Km 7 Route d'El Jadida, Casablanca, Maroc*
[3]*Ecole Nationale des Travaux Publics de l'Etat (ENTPE), Rue Maurice Audin, Vaulx-en-Velin, France*

Abstract. Casablanca is the largest metropolis in Morocco and the Maghreb and one of the largest cities in Africa. However, like most cities in developing countries, it faces major problems in terms of planning and transport. The problems in transport are of different types and affect all areas of the city, especially the peripheral areas of Casablanca, which are experiencing significant difficulties in terms of accessibility, knowing that these areas are the most precarious. The aim of this work is to confront on the one hand the global GIS analysis of the transport supply and its relation with the precariousness of the populations, and on the other hand the real problem of access to economic opportunities because of the exorbitant costs of public transport in relation to the income of these peripheral populations, taking into account the case of the peripheral city of Errahma. This second analysis was carried out through a field survey, on the area of the new city of Errahma. Initial analysis of this work has confirmed the existence of a very strong relationship between transport supply, loss of economic opportunity and precariousness in these peripheral areas.

1 Introduction

Transport development has always been linked to economic development. It's in the same time factor and result of economic activity.

Indeed, firstly transport is considered as a factor of production of goods and services. Transportation participates in the added value to goods and services. On the other hand, the development of transport is an implication of the changing economy.

Like most large cities in developing countries, Casablanca has many transportation problems: both in terms of quantity and quality. These problems can dampen the economic activity of the population, especially the most fragile one [1, 2, 3].

The question of the equity of access to the city and urban services has been the subject of a lot of researches in the field of urban studies, but the consideration of social inequalities and the use of the notion of accessibility in the field of transport can still today be considered undeveloped. Mobility can be seen as a condition of social integration. [4]

Besides, "The fit between transport alignments and contemporary poverty mappings is rarely considered in the design of transport systems funded or leveraged through the public purse" [5]. In this context, known by low supply of urban public transport services, there is the question of the confrontation of this offer with the socioeconomic characteristics of the population of the various districts of Casablanca, especially the question of the confrontation of the characteristics of the urban transport supply and the poverty data in Casablanca, to test the hypothesis where poorest neighborhoods are the least served by public transport. This hypothesis has been the subject of several studies carried out on cities in developing countries [6, 7, 3].

Another research has shown the causal link between public transport supply and urban distribution of poverty: "access to public transportation plays a significant role in determining the spatial distribution of poverty in a metropolitan area, and if expansion of bus public transit is a factor contributing to the suburbanization of poverty" [8]. In Casablanca, this link is not proved because of the weakness of the urban transport offer.

Transport and poverty has also been dealt with in a broader context evoking the notion of "transport poverty" or "energy poverty". [9, 10, 11].

[a] Corresponding author: ishak.hbiak@gmail.com

2 Study area and used data

2.1 Study area

Casablanca is the first metropolis of Morocco and North Africa, and one of the largest cities in Africa. The area of Casablanca agglomeration spreads over 1615 km².

According to the High Commission for Planning (HCP), Grand Casablanca has more than 4.3 million inhabitants, and Casablanca city alone totals 3.4 million inhabitants, with a large part of young population (31% are under 15). Casablanca is also a very active city with an activity rate of 51%. It hosts 39% of the production units in Morocco and 60% of the industrial workforce [12].

The Grand Casablanca region still has about 500 shanty towns in which nearly 111,500 families are crowded (according to statistics of 2012), despite all the efforts of the government in the program "Cities without slums" [13].

This demographic pressure has led to a still very strong urban sprawl in action, and strong demand on the various urban services.

2.2 Used data

In our work, we used first, the data of the supply of urban public transport in Casablanca. These data are composed of the mapping of the bus network, as well as the route of the first tramway in operation since late 2012 [14-15].

In addition to these offer transportation data, we used data on poverty in the region of Casablanca in 2014. These data were produced by an investigation of the territorial authorities of Casablanca in collaboration with the High Planning Commission and other actors, such as the social development Agency (ADS) [4-16].

Figure 1 presents the results obtained from this survey for the different districts of Casablanca and the province of Nouacer.

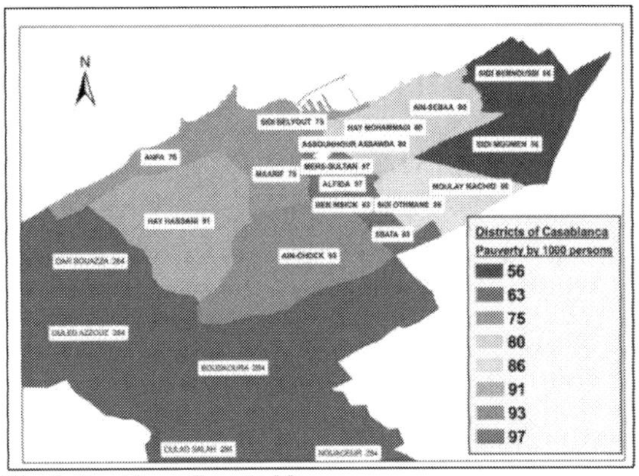

Fig. 1. Casablanca Poverty Map

We have also exploited spatial data related to the territory of Casablanca, including the administrative division, as well as a source of very high resolution satellite imagery (up to a resolution of 30 centimeters).

We also used a layer showing the main economic activity zones in the Casablanca region territory in order to analyze the relationship between economic opportunities and transport.

We have also conducted a survey of households in certain peripheral areas of the city of Casablanca. Here after, we will present some preliminary results of this survey concerning the peripheral city of Errahma.

3 Results and analysis

3.1. Development of a spatial database

Before starting our analytical work we have proceeded with the compilation of all collected data in a unified spatial database and in a GIS environment.

Figure 2 illustrates Casablanca's public transport supply network, consisting of bus lines and the first tram line, based on the developed database.

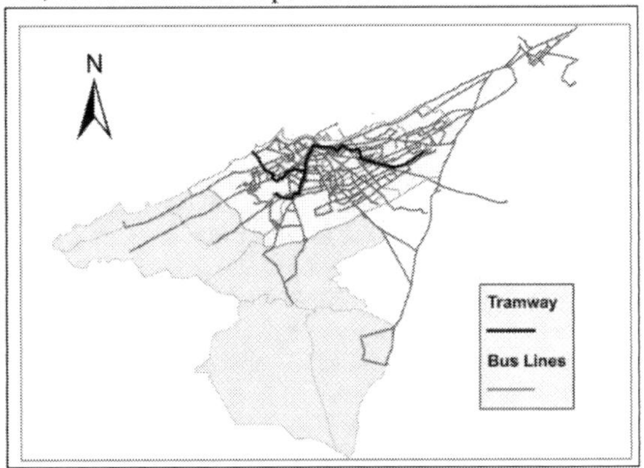

Fig. 2. The Casablanca public transport supply network

The spatial database has helped us perform spatial analysis relative to our original problem using a GIS tool.

3.2 Density of public transport supply in Casablanca

After mapping the network of public urban transport in Casablanca, we conducted an analysis of its density in space. So we defined an indicator to highlight the concentration of supply in certain districts of Casablanca. The map in figure 3 shows the results.

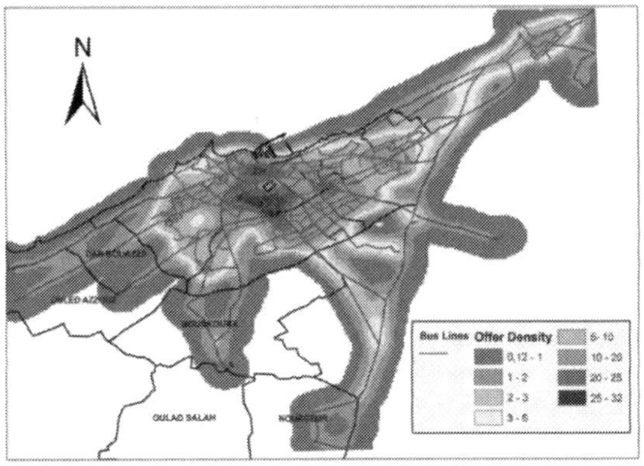

Fig. 3. The density of the supply of urban public transport

The analysis of this map shows the concentration of supply in the central districts of the city, with a gradual decrease towards the periphery. Some peripheral areas experiences almost total absence of the network.

3.3 Urban planning and transportation in Casablanca

We looked after to the relationship between the distribution network and urban form in Casablanca.

The analysis of the directional distribution of the urban public transport network in Grand Casablanca confirms the historical link between supply of transportation and urban form concentrated around the port. Figure 4 illustrates directional distribution of the urban public transport.

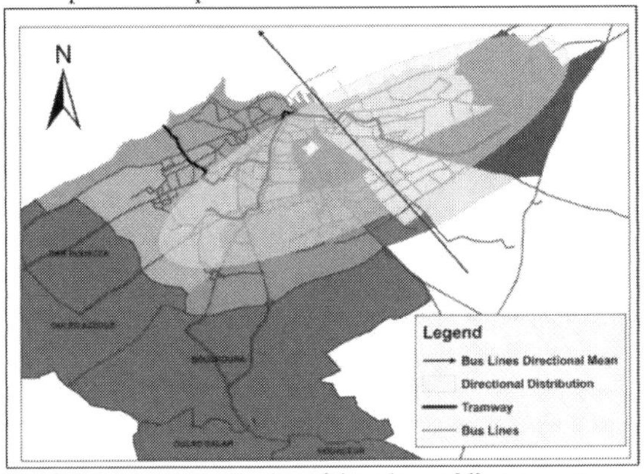

Fig. 4. Directional distribution of the urban public transport

3.4 Precariousness and accessibility to the urban transport network in Casablanca

In this step of our analysis, we have calculated another indicator to compare access to the network of urban public transport and the precariousness of the populations served by this offer.

The result is shown in the figure 5, and it clearly shows that the poorest populations, in the majority of cases, are also the least served by public transport offer.

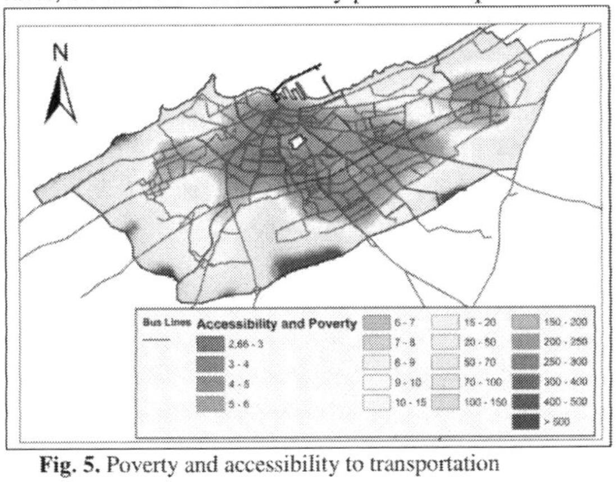

Fig. 5. Poverty and accessibility to transportation

3.5 transport and economic opportunity of the peripheral populations of Casablanca

Finally, we were interested in verifying the hypothesis of the negative effect of transport supply on the economic opportunity of the peripheral populations of Casablanca. We took the example of the peripheral city of Errahma.

Figure 6 presents the main areas of economic activity in the Casablanca region. It shows that the most important economic zones of the Casablanca region are concentrated in the North and it is the opposite of the city of Errahma which is south of the city of Casablanca.

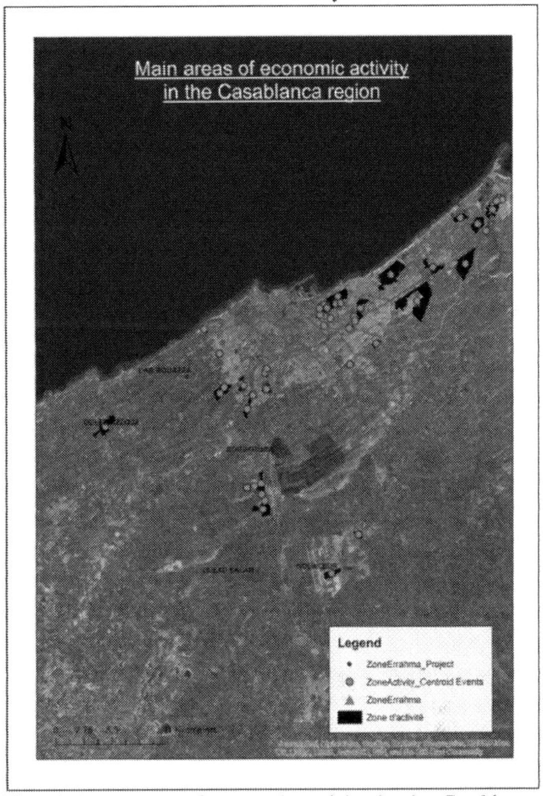

Fig. 6. Main areas of economic activity in the Casablanca region

We have analyzed the distance between Errahma and the economic activity zones of Casablanca. Figure 7 shows the proximity of these zones along concentric rays between 5 and 40 kilometers.

It can be seen that the main economic activity zones of Casablanca are at distances between 10 km and 20 km from Errahma, and even up to 40 km for certain areas. This is equivalent, in terms of time, to trips lasting between 1h and 2h20min.

To confirm this result, we conducted a survey of households in the city of Errahma. The purpose of this survey was to answer essentially to the questions relating to the mobility of people living in Errahma. The results of this survey relate to a sample of 60 households.

Preliminary results show that the average of the ratio of the monetary cost of travel compared to the income of the outlying households of Errahma is much higher than the national average.

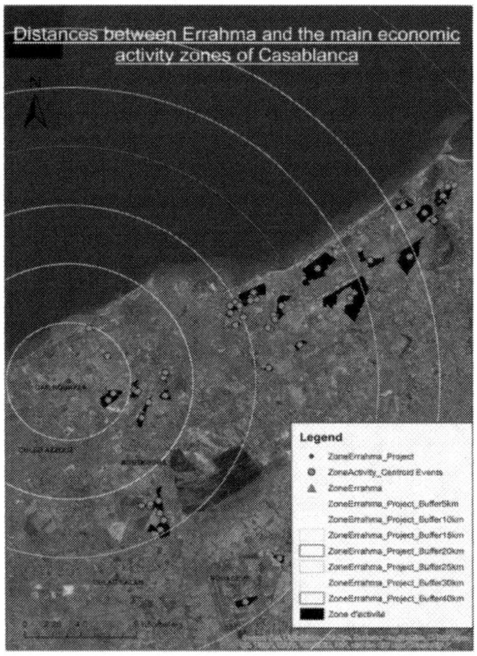

Fig. 7. Distance between Errahma and the main economic activity zones of Casablanca

We note also that Errahma's household travel time is very high and can in some cases reach 4 to 5 hours a day.

We notice that more than a quarter of households say they have lost an economic opportunity mainly because of the high generalized cost (Time, Cost, Security ...) of the trips to and from the city of Errahma.

4 Conclusion

Our study shows that the new urban sprawl has concentrated precarious population on the peripheral areas of the city of Casablanca without supporting this extension by an offer of adequate transport and thus decreased its potential access to economic opportunities.

This conclusion raises questions about the equity in the supply of urban public transport in Casablanca and the choice of the proposed urbanization model for the Casablanca region and Morocco as a whole.

The results of the survey showed that the transport supply of a peripheral district (case of Errahma) can be a serious obstacle to access to economic opportunities. A low supply of transport therefore in this sense appears as a cause of impoverishment of households in peripheral areas.

References

1. Godard, X. (2000). Urban mobility in developing cities: Difficulties of measures, uncertainty of trends and of sustainability appraisal. In *Urban transportation and environment. Proceedings of the international conference CODATU IX, Held Mexico City, 11-14 April 2000.*
2. Diaz Olvera, L., Plat, D., & Pochet, P. (2005). La ville hors de portée? Marche à pied, accès aux services et ségrégation spatiale en Afrique subsaharienne. *Espace populations sociétés. Space populations societies*, (2005/1), 145-161.
3. Rebouha, F., & Pochet, P. (2009). Mobilité et accès à la ville pour le travail des populations pauvres du périurbain d'Oran. *Cahiers de la Méditerranée*, (78), 335-353.
4. Sylvie, F., & Gallez, C. (2013, January). Mobilité, accessibilité et équité: pour un renouvellement de l'analyse des inégalités sociales d'accès à la ville. In *Colloque International Futurs urbains: Enjeux interdisciplinaires émergents pour comprendre, projeter et fabriquer la ville de demain*.
5. Grieco, M. (2015). Poverty mapping and sustainable transport: A neglected dimension. *Research in Transportation Economics*, *51*, 3-9.
6. Olvera, L. D., Plat, D., & Pochet, P. (2005). Marche à pied, pauvreté et ségrégation dans les villes d'Afrique de l'Ouest. Le cas de Dakar.
7. Olvera, L. D., Plat, D., Pochet, P., & Audin, R. M. (2011). Pauvreté, mobilité quotidienne et accès aux ressources dans les villes subsahariennes. *Dynamiques de Pauvretés Et Vulnérabilités en Démographie Et en Sciences Sociales: Actes de la Chaire Quetelet 2007*, 281.
8. Pathak, R., Wyczalkowski, C. K., & Huang, X. (2017). Public transit access and the changing spatial distribution of poverty. *Regional Science and Urban Economics*, *66*, 198-212.
9. Lucas, K., Mattioli, G., Verlinghieri, E., & Guzman, A. (2016). Transport poverty and its adverse social consequences. In *Proceedings of the institution of civil engineers-transport* (Vol. 169, No. 6, pp. 353-365). Thomas Telford (ICE Publishing).
10. Berry, A., Jouffe, Y., Coulombel, N., & Guivarch, C. (2016). Investigating fuel poverty in the transport sector: toward a composite indicator of vulnerability. *Energy Research & Social Science*, *18*, 7-20.
11. Mattioli, G., Lucas, K., & Marsden, G. (2018). Reprint of Transport poverty and fuel poverty in the UK: From analogy to comparison. *Transport Policy*, *65*, 114-125.
12. au Plan, H. C., Direction de la Statistique (2016). Le Maroc des régions 2014.
13. Ministry of Housing and Urban Policy, Morocco (2012). "Cities without slums" program.
14. Hbiak, I., Adidi, A., Nicolas, J.P., El brirchi, E. (2016). Choix d'indicateurs de mobilité urbaine durable pour le territoire de Casablanca. In *Proceedings of the 1st AMSR Congress and 23rd APDR Congres* (pp 76-85).
15. Hbiak, I., Adidi, A., Nicolas, J.P., El brirchi, E. (2016). Analyse de la desserte des transports publics urbains sur le territoire de Casablanca par les systèmes d'information géographique. In *3ème Colloque International sur le Monitoring des Systèmes Industriels, CIMSI'2016, Fès- Maroc.*
16. Wilaya de Casablanca (2015). Carte de pauvreté de la région de Casablanca.

Wireless Sensors Network for Traffic surveillance and management in Smart Cities

Anass El Mrini[1]* and Abdellatif Ghacham Amrani[1]

[1]Dept of Physics, Laboratory of Science and Advanced Technology, Multidisciplinary Faculty of Larache, Morocco

Abstract. The efficiency of a city's own transportation system heavily influences its economic growth due to the increasing need to transport labour force, consumers and cargo. The use of the latest technologies to solve traffic problems is the main theme of ITS "Intelligent transportation system", several solutions are already in use in many cities around the world and it is an indispensable part of the smart cities project. Our research focuses on developing a solution to manage road traffic in a smart city environment by utilizing wireless sensor network as medium to surveille the traffic. In this paper, we propose a set of countermeasures to avoid congestion and improve the flow of the traffic. The proposed countermeasures were simulated, and the results showed that our plan reduced the traffic congestion, minimized gases emission and increased fuel usage efficiency.

1 Introduction

There is a long-term dynamic interaction between transport and the nature of the city economy. Cities have economic cores explained by various forms of agglomeration economies, which are often based on a traditional industrial or trading base [1]. Cities are changing with the movement of people and businesses out from the centre, increased suburbanization, and the desire for lower residential and job densities. Suburbanization of employment has followed, and the simple local journey-to-work pattern or the public transport-based movement to the strong central area employment location has been replaced by more complex longer distance car-based movements. Traffic problems occur when the city own transport system can't cope with its transportation demands, and the bigger the city, the more it is vulnerable to traffic problems.

ITS, Intelligent Transportation Systems is a set of solutions and tools to improve transportation efficiency and safety, it also aims at improving energy saving and reducing vehicles emissions [2]. Generally, ITS can be classified into several subsystems according to their functions: Travel Management and Traveller Information System (TMTIS), Advanced Vehicle Control and Safety System (AVCSS) and Commercial Vehicle Operation (CVO). TMTIS is responsible for providing real-time traffic information, and for managing and controlling traffic condition and traffic lights.

Several WSN-based traffic surveillance systems have been proposed, Cheung *et al.* [3] built a traffic surveillance system based on wireless sensors. The proposed system is deployed along roadsides and at intersections, it uses magnetic sensors to detect vehicle detection through disturbance in the earth's magnetic field. The system can collect traffic data such as vehicle count, occupancy, speed, vehicle classification and reidentification.

O. Karpis *et al.* [4] tested the applicability of WSN in traffic monitoring, two magnetic sensors nodes were placed on the verge of the road with spacing of 2 m. Traffic parameters were accurately measured including the number and the speed of vehicles, initial experiments about estimating the length of the vehicles were also conducted, showing the possibility of the classification of vehicles.

Chen *et al.* [5] propose a system of WSN for Intelligent Transportation System (WITS), used for the information gathering and data transferring. In this system, each vehicle is equipped with a wireless sensor unit that measures the vehicle parameters, the *roadside unit* collect the data from the passing vehicles units and transfers it to the *intersection unit*. The *roadside unit* gathers the information of the vehicles around and transfers it to the *intersection unit*. The *intersection unit* analyses the received information, and forward them to the *strategy sub-system*, which in turn calculates an appropriate scheme according to the pre-set optimization target (such as maximum throughput, minimum waiting time, etc.). The drawback of this approach is the necessity of equipping every vehicle with a sensor node, a requirement that is very unlikely to be met in a real-life application.

M. Franceshinis *et al.* [6] proposes a solution, within the SAFESPOT Project, using WSN to obtain traffic related data in order to generate automatic safety warnings along the road network. The architecture follows a Master-Salve paradigm in a linear topology, the system consists of several *sensor nodes* that are connected to a single *gateway node GN*. The sensor nodes are installed on the roadside and are responsible for estimating vehicle count, speed and direction. the gateway node forwards the data to the *road side unit RSU* via a wired connection. The linear topology causes delay and scalability problems, also the wired connection between the GN and RSU becomes costly and inefficient in a large-scale application (i.e. smart city).

J.J. Fernandez *et al.* [7] proposes a platform called Urban Information System based on WSN in order to characterize urban traffic parameters. The platform collects traffic data by taking advantage of Bluetooth devices in the vehicles. Furthermore, it can collect environmental information such as noise levels, light intensity, pressure, temperature, humidity, airborne dust and gases. Even though the system can generate Origin-Destination matrix in real-time with acceptable accuracy, the system assumes all vehicles to have a functioning Bluetooth device.

The literature shows that applying the WSN technologies in the field of intelligent transportation systems, covers the need for real-time and accurate traffic parameters. A network of wireless sensors offers flexibility, scalability, and lower installation and maintenance costs than the traditional surveillance systems. However, these studies mainly focus only on effective traffic information acquisition, but do not propose a whole system capable of traffic monitoring and management at the same time.

The remainder of this paper is organized as follows, Section II exposes the motivation for using WSN to monitor urban traffic, section III defines the functionalities of the proposed traffic management system, in section IV the results of a basic simulation of the proposed system is evaluated. Section V concludes the paper and outlines our future work.

2 APPLICATION OF WSN IN TRAFFIC SURVEILLANCE

ITS rely on accurate, reliable real-time traffic data. And a successful application of ITS technologies in a smart city means a large-scale deployment of surveillance systems on all major highways and local streets. Surveillance systems nowadays rely heavily on intrusive sensors - inductive loops specifically, which has many undesirable features. That's why the biggest motivation for using WSN for traffic surveillance is the advantages that it presents in comparison to the other sensors technologies:

- **Cost:** a cost comparison between different surveillance technologies - including device cost, installation and maintenance cost – shows that the cost of these networks is about half that of inductive loops [8], and it is expected to rapidly decline with volume.
- **High accuracy:** wireless sensors proved to be the most accurate sensor besides the inductive loops when it comes to vehicles count or speed measurement [8]. Also, by using the collaborative approach in a WSN the detection accuracy and reliability can be improved greatly. Other test also shows that vehicle classification is also possible with wireless sensors with an acceptable accuracy [9].
- **Maintenance:** maintenance of wireless sensors nodes is made easier as they are smart and can assist automatic diagnosis, so any arising problems can be resolved quickly. Repairing or replacing the wireless nodes is easier because of the non-intrusive nature of WSN.
- **Flexibility:** the biggest advantage of wireless sensors is that they can be easily deployed and withdrawn anywhere close to the road (street furniture like: traffic signs, street lights, traffic lights etc.), which make them useable for permanent or temporary traffic measurement. Wireless sensors configurations can be customized to adapt to different purposes and environments, for example environmental sensing can be done alongside traffic surveillance.
- **Wireless communication:** wireless communication capability is indispensable for road-to-vehicle communication; With the emergence of smart vehicles, the origin destination matrix can now be generated without relying on surveys, as it can be done by retracing the log of the communication between the sensing nodes and the vehicle during its trip.

In summary, the new generation of wireless sensors overcome the disadvantages of the traditional sensors. They are generally small in size which make them easy to install and maintain, last longer on small batteries and they generally cost less. In addition, their wireless capabilities make the sensing network scalable and more flexible.

3 Traffic management

The system we propose consists of two subsystems:

A. Data Acquisition Subsystem:

The WSN technology is the main actor in this subsystem. The sensor nodes installed by all road intersections need to collect and send real-time traffic data including:

- Vehicle count.
- Vehicles speed.
- Vehicle classification.
- Occupancy.
- Environmental data.
- Vehicle re-identification, the nodes also need to be able to reidentify vehicles for origin-destination traffic map.

The communication within the subsystem must be reliable and fail-safe to maintain a degree of Quality of service. Because of its low-cost, low-power and low complexity, the standard IEEE 802.15.4, adapted by the ZigBee Alliance [10], will be used in the deployment of our Sensor Network. While other wireless technologies, like Bluetooth or WIFI, provide high data rates, they are disadvantageous due to their high-power consumption, application complexity, and cost [11].

B. Control and data analysis subsystem:

The Control Station "CS" collects and processes the data, then takes appropriate sets of predetermined actions to solve or to prevent traffic problems. The CS has three main functions:

- The CS should collect, compile then store the environmental and the relevant traffic data in a database accessible for further studying and reviewing purposes.

• This system will use the real time traffic data to give the macroscopic or microscopic evaluation of the traffic flow in the city.

• The CS will assess traffic condition based on the collected data, then if a traffic problem is detected it will take an automated response to resolve the situation.

4 Simulations results

The targeted environment of our research is a smart city, which by theory will have full control over intersections lights and electric traffic signs. LED lights replace paint to make road lines and signage, which opens the possibility of changing the traffic flow by opening/blocking some roads or exchanging the number of lanes used between the two opposite directions in a specific street.

The proposed actions were simulated in VISSIM [12] software to test their feasibility and effectiveness. The subject test bed in this study is a 4 km stretch of national no1 going through LARACHE city from the city centre to the highway entrance, the road has 2-lanes in each direction and only 4 major intersections were simulated, the street serves all kinds of vehicles. The simulation parameters were determined after observing the traffic in the simulated area daily for the period of one week. The traffic composition was set at the default setting by the VISSIM program. The study period encompassed the evening peak hour between 5 p.m. and 6 p.m. The main inbound traffic flow was set to 1200 vehicle per hour and to 100 vehicles per hour for every other vehicle input. The inbound traffic has a 20% chance to change direction at every intersection. And every vehicle coming from a branch road to an intersection has a 90% chance to go to the inner city.

Two scenarios were simulated: the first scenario we simulated a standard situation without any change to the infrastructure or traffic signal behaviour (standard city), in the second scenario we implemented the proposed counter-actions in the same network (smart city): we changed the number of lanes for inbound direction to 3 and the opposite direction was left with one lane. We also changed the priority rules at every intersection as to prioritise the bigger flow, the traffic lights were also programmed to give more green-light time to the inbound traffic.

Each scenario was simulated several times and several traffic parameters were exported from the VISSIM evaluation lists such as average delay time, number of stops per vehicle and the amount of harmful gases emitted by the vehicles. Fig 1. to Fig.5 show the average results of each scenario.

Fig. 1 compares the delay parameters between the two scenarios, the average delay per vehicle and average standstill time per vehicle and the latent delay (total delay time of all vehicles that could not enter the network immediately), the second scenario shows less delay time across all the parameters. Fig.2 compares the average vehicle stops per vehicle, the vehicle in the first scenario were forced to stop in more occasions than in the second scenario. Fig 3. Compares the average traffic queue length and the maximum queue length detected in the two scenarios, our counter-measures minimized the traffic queue length.

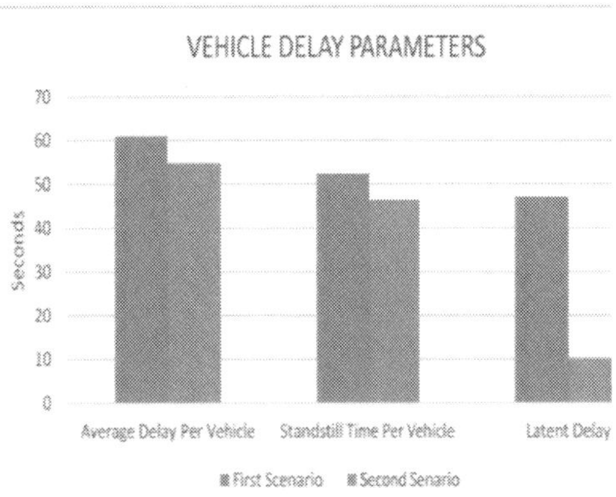

Figure 1. Vehicle delay parameters in the network.

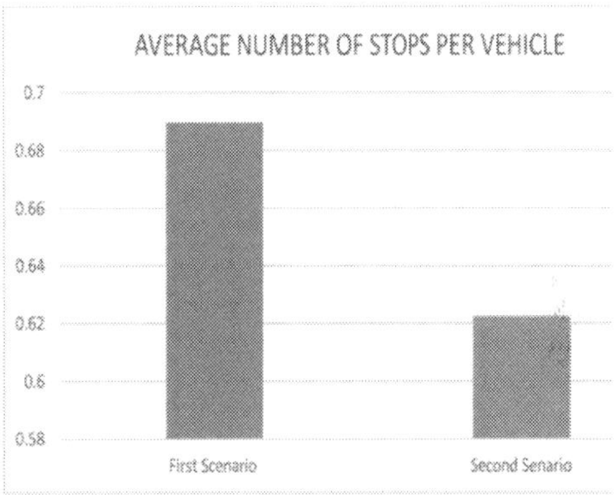

Figure 2. Average number of stops per vehicle.

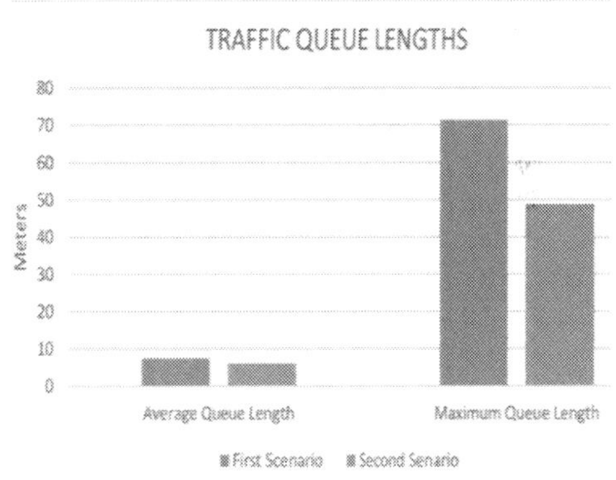

Figure 3. Average and maximum vehicle queue length measured.

The effect of increasing the number of lanes for the inbound traffic minimized the delay time and reduced traffic congestion, Also, the changing of traffic light behaviour and the priority rules for the benefits of the dense inbound traffic produced decreased the number and the duration of vehicle stops. These results show the effectivness of the proposed solutions against the traffic problems.

To complete the evaluation, we compared the environmental data extracted from the simulation to evaluate the impact of our counter-actions on the environment. Fig. 4 shows the total mass of the gases emitted by all vehicles in the network. The smart city had relatively less gases emissions than the standard city. The difference in speed between the two simulations is also relatively small: 36 km/h for the first scenario, and 38 km/h for the second. Fig. 5 compares the fuel consumed by the vehicles that travelled in the network. the second scenario

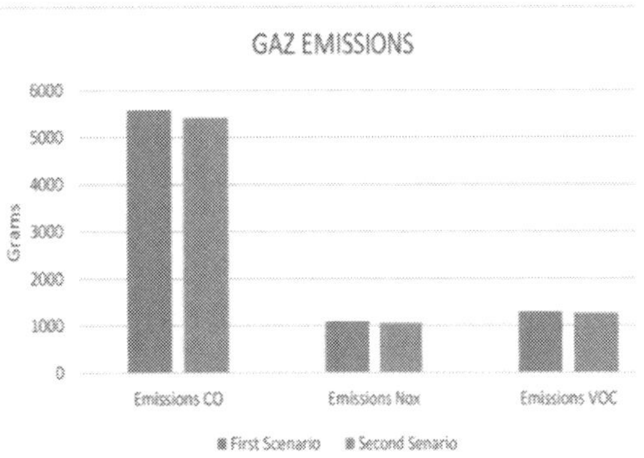

Figure 4. Vehicle emissions of CO, NOx, and Volatile Organic Compounds

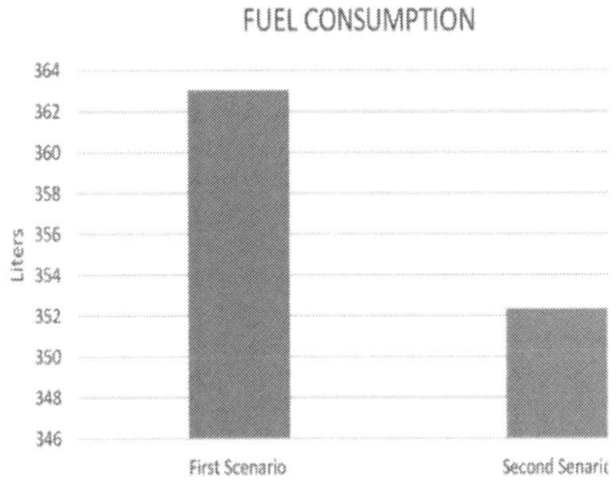

Figure 5. Vehicle emissions of CO, NOx, and Volatile Organic Compounds

had less fuel consumption which can be translated to higher fuel usage efficiency.

5 Conclusion and future work

In this paper we have proposed a traffic surveillance and management system based on WSN. The surveillance system is based on a network of wireless sensor nodes installed across all the intersection of the traffic network. The WSN will detect traffic parameters across the city to give an overview of the traffic status in the city. The central Unit will analyse the data and take automated counter-measures to arising traffic problems thanks to remote controlled traffic lights, road signs and lane assignment. We evaluated the counter-measures in traffic simulation software VISSIM and results showed that our plan reduced the traffic congestion, minimized gases emission and increased fuel usage efficiency.

In a future study, we intend to determine the WSN topology and routing protocols. The performance of the chosen WSN architectures will be simulated by the CUPCARBON software to test their effectiveness and quality of service. We will re-test our smart city solutions in a bigger scale and a more complex network, we will also try to implement a customer service system that sends real-time traffic status to drivers which will help them choose faster, less congested streets, toward their destination.

Acknowledgement

We would like to express our gratitude to the PTV Group for granting a free licence to use the VISSIM software.

References

1. K. Gwilliam, *World Bank. Cities on the Move : A World Bank Urban Transport Strategy Review*. Washington, DC. © World Bank. (2002)
2. Z. BINBIN, *Adaptive Traffic Light Control in Wireless Sensors Network-based Intelligent Transportation System*, The Hong Kong Polytechnic University, (2010).
3. P. Varaiya and S.-Y. Cheung, *Traffic Surveillance by Wireless Sensor Networks*, UNIVERSITY OF CALIFORNIA, BERKELEY, final report UCB-ITS-PRR-2007-4, (2007).
4. O. Karpis, *Wireless sensor networks in intelligent transportation systems*, Int. J. Mod. Eng. Res. IJMER, **vol. 3**, pp. 611–617, (2013).
5. W. Chen, L. Chen, Z. Chen, and S. Tu, *Wits: A wireless sensor network for intelligent transportation system*, in Computer and Computational Sciences, 2006. IMSCCS'06, **vol. 2**, pp. 635–641,(2006).
6. M. Franceschinis, L. Gioanola, M. Messere, R. Tomasi, M. A. Spirito, and P. Civera, *Wireless sensor networks for intelligent transportation systems*, in Vehicular Technology Conference, 2009. VTC Spring 2009. IEEE 69th, pp. 1–5 , (2009).
7. J. J. Fernández-Lozano, M. Martín-Guzmán, J. Martín-Ávila, and A. García-Cerezo, *A Wireless Sensor Network for Urban Traffic Characterization and Trend Monitoring*, Sensors, **vol. 15**, no. 10, pp. 26143–26169, (Oct. 2015).
8. P. T. Martin, Y. Feng, X. Wang, and others, *Detector technology evaluation*, Citeseer, (2003).
9. D. P. Mishra, G. M. Asutkar, and S. S. Dorale, *An Application of Wireless Sensor Network in Intelligent Transportation System*, pp. 90–91, (2013).
10. 'zigbee alliance'. Bishop Ranch, CA, http://www.zigbee.org/
11. K. Sohraby, D. Minoli, and T. F. Znati, *Wireless sensor networks: technology, protocols, and applications*. Hoboken, N.J: Wiley-Interscience, (2007).
12. 'PTV Vissim'. [Online]. Available: http://vision-traffic.ptvgroup.com/en-us/products/ptvvissim

AUTHOR INDEX